Fused Deposition Modeling

This book covers fused deposition modeling (FDM) from its fundamentals to research advances including developments in FDM machines, development of polymer materials, novel applications, process modeling, simulation, and optimization backed by case studies and research results.

It explains the effect of important FDM parameters on part quality, appropriate methods for testing and characterization of FDM-fabricated parts, FDM of composite materials, process monitoring and control. It also covers applications of FDM process in automobile, aerospace, biomedical, pharmaceutical, and tissue engineering.

Key features:

- Provides all the necessary fundamental details and advancements in FDM
- Details novel ways of fabrication of complex and intrinsic parts by using different types of plastics and polymer materials
- Covers fabrication of miniature devices by FDM
- Discusses finite element analysis, modeling, optimization, and sustainability rating, for FDM
- Includes monitoring and control, intelligent processing, and application of artificial intelligence (AI) and machine learning (ML)-based part fabrication

This book is aimed at graduate students and researchers in materials science and manufacturing.

Advanced Materials Processing and Manufacturing

Series Editor: Kapil Gupta

The CRC Press Series in Advanced Materials Processing and Manufacturing covers the complete spectrum of materials and manufacturing technology, including fundamental principles, theoretical background, and advancements. Considering the accelerated importance of advances for producing quality products for a wide range of applications, the titles in this series reflect the state-of-the-art in understanding and engineering the materials processing and manufacturing operations. Technological advancements for enhancement of product quality, process productivity, and sustainability, are on special focus including processing for all materials and novel processes. This series aims to foster knowledge enrichment on conventional and modern machining processes. Micro-manufacturing technologies such as micro-machining, micro-forming, and micro-joining, and Hybrid manufacturing, additive manufacturing, near net shape manufacturing, and ultra-precision finishing techniques are also covered.

Novel Applications of Functionally Graded Materials
Edited by Neeraj Kumar Bhoi, Harpreet Singh and Himansu Sekhar Nanda

Advanced Functional and Composite Materials: Fabrication, Characterization, and Applications
Edited by Neeraj Kumar Bhoi, Harpreet Singh and Himansu Sekhar Nanda

Advanced Welding Technologies: Bridging Theory, Practice and Innovations in Fusion, Solid-State and Beyond
Edited by Bappa Acherjee, Dawei Zhao and Fábio A.O. Fernandes

Process Modeling and Optimization in Modern Manufacturing
Edited by Ravi Pratap Singh, Narendra Kumar, Vishal Francis and Ankur Jaiswal

Additive Manufacturing for Biomedical Implants and Prosthetics
Edited by Pawan Kumar Rakesh and Pravin Kumar

Manufacturing in the Digital Age
Edited by Mohan Kumar Pradhan, Mahendra U. Gaikwad, and Ragavanantham Shanmugam

Fused Deposition Modeling: Technology, Applications, and Developments
Edited by Anshuman Kumar Sahu and Siba Sankar Mahapatra

For more information about this series, please visit: www.routledge.com/Advanced-Materials-Processing-and-Manufacturing/book-series/CRCAMPM

Fused Deposition Modeling

Technology, Applications, and Developments

Edited by

Anshuman Kumar Sahu and Siba Sankar Mahapatra

CRC Press
Taylor & Francis Group
Boca Raton London New York

CRC Press is an imprint of the
Taylor & Francis Group, an informa business

First edition published 2026
by CRC Press
2385 NW Executive Center Drive, Suite 320, Boca Raton, FL 33431

and by CRC Press
4 Park Square, Milton Park, Abingdon, Oxon, OX14 4RN

CRC Press is an imprint of Taylor & Francis Group, LLC

ISBN: 9781032798394 (hbk)
ISBN: 9781041169024 (pbk)
ISBN: 9781003686941 (ebk)

DOI: 10.1201/9781003686941

Typeset in Times
by Newgen Publishing UK

Contents

Preface

Additive manufacturing (AM) is one of the emerging areas in manufacturing field due to its capability of producing complex and intrinsic parts which are difficult to produce by conventional manufacturing route. AM is a rapidly growing technology, through rapid innovation, with unprecedented freedom of design and production of complex-shaped products. Furthermore, the development of AM technologies bears practical implications for economic growth, healthcare, national security, space exploration, and sustainability.

Recent advancement in the field of fused deposition modelling (FDM), one of the filament-based AM processes, opens a new field in the area of plastic and polymer-based AM. FDM uses a wide range of materials including plastics, polymers, and composite materials to build complex parts in a layer-by-layer addition of materials. The versatility along with low machine and feedstock material cost and less processing time has enabled filament-based AM processes to be acceptable in today's competitive manufacturing scenario. FDM has transformed the way designers, engineers, and innovators approach product development and prototyping. Along with the materials like acrylonitrile butadiene styrene (ABS), polylactic acid (PLA), thermoplastic polyurethane (TPU), polyethylene terephthalate glycol (PETG), nylon, etc. composite materials like carbon fiber-reinforced polymers, ceramic-reinforced polymer composites, and natural fiber composites can also be easily printed by FDM process, increasing its applications in different fields. Recent advancements in FDM enhance it applications in pharmaceutical production processes with focus on mass production of dosage forms, suiting the average patient in the population and moving generalized healthcare toward personalization with versatile production techniques that create high-quality unique dosage forms developed for treatment of patients. Similarly, FDM demonstrated the fabrication of implant materials for patients' specific applications.

An attempt has been made in this book to provide a thorough understanding of the three-dimensional (3D) printing process like FDM with recent studies on experimental investigations, specific applications as well as modeling and optimization of the process. This book aims to provide a glance on historical development of this unique process, overview of the process, detailed discussion on different field applications like automobile, biomedical, aerospace, and use of a wide range of feedstocks. It also addresses the aspects of product development, processing, part fabrication, and characterization of build parts.

Acknowledgments

We would like to express our sincere thanks to Prof. Marco Leite and Prof. Paulo Peças, IDMEC, Instituto Superior Técnico, Universidade de Lisboa, Lisbon, Portugal, for their continuous support. We are also thankful to Arpan Paul and Nikhil Aditya, Research Scholars, Department of Mechanical Engineering, National Institute of Technology Rourkela, India; Mukesh Yadav, Research Scholar, Department of Metallurgical and Materials Engineering, Indian Institute of Technology, Kharagpur, India; Piyush Mishra, Research Scholar, University of Illinois Chicago, USA.

We acknowledged the support provided by Prof. Kapil Gupta, Department of Mechanical and Industrial Engineering Technology, University of Johannesburg, South Africa, for completion of this book. We want to express our sincere gratitude to everyone involved in the production of this book.

We would like to acknowledge the support provided by the European Union under the Next Generation EU, through a grant of the Portuguese Republic's Recovery and Resilience Plan (PRR) Partnership Agreement, within the scope of the project PRODUTECH R3 – "Agenda Mobilizadora da Fileira das Tecnologias de Produção para a Reindustrialização."

Anshuman Kumar Sahu
Siba Sankar Mahapatra

About the Editors

Anshuman Kumar Sahu is a post-doctoral researcher at IDMEC, IST, University of Lisbon, Lisbon, Portugal. His areas of research are additive manufacturing, rapid prototyping and rapid tooling, non-traditional machining, powder metallurgy, and optimization of manufacturing processes. He is presently working on Lean and Green Manufacturing in Industry 4.0.

Siba Sankar Mahapatra is working as a professor (HAG) in the Department of Mechanical Engineering, National Institute of Technology, Rourkela, India. He has more than 30 years of experience in research and teaching. He has published 300 research articles in reputed peer-reviewed international journals. His h-index in Google scholar is 62 with more than 16,068 citations. He has three Indian patents. He has authored four books. His academic interest includes rapid prototyping, non-traditional machining, modeling, and simulation. He is busy carrying out many projects with national and international partners on additive manufacturing, modern machining, and manufacturing optimization.

Contributors

Sateesh Kumar Adapa
National Institute of Technology Raipur
Raipur, Chhattisgarh, India

Nikhil Aditya
National Institute of Technology Rourkela
Rourkela, Odisha, India

Rohit Anand
Indian Institute of Technology Kharagpur
Kharagpur, West Bengal, India

Ch. Bharath Kumar
Vignan's Institute of Information Technology(A)
Visakhapatnam, Andhra Pradesh, India

Roop Sandeep Bommidi
Department of Mechanical Engineering, Aditya Institute of Technology and Management (A)
Tekkali, Andhra Pradesh, India

Neha Choudhary
National Institute of Technology Karnataka
Surathkal, Karnataka, India

Manik Chandra Das
Maulana Abul Kalam Azad University of Technology
Kolkata, West Bengal, India

Borra N. Dhanunjayarao
Vignan's Institute of Information Technology(A)
Vishakapatnam, Andhra Pradesh, India

R. Durga Prasad Reddy
Indian Institute of Technology Bombay
Mumbai, Maharashtra, India

Arka Ghosh
C V Raman Global University
Bhubaneswar, Odisha, India

Kumar Aviseck Jati
Biju Patnaik University of Technology
Rourkela, Odisha, India

Sally AbdulHussain Kadhum
University of Technology
Baghdad, Iraq

Karthi S.
Central Institute of Petrochemicals Engineering & Technology (CIPET): Institute of Petrochemicals Technology (IPT)
Bhubaneswar, Odisha, India

Gill Santosh Kumar
Vignan's Institute of Information Technology(A), Vishakapatnam, Andhra Pradesh, India

Mohit Kumar
Indian Institute of Technology Bombay
Mumbai, Maharashtra, India

Pankaj Kumar
Indian Institute of Technology Kharagpur
Kharagpur, West Bengal, India

Yong Hoon Lee
University of Memphis
Memphis, TN, USA

Marco Leite
Universidade de Lisboa
Lisboa, Portugal

Siba Sankar Mahapatra
National Institute of Technology Rourkela
Rourkela, Odisha, India

Sanhita Mandal
Indian Institute of Technology Kharagpur
Kharagpur, West Bengal, India

Ganesh Sahadeo Meshram
Indian Institute of Technology Kharagpur
Kharagpur, West Bengal, India

Pradeep Kumar Mishra
Biju Patnaik University of Technology
Rourkela, Odisha, India

Afonso Oliveira
Universidade de Lisboa
Lisboa, Portugal

Pratap Chandra Padhi
Central Institute of Petrochemicals Engineering & Technology (CIPET): Institute of Petrochemicals Technology (IPT)
Bhubaneswar, Odisha, India

Parth Patel
National Institute of Technology Rourkela
Rourkela, Odisha, India

Albert E. Patterson
Texas A&M University
College Station, TX, USA

Arpan Paul
National Institute of Technology Rourkela
Rourkela, Odisha, India

Anuj Rajoriya
National Institute of Technology Rourkela
Rourkela, Odisha, India

Vinod Singh Rajput
Government Engineering College Nowgong
Chhatarpur, Madhya Pradesh, India

Ravindra Reddy, N.
Central Institute of Petrochemicals Engineering & Technology (CIPET): Institute of Petrochemicals Technology (IPT)
Bhubaneswar, Odisha, India

Uzwalkiran Rokkala
Aditya Engineering College (A)
Surampalem, Andhra Pradesh, India

Rakesh Roshan
National Institute of Technology Rourkela
Rourkela, Odisha, India

P. V. N. Sai Kumar
Vignan's Institute of Information Technology(A)
Visakhapatnam, India

Usha Kiran Sanivada
University of Minho
Braga, Portugal

Varun Sharma
Indian Institute of Technology Roorkee
Roorkee, Uttarakhand, India

Aswani Kumar Singh
Indian Institute of Technology Bombay
Mumbai, Maharashtra, India

Ramandeep Singh
Banasthali Vidyapith
Banasthali, Rajasthan, India

N. V. Swamy Naidu
National Institute of Technology
Raipur
Raipur, Chhattisgarh, India

1 Exploring Fused Deposition Modeling Processes

Mechanics, Working Principles, and Applications in Research and Development

Parth Patel, Anuj Rajoriya, Arka Ghosh, Rakesh Roshan, and Sally AbdulHussain Kadhum

1.1 INTRODUCTION

Fused deposition modeling (FDM) represents an additive manufacturing (AM) technique wherein the material, once melted, is methodically deposited in successive horizontal strata to fabricate a three-dimensional (3D) entity. The process initiates with a digital model, specifically a computer-aided design (CAD) model, serving as the input. The interplay of hardware, software, and materials is integral to achieving the desired final product. Initially, FDM was predominantly utilized for the creation of prototypes rather than fully functional end products. However, advancements in material science and precision engineering have enabled the production of objects possessing mechanical and material properties that closely mirror, or even match, those of end-use functional products. The standard pre-processing stages of the conventional FDM process are delineated in Figure 1.1.

The first step in the FDM process involves creating a CAD model. This can be done by designing a new model from scratch or by utilizing reverse engineering techniques to scan an existing object. Once the CAD model is created, it is converted into the STL file format, which stands for "Standard Triangle Language" or "Standard Tessellation Language." The STL format is widely accepted in the field of AM. The STL format approximates all surfaces of the object with triangles, facilitating the slicing of the object into discrete layers.

The slicing process is performed using a horizontal plane parallel to the printer's base plate (bed), commencing from the bottom surface of the object and progressing incrementally until the top surface is reached. The increment is equivalent to the layer thickness. For generating the initial layer at the base, the horizontal plane intersects the triangles to derive intersection points. By connecting these points, the boundary

DOI: 10.1201/9781003686941-1

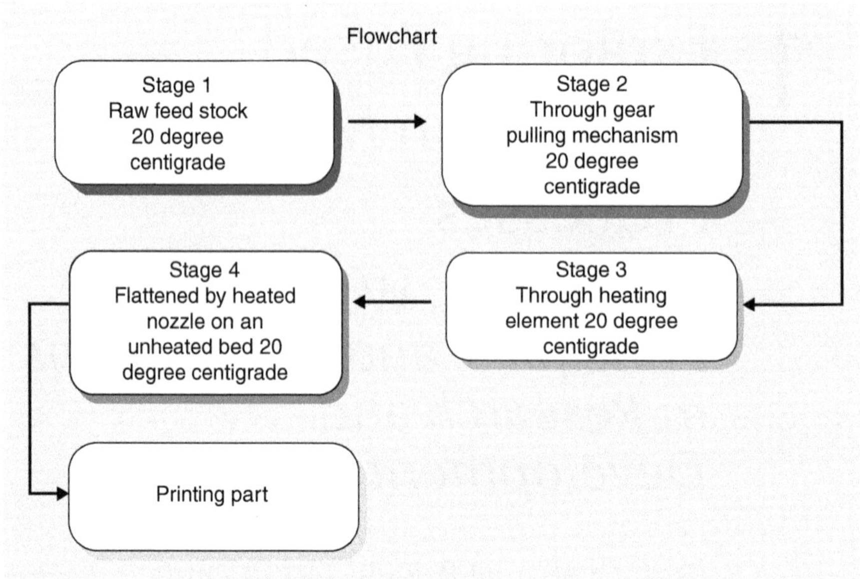

FIGURE 1.1 Processing steps in the FDM process.

of the bottom layer is delineated. The slicing operation then proceeds incrementally, with each increment corresponding to the specified layer thickness, until the top surface is reached, resulting in a complete set of layers.

After slicing each layer, the toolpath is generated using G and M codes. This process creates a G-code file that contains all the necessary instructions. The G-code file is then transferred to the FDM machine to begin the printing of the item. The machine utilizes G and M codes to control the movement of the extruder and manage material deposition. Figure 1.2 presents a schematic diagram of the FDM machine.

As depicted in Figure 1.2, the FDM machine primarily comprises a movable extruder and a printing base. Typically, the FDM extruder is equipped with two nozzles: the primary nozzle for depositing the build material and the secondary nozzle for dispensing support material. In instances where the part includes overhanging features, the secondary nozzle deploys support material to facilitate proper construction. The printing material, supplied in filament form with a standard diameter of 1.75 mm, is fed through these nozzles. Thermoplastic materials are predominantly utilized in the FDM process. Each nozzle is precisely maintained at a specific temperature to ensure the extrusion of the material. The filament material once passed through the heat block of the extruder melts and is subsequently deposited through the nozzle. Upon exiting the nozzle and being exposed to the atmosphere, the material solidifies almost instantaneously, within approximately 0.1 seconds, due to the inherent properties of thermoplastic materials. The most commonly employed printing materials are polylactic acid (PLA) and acrylonitrile butadiene styrene (ABS). The extruder maneuvers within the X–Y plane to deposit material without altering its orientation. After completing each layer, the printing bed lowers in the Z-direction by an amount

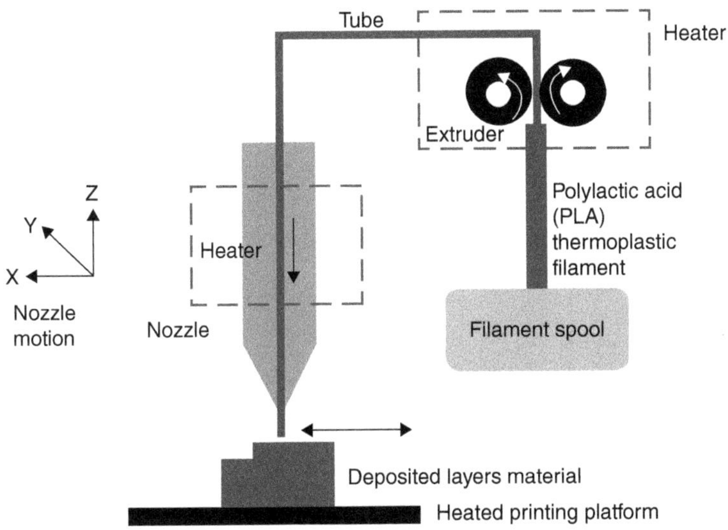

FIGURE 1.2 Working principle of the FDM process.

equal to the thickness of that layer, allowing for the deposition of the next layer. This process continues until the entire object is printed. An important post-processing step in the FDM process involves removing the printed support structures and, if necessary, polishing the object to enhance its surface quality.

FDM is a widely used manufacturing process that involves applying constant pressure as material is extruded through a nozzle. The extruded material is deposited on a substrate, where it solidifies. This deposition process ensures that each new layer adheres to the one beneath it, allowing for the creation of 3D objects. The material extrusion technique is commonly employed for various polymers, including polyphenylsulfone (PPSF/PPSU), ABS, thermoplastic polyurethane (PU), and polycarbonate (PC PLA). The rise in popularity of extrusion-based AM technology can be attributed to its accessibility, simplicity, affordability, and ability to handle multiple materials [1–3].

FDM technology is compatible with various materials, including PPSU, PC, PC-ABS blends, and ABS. The technique utilizes a computer-controlled deposition nozzle to create 3D objects with specified structures and compositions [3, 4]. In FDM, a thin plastic filament is melted through the print head and extruded in a specified thickness. The primary components of the FDM process, as illustrated in Figure 1.2, include the extrusion head (responsible for heating and melting the material) [5], the nozzle or extruder (for material exit in a specified size) [6], the bed (the base for material deposition), and the Ram (for vertical bed movement). A gear mechanism controls and drives the filament from the spool to the liquefier. Typically, FDM systems feature one nozzle for part creation and another for generating support structures.

In 1988, Scott Crump created, developed, and patented a type of AM technology known as fused deposition modeling through Stratasys shown in Figure 1.3 Inc., USA. Recently, there has been growing interest in using FDM technology for the

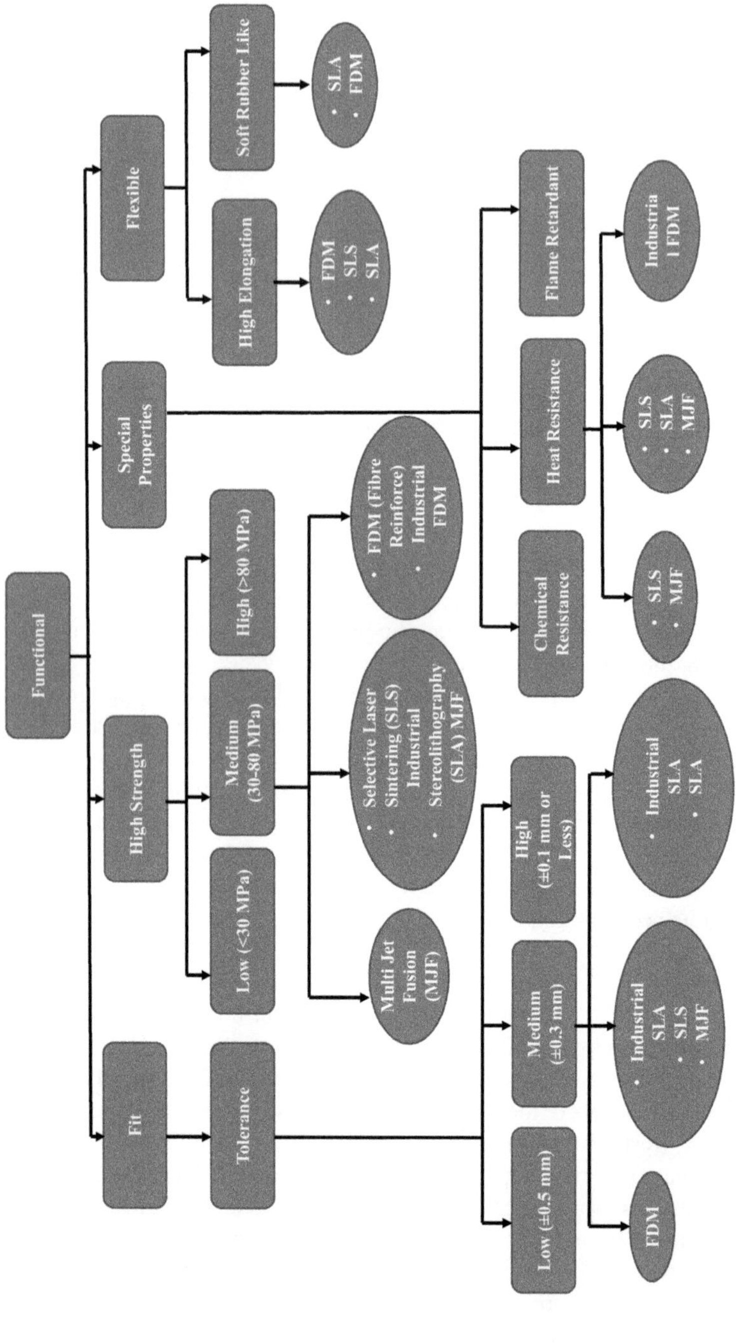

FIGURE 1.3 Different properties of AM processes.

customized production of small industrial and home components. Several countries have integrated FDM technology into pre-secondary education curricula, and it is increasingly utilized domestically [7]. In India, the NITI Aayog government initiative is establishing FDM-type 3D printers in secondary schools under the Atal Innovation Mission (AIM)-Atal Tinker Labs (ATLs) scheme. FDM is recognized as the most economical and user-friendly AM process for producing thermoplastic domestic application parts. Compared to traditional manufacturing processes, FDM offers greater flexibility in producing intricate and customized parts. The advantages of waste reduction, minimal capital investment, high speed, and low cost set FDM apart from other manufacturing techniques [8]. FDM enables the rapid creation of parts in an office environment without any toxicity risks.

FDM technology has a variety of uses, including automobile parts, domestic accessories, biomedical components, snap-fits, conceptual models for final manufacture, casting, molding, and other plastic parts [9]. Stratasys was the pioneering manufacturer of extrusion-based AM systems. Today, numerous companies produce these systems, including MakerBot, Ultimaker, Formlabs, Stratasys, Aleph Objects, 3D HUBS, 3D Systems, Shapeways, Materialise NV, Sculpeto, Zortrax, HP Inc., The UPS Store, 3D PI, Carbon, Protolabs, Airwolf 3D, Solidoodle, Voxeljet, Fusion 1.3, Envision Tec, Sinodh, Prusa, and Rapide 3D.

1.1.1 Process Parameters

FDM technology facilitates the creation of customized products across various applications by adhering to a well-defined set of process parameters. The efficacy of the FDM process is highly contingent upon these parameters, which significantly influence the part properties, material consumption, and fabrication duration. To achieve an optimal balance between mechanical properties, printing time, and material usage, it is imperative to determine the most suitable range of parameters. Manufacturing parts under these optimal conditions ensures the highest quality output. According to a comprehensive literature survey, FDM process parameters can be categorized into four primary groups, as illustrated in Figure 1.4. Operation-specific parameters, in particular, have a substantial impact on material conservation and mechanical properties [10].

The frequently examined parameters that affect the properties of FDM-produced parts include nozzle diameter, raster size, raster orientation, raster-to-raster gap, infill design and density, layer thickness, and printing speed.

Nozzle Diameter: The nozzle diameter refers to the size of the through-hole used for extruding both model and support materials. This parameter is crucial as it influences other significant process parameters such as road width, air gap, and layer thickness, which are critical in the AM process. Road width, also known as raster width, denotes the lateral dimension of the extruded material from the nozzle to complete each layer of the 3D part. This parameter directly impacts the build time of the final object [11].

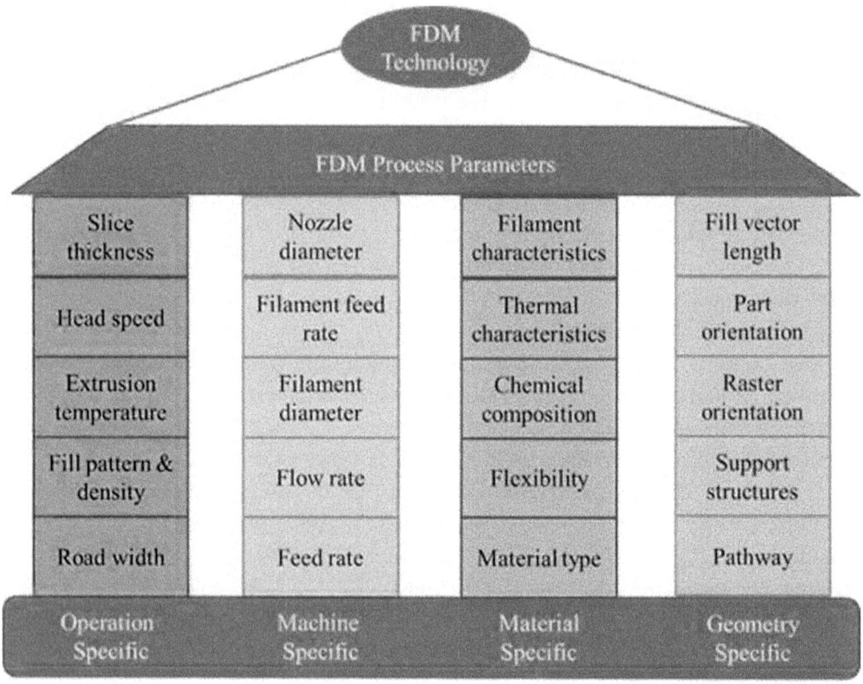

FIGURE 1.4 Categories of process parameters.

Raster Width: Raster width is defined by the size of the deposition path corresponding to the nozzle size [12]. Generally, a larger raster width reduces build time [11–13]. Conversely, reducing the road width increases the printing time by necessitating a greater number of perimeters.

Raster Angle: The raster angle refers to the angular position of the raster on the bottom layer concerning the X-axis. This angle, which can vary from 0° to 90°, significantly influences the mechanical properties of the printed pieces [13–15].

Air Gap: The space between two adjacent rasters is called the air gap. If there is zero or negative air gap, raster overlap results in bumps [13]. A negative air gap makes the part denser, stronger, and stiffer [11]. This parameter directly influences the material usage and the time required for fabricating the components.

Filling Style: The filling style refers to the raster pattern used to fill the interior regions of parts [14]. Common raster patterns include long raster (0°), short raster (0°/90°), and offset raster, which combine the two. The filling pattern affects material consumption [16] and can provide high strength while reducing the overall weight and printing time. Researchers have explored various infill patterns such as rectilinear, grid, honeycomb, and gyroid to optimize these properties [17].

Layer Thickness (LT): Layer thickness denotes the vertical size of each layer. For most FDM machines, the standard layer thickness is 0.254 mm. This value is influenced by the type of material and nozzle size [13,18,19].

Speed of Deposition: The deposition speed refers to the rate at which semi-molten material is extruded through the nozzle. Lower slice thickness reduces deposition speed, thus increasing the time and energy required [20]. Optimal selection of process parameters is essential to achieve the desired product quality with minimal resource utilization.

1.1.2 Factors to be Considered in FDM Printing

The physical and mechanical characteristics of 3D-printed components made using FDM are significantly influenced by various manufacturing factors and the properties of the structural polymers used. FDM involves modifying several processing parameters during the printing process, such as build platform temperature, nozzle temperature, extrusion speed, layer thickness, raster angle, layer count, infill density, and cooling fan speed. Optimizing these parameters is crucial for creating stable models with desirable characteristics. These factors directly impact the material properties, and even minor adjustments to operational and structural parameters can lead to considerable changes in mechanical strength and overall quality. The quality of FDM components is typically assessed based on their dimensional accuracy and surface texture. Build direction, or build orientation, refers to how the component is positioned on the printing platform. FDM printing can be done in three construction orientations: horizontal, vertical, and perpendicular.

To enhance the mechanical strength, it is essential to regulate the temperature. The viscosity of the filament is closely linked to the temperature in the FDM process. If the temperature is too high, the polymer can become overly viscous, potentially resulting in poor surface finish and dimensional issues. Research indicates that temperatures between 200°C and 220°C enhance tensile strength, while temperatures above this range can weaken the final product. In FDM components, higher temperatures can increase melt flowability, which may improve layer adhesion and enhance fracture resistance.

In addition to temperature, printing speed and extrusion speed are also critical factors in FDM. Incorrect speed settings can lead to excessive or insufficient material extrusion.

Raster orientation significantly affects a part's strength by altering its anisotropic properties. The mechanical strength of FDM products is influenced by both raster orientation and infill geometry. The density and strength of a part are determined by its infill percentage. Additionally, the thickness and height of the layers play a role in mechanical strength. Some studies suggest that variations in layer height do not significantly affect strength. However, several challenges arise when creating composites through 3D printing. Compared to traditional methods, 3D-printed fiber composites tend to be less strong due to their greater heterogeneity and anisotropic characteristics.

In FDM, selecting the appropriate thermoplastic material is essential. High-quality thermoplastics with either semi-crystalline or amorphous properties are preferred. Filaments are typically manufactured with a diameter of 1.75 mm and a tolerance of ±0.1 mm. The specifications of the machine also influence the nozzle diameter. Small internal stresses can develop during the solidification of FDM components, leading to

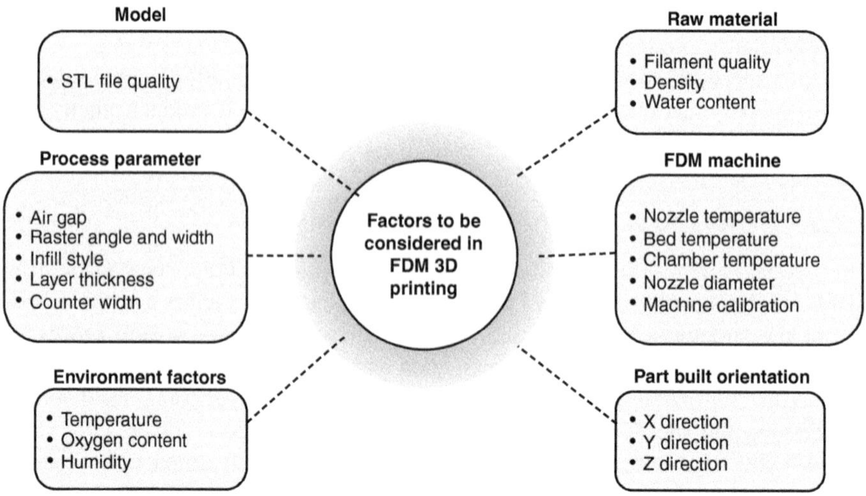

FIGURE 1.5 Factors considered in FDM 3D printing.

potential delamination or distortion. Important considerations for FDM 3D printing are shown in Figure 1.5.

One crucial factor that significantly impacts the strength of 3D-printed parts is layer bonding. For printed components to possess improved mechanical properties, strong connections between layers must be established. The process of bonding layers in FDM technology is similar to welding. As the filament is deposited layer by layer during printing, the polymer layers diffuse and fuse together. Ineffective bonding can lead to the formation of pores, which may increase part inhomogeneity. Several factors influence the effectiveness of layer bonding, including the melt temperature, viscosity, and surface energy. To ensure optimal bonding and, consequently, excellent mechanical properties of the printed components, careful management of these parameters is essential.

1.1.3 Advantages of FDM Process

- The FDM process is a relatively low-cost AM process compared to the other processes.
- A wide variety of thermoplastic materials in different colors are available for the FDM process. These materials are environmentally and mechanically stable. Material change can be done easily one after another. Minimum wastage of material is also one major advantage from a cost point of view.
- FDM process is capable of printing the prototypes with almost 85% strength of the actual part achieved by using materials like ABS.
- The FDM process is also called a desktop prototyping facility because it is simple to use and easy to clean. Hazardous rays or toxic fumes will not get generated in this process, making it a user-friendly and environmentally safe process.

- The support structures for the overhanging features in the object can be easily removed. Hence, post-processing is minimized or eliminated depending upon the requirement.
- A high skilled operator is not required.

1.1.4 Limitations of FDM Process

- The traditional FDM machine has limited freedom, only able to print horizontal flat layers.
- The layer-by-layer deposition of material in the FDM process causes the warping of the object. The thermoplastic material rapidly cools once it gets extruded through the hot nozzle of the extruder. This rapid cooling causes the development of stresses within the object. This leads to shrinkage and affects the dimensional accuracy of the printed object.
- The objects printed through the FDM process tend to have restricted accuracy because the printing material supplied to the extruder is in filament form.
- In the FDM process, material is extruded through the nozzle's circular hole, making it particularly challenging to print sharp corners and edges. These features tend to have a rounded appearance.
- The thermoplastic build materials used in the FDM process tend to have high viscosities. Hence, the deposition through the extruder takes place at a slower rate. This ultimately makes the FDM process slower.

1.2 RESEARCH APPROACH

Over the past century, printed scriptures have largely been limited to 2D printing, making it difficult to accurately represent and visualize real-life 3D models. However, the development of 3D printing has opened up new possibilities in this realm. The main objective of 3D printing is to directly fabricate 3D designs utilizing FDM technology. Many companies are now adopting this method for their manufacturing and product development processes. The industrial adoption of 3D printing is experiencing rapid growth. In 2020, FDM emerged as the predominant 3D printing technique, capturing 46% of the market share. This method utilizes a continuous filament composed of thermoplastic material. Other widely utilized techniques include stereolithographic (SLA) and selective laser sintering (SLS). A recent survey revealed that 17% of respondents currently own a 3D printer, while 38% are considering purchasing one in the near future [21]. FDM technology generally uses thermoplastic materials, with ABS, PLA, and polyethylene terephthalate glycol (PETG) being the most commonly used due to their appropriate thermal and rheological properties [22]. For specialized engineering applications, polymers like polyether ketone (PEEK), nylon, and polyetherimide (PEI) with enhanced mechanical and thermal properties are also used [23]. Researchers are actively developing new materials aimed at producing components with improved mechanical, electrical, and thermal properties.

Although FDM technology offers benefits including the capability to produce intricate shapes and internal spaces, 3D-printed components frequently exhibit diminished mechanical characteristics in comparison to traditionally made ones. The anisotropic qualities and lower interlayer adhesion in the FDM process are a result of its layered nature. Utilizing various materials in FDM, such as filled filaments reinforced with carbon fibers or nanoparticles, can significantly enhance the mechanical and functional properties of printed parts [24–26]. However, the incorporation of fillers in FDM presents challenges, such as potential processing difficulties and compatibility issues with the base polymer. Despite optimizing printing parameters like layer height, printing speed, and raster angle, achieving a 100% infill density and eliminating internal porosity in FDM objects remains challenging. The presence of pores and the stratified composition of FDM components can result in diminished mechanical properties, particularly in terms of tensile and flexural strength. To address these limitations, researchers are exploring methods such as vacuum-assisted FDM and post-processing technologies to enhance the strength and performance of FDM-printed products.

One effective method to enhance the mechanical properties of FDM-printed components is to reinforce them with functional fillers. Selecting a suitable polymer matrix and reinforcing phase that complement each other is crucial [27–29]. The mechanical and rheological characteristics of the filament are primarily determined by the polymer matrix. Examples of polymers used include PLA, ABS, nylon, and PEEK. To improve mechanical performance, researchers have incorporated both short and continuous fibers, such as carbon fiber (CF) and glass fiber (GF), as well as Kevlar fiber (KF) [30, 31]. Carbon nanotubes (CNTs) and other nanocomposite materials have also been employed to improve mechanical and electrical qualities [32–34]. The choice of reinforcing phase is critical, as it needs to improve mechanical properties while maintaining good process ability. Continuous fibers generally provide greater strength and stiffness improvements compared to short fibers [35–37]. However, the fiber type, length, and orientation must be carefully considered to optimize the composite's performance. With the development of specialized composite filaments, it is now possible to 3D print parts with enhanced mechanical and even electrical properties [38–40].

Researchers have developed design and manufacturing strategies to address the limitations of current FDM technologies and systems. They have proposed a comprehensive set of design guidelines that take into account the technical drawbacks of FDM, such as:

1. Anisotropic mechanical properties due to layer-by-layer fabrication
2. Stair-stepping effect on inclined and curved surfaces
3. Dimensional inaccuracies and geometric deviations
4. Surface roughness and poor surface finish
5. Limited material selection and compatibility

The researchers have created a set of design techniques that are intended to maximize the effectiveness, efficiency, and performance of parts that are printed using FDM technology by taking these elements into account. These methodologies give a framework for designers and engineers to construct parts that harness the advantages of FDM while limiting its constraints [41].

Process parameters in FDM can be categorized into controllable and uncontrollable variables. The most critical controllable variables include nozzle diameter, build orientation, infill pattern, layer thickness, and infill density [42]. These factors significantly influence the mechanical properties of the final items produced by FDM [43]. Specifically, layer thickness, nozzle temperature, bed temperature, infill density, nozzle diameter, infill pattern, build orientation, and printing speed are essential for achieving high accuracy in FDM. Additionally, other factors, as shown in Figure 1.6, also affect surface quality and mechanical properties. These include raster angle, air gap wire width, raster/road width, and the number of contours [44].

FDM is an efficient procedure for the quick and cost-effective manufacture of samples. Nonetheless, certain issues must be solved. The layer-by-layer printing approach typically results in staircase flaws, which can severely influence the mechanical qualities and surface texture of the final product [45]. Issues such as warping and shrinkage are widespread in FDM, generally stemming from non-optimized process settings or inadequate temperature control [46]. Furthermore, correctly defining the thermoplastic qualities might be problematic due to substantial deformation, leading to lower strength and stiffness compared to metal counterparts. The surface roughness quality achieved through FDM is inferior to that of other processes, such as selective laser melting (SLM). Achieving tight tolerances in FDM is challenging,

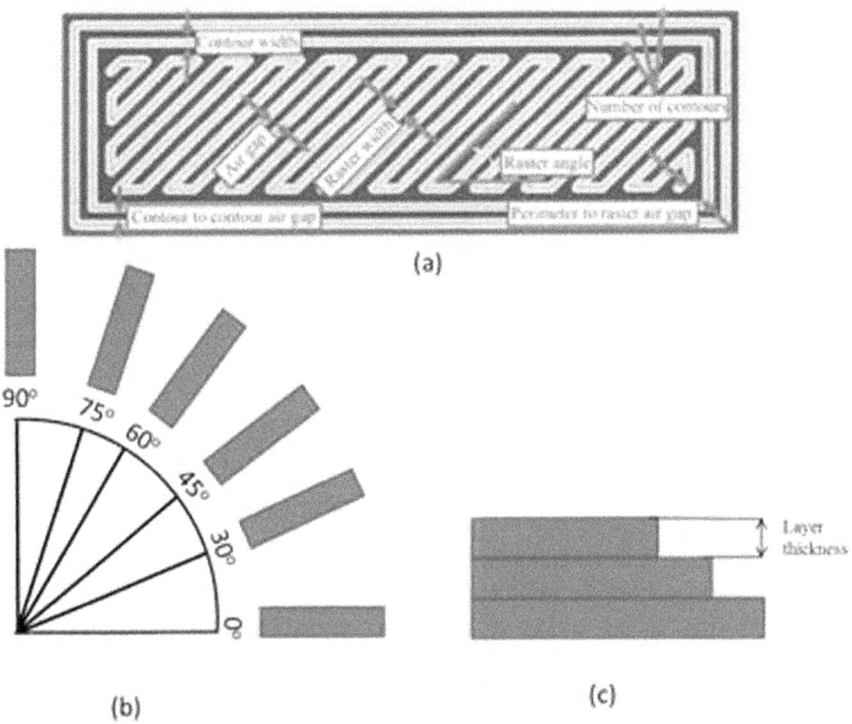

FIGURE 1.6 (a) FDM path parameters, (b) build orientations, and (c) layer thickness.

with dimensional accuracy typically ranging from ±0.2 mm to ±0.5 mm, which is less precise compared to machining processes. Consequently, creating a smooth surface texture can be difficult, often necessitating post-processing techniques to enhance the quality of 3D-printed objects. Additionally, technical polymers used in FDM can be classified as isotropic or anisotropic, making it crucial to accurately assess parameters such as tensile strength, Young's modulus, and elongation at break to ensure precise results [47].

1.2.1 3D Printing by FDM

Mitch Heynick and Ivo Stotz conducted a study that determined that FDM is the second most extensively deployed fast prototyping process, following SLA. The FDM technique incorporates many materials, including ABS and investment casting wax. ABS is recognized for delivering robust strength, but materials such as PPS and PC offer higher strength and a broader temperature range. In FDM, support structures for overhanging components are automatically built and then deleted. Additionally, a "water-soluble" support substance is provided for ABS parts, boosting ease of removal. FDM is recognized for being a user-friendly and quiet method, particularly suitable for the rapid production of small parts, though it may be slower for larger components. Depending on the shape and orientation of the part, support material may not always be necessary in FDM, and in certain cases, it may even be more necessary than the part itself. The final pieces created by FDM show anisotropic mechanical properties, displaying distinct features in different orientations. While the resolution may not be as fine as that attained by SLA, the items created by FDM are often sturdier.

1.2.2 Effects of Different Parameters on the Traditional FDM Process

The FDM process is governed by various parameters that exert direct influence on the mechanical characteristics, overall quality, and duration of printing for the produced object. Each layer is defined by its outer perimeter, termed as a shell or contour, which may comprise multiple such boundaries within a single layer. Infill, on the other hand, constitutes the repetitive internal structure that occupies the space enclosed by the outer shell layers.

There are various infill patterns available for choosing in FDM including concentric, lines, triangular, zigzag, grid, and others, each directly affecting the mechanical qualities of the printed product. Figure 1.7 shows an illustration of these varied infill patterns often applied in the FDM process.

Infill density refers to the amount of material used within the infill structure, which significantly affects both the overall strength and weight of the printed product. The selection of infill density is largely dependent on the intended application of the printed product. Lower infill densities are often preferred for visual models to reduce material consumption, while higher densities are preferred for functional parts to enhance strength, despite longer print times. The road also referred to as a raster or bead, which represents a narrow strip of thermoplastic material extruded through the

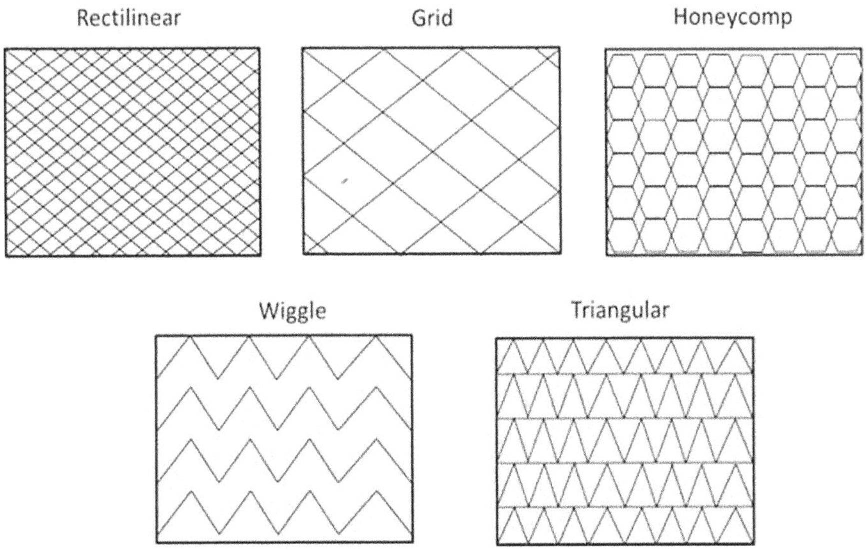

FIGURE 1.7 Infill patterns.

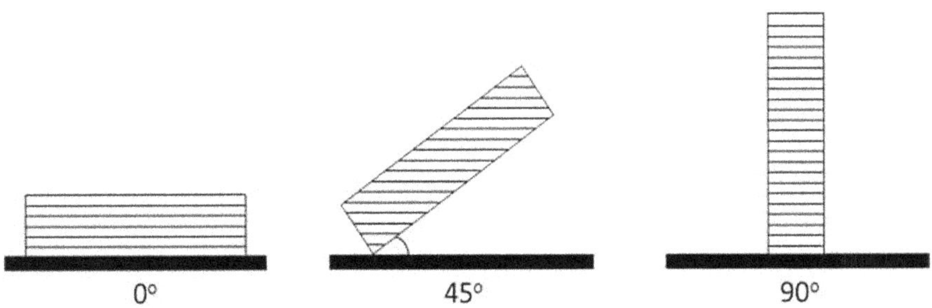

FIGURE 1.8 Different build orientations.

printer's nozzle. The width of the raster is determined by the nozzle diameter, while the raster angle refers to its orientation relative to the X-axis. To achieve mechanical integrity and robust bonding adjacent layers often include raster angles varied by 90°. The air gap represents the space between adjacent highways inside a single stratum.

Layer thickness defines how thick each deposited layer is, significantly affecting the final quality and mechanical properties of the printed product. Choosing a smaller layer thickness reduces surface roughness on curved surfaces but increases printing time. Conversely, choosing a larger layer thickness reduces surface roughness but may compromise mechanical strength. Build orientation is another critical consideration influencing mechanical performance and the need for support structures during printing, thereby impacting both time and material costs. Selecting the appropriate build orientation is crucial to achieving optimal printing outcomes. Figure 1.8 illustrates examples of different build orientations commonly utilized in FDM printing.

Various aspects of the FDM process are influenced by different factors that directly affect the mechanical properties, quality, and efficiency of printed items. The extruder's velocity during material deposition on the X–Y plane is known as the printing speed, and it influences how long the printing process takes in total. The rate of material extrusion via the extruder's nozzle is indicated by the deposition rate, on the other hand. Precise control over both printing speed and deposition rate is essential to maintain consistent raster widths during printing. Furthermore, maintaining a consistent nozzle temperature is crucial to ensure the proper melting of the filament material for extrusion, with this parameter varying based on the type of material used. Bed temperature is another key factor that must be carefully controlled during the printing process to promote strong adhesion between the base layer and the printing bed. This adhesion helps mitigate issues such as residual stresses and unexpected shrinkage that may occur during printing.

The following research by different authors focused on various features of FDM and 3D printing processes: Masood and Song [48] studied the manufacturing of quick tooling utilizing composite materials in FDM, proving the possibility of making injection molding tool inserts with appropriate mechanical qualities. Pennington *et al.* [49] examined the dimensional accuracy of FDM-printed items using ABS material and determined that bed temperature significantly affects dimensional precision. Lee *et al.* [50] explored optimal FDM parameters to enhance the elastic properties of ABS prototypes, while Wang *et al.* [51] developed a mathematical model to analyze heat stress-induced deformations in FDM-printed prototypes. Zhang and Chou [52] investigated distortion caused by residual stresses in FDM, evaluating the influence of process factors on residual stress levels. Saravanan [53] focused on minimizing thermal stress-induced deformation in FDM parts, providing suitable parameters to reduce deflection.

1.2.3 Role of FDM in Investment Casting

Developing a molding pattern is a critical phase in casting procedures. In methods like investment casting (IC) or sand casting, the preparation of core boxes and gating systems for intricate castings is labor-intensive and expensive. Rapid prototyping (RP) techniques (RPTs) offer a solution by allowing foundries to generate patterns without requiring tooling, particularly advantageous for small-scale production batches [54]. These RPT-based designs efficiently cut costs and lead times for low-volume manufacturing runs, removing the demand for machining and welding in the IC process.

FDM, a popular AM technology, achieves a large 82–93% decrease in lead times relative to conventional wood and metal forms [55–57]. FDM is widely employed for creating patterns used in IC and sand-casting processes [58, 59], and offers advantages such as decreased cycle times, reduced material costs, and simplified capital equipment requirements for medium-volume manufacturing. It has been identified as the most cost-effective method for manufacturing core boxes compared to conventional machining techniques [60]. FDM patterns can effectively substitute traditional wax patterns in IC processes [61], supporting the manufacture of functioning prototypes and end-user products employing diverse customized materials [62].

TABLE 1.1
Commercially Available Thermoplastic Composite Filaments

Company	3D Printing Filament Composition
Black Magic 3D	G6—impact HIPS—carbon fiber—graphene filament
	Ferro-magnetic PLA
	Conductive graphene filament
Filabot	MWCNT pellets
3DXTech	Carbon fiber-based filament (matrix: PETG, PLA, PC, ABS)
	Carbon X™ carbon fiber 3D printing filament

In IC applications utilizing ABS, wax gates and vents can be affixed to ABS patterns using adhesives or integrated directly into the pattern using the FDM system. Following the creation of an ABS pattern, it undergoes burn-out in a flash fire furnace at 1050°C, leaving minimal ash residue, after which molten metal is poured into the resultant mold. Singh's review underscores the suitability of FDM-based plastic components for rapid investment casting (RIC) scenarios [63], advocating the use of mathematical modeling and simulation software to optimize gating and riser design for defect-free castings. Gouldsen and Blake's [64] extensive study (Gouldsen and Blake) across six foundries demonstrated that ABS parts fabricated via FDM are conducive to RIC solutions, offering benefits such as dimensional stability and robustness despite the inherent rough surface finish associated with FDM build styles.

1.2.4 Recent Materials Used for Fused Deposition Modeling

Thermoplastic polymers, such as ABS, PLA, polyvinyl alcohol, polyamide, and polycaprolactone, are among the most commonly used materials for FDM filaments. Additionally, there are high-performance engineering thermoplastics available on the market, including PEEK and PEI. Companies like 3D4Makers and 3DXTech offer PEEK and PEI filaments, respectively. While these high-performance engineering polymers tend to be more expensive, they provide superior mechanical and thermal properties. It is important to note that not all 3D printers can achieve the high temperatures necessary for these advanced materials [65]. An overview of commercially available thermoplastic composite filaments can be found in Table 1.1.

In FDM 3D printing using thermoplastic filaments, it is typical to incorporate particles, fibers, and nanomaterials for reinforcement. To prevent clogging, it is crucial to carefully control the size of these fillers in relation to the nozzle diameter. Aggregation of fillers can exacerbate clogging issues, necessitating precise control over filler dispersion to maintain printing reliability.

1.2.5 Literature Survey for Commercially Available Feedstock Filaments of FDM

Various polymeric materials are frequently utilized in FDM, each offering distinct properties suited for specific applications. Common materials include PA, PLA, ABS,

PP, PC, and others. ABS, available in filament form for FDM, exhibits anisotropic qualities with 65–72% tensile strength and 80–90% compressive strength compared to injection-molded parts. Despite its reduced strength, FDM's capability for complex design makes it preferable over injection molding for certain applications. ABS is extensively used for transparent parts in mechanical and electrical components. Post-processing procedures, such as temperature treatment and surface sanding, can enhance translucency by up to 16% without causing dimensional shrinkage [66]. Dynamic mechanical analysis (DMA) indicates that ABS parts with standard build styles exhibit greater strength than dense or sparse structures, with viscosity decreasing as temperature increases. Optimal process parameters for ABS include raster angle selection to maximize tensile or flexural strength [67].

Research has also emphasized wear resistance, highlighting optimal conditions such as a layer thickness of 0.127 mm, a raster angle of 0°, an air gap of 0.008 mm, a raster width of 0.5064 mm, and a build orientation of 0° to minimize wear during FDM [68]. Electrical conductivity enhancements were achieved through chemical deposition of copper (Cu) particles on ABS parts, with chromic acid etching proving most effective in transforming the insulator into a conductor [69]. Further studies have explored mechanical performance and processing parameters specific to ABS, underscoring its diverse applications [50, 70, 71].

Biodegradable PLA has garnered attention for its mechanical properties, with studies showing optimal tensile modulus at 225°C nozzle temperature and geometry's role in influencing mechanical behavior [72]. PLA's suitability for biomedical applications is enhanced by its strength and biodegradability. Researchers have explored various FDM printers like Divide by Cube-2, Stratus's, and Zero FDM, focusing on mechanical properties under different parameter sets [73]. PLA's surface has been enhanced with silver particles for antimicrobial applications, leveraging voids within and between layers to facilitate particle synthesis [74].

PA, a semi-crystalline thermoplastic, offers flexibility and high strength, making it ideal for aerospace and consumer goods in FDM applications. Studies on PA's mechanical strength emphasize the role of build orientation in anisotropic tensile properties and strategies to mitigate high shrinkage stress during printing. Infusing additives like POE-g-MAH into nylon-6 improves dimensional stability, expanding its applications. Optimization studies have explored PA's feedstock filaments to refine printing conditions and output properties [75].

PC, known for toughness and optical transparency, is studied for its mechanical properties under various FDM parameters. Research highlights the impact of build parameters on tensile modulus and failure analysis, supporting predictive modeling of isotropic material behavior [76, 77]. Optimization efforts for PC-based filaments include input parameters like build angle, orientation, raster width, and infill speed to improve print quality and mechanical performance [47, 78, 79].

PP, another thermoplastic, shows promise in FDM for its mechanical and flexural properties [80]. Optimization studies have identified parameters like orientation angle, layer thickness, and infill percentage crucial for achieving desired mechanical performance [81]. PP's versatility extends to applications such as nonreactor fabrication with integrated components using FDM technology [82]. Commercially

TABLE 1.2
The FDM Platform Uses a Variety of Widely Accessible Polymeric Materials

Materials	Studies Performed
PLA	(a) Characterization of the adhesiveness of an FDM-fabricated prototype
	(b) Fusion of PLA prototype with silver particles for antibacterial properties in the biomedical field
	(c) FDM process improvement for 3D-printed component of PLA
	(d) Comparing several FDM machine configurations to assess the repeatability of 3D-printed components
	(e) Printing designs based on textiles using FDM setup for soft PLA matrix
	(f) Analysis of PLA characteristics for FDM 3D-printed component for potential biological use
	(g) A comparative analysis of 3D-printed ABS and PLA prototypes
ABS	(a) Wear analysis and functional prototype process optimization
	(b) Improving the electrical conductivity of working prototypes for use in electronics and electricity
	(c) Printing prototypes to package electrical and mechanical components
	(d) Optimization of the FDM process
	(e) Dynamic behavior study of an item made using FDM
PC	(a) Creepiness for a 3D-printed prototype
	(b) Analysis of isotropic PC-based prototype using finite elements (FEA)
	(c) DMA examination of a 3D-printed prototype to assess the PC feedstock filament's characteristics
PA	(a) Optimization of the FDM process for 3D-printed parts
	(b) Reduced shrinking of the 3D-printed polyamide component
	(c) Anisotropic characterization of a 3D-printed specimen made of nylon for tensile and compression testing
PP	(a) Optimization of the FDM process and assessment of the mechanical characteristics for PP polymeric feedstock filament

available feedstock filaments for FDM are summarized in Table 1.2, highlighting their specific applications and advantages in various industries.

1.2.6 Literature Survey for Non-Commercial Feedstock Filaments of FDM Developed In-House

Commercially available virgin polymeric feedstock filaments have been the subject of much investigation; yet, they are known to have some drawbacks, including low electrical and thermal conductivity, diamagnetic features, and inadequate mechanical and surface qualities. To address these challenges, researchers have explored methods for enhancing virgin polymers by incorporating foreign particles. This reinforcement can be achieved through two primary approaches: mechanical and chemical routes. The foreign particles used for reinforcement are typically categorized based on their size into micro particles and nanoparticles. Consequently, the resulting composite materials are classified into two main types: nanocomposites and micro composites.

1.2.6.1 Micro Composites

Several types of reinforcements have been studied to enhance the properties of polymeric matrices used in manufacturing laboratory-scale feedstock filaments for FDM applications shown in Table 1.3 [83]. For example, low-density polyethylene (LDPE) has been reinforced with SiC/Al_2O_3 particles to produce composite feedstock filaments suitable for FDM. Optimal mechanical performance was achieved with LDPE at 50 wt%, SiC at 25 wt%, and Al_2O_3 at 25 wt%. Similarly, nylon polymer was extruded with Al-Al_2O_3 reinforcement to make FDM feedstock filament, although it was noted that the inclusion of Al/Al_2O_3 particles decreased the tensile characteristics of the nylon matrix compared to virgin nylon [99]. DMA demonstrated that increasing Cu filler content (22–26% by volume) in ABS-Cu blends significantly enhanced both storage and loss moduli [84].

Various types of fibers, including KF, GF, and CF, have been studied as reinforcements in nylon matrices. CF, for instance, significantly increased the tensile strength of nylon composites by a factor of six compared to the unreinforced matrix [85]. Incorporating thermochroic liquid crystalline polymer (TLCP) fibers, up to 40 wt% in polypropylene (PP), notably enhanced mechanical properties. In bone graft applications, PP blended with tricalcium phosphate (TCP) demonstrated promising

TABLE 1.3
Utilizing a Polymeric Micro-Composite, Feedstock Filament Is Prepared for Use in FDM Applications

Polymer	Reinforcement	Studies Performed
HDPE	UHMWPE	Applications for biomedical implants with superior mechanical properties
ABS	(a) Cu	Analysis of the dynamic behavior of a polymeric mix
	(b) TLCP	Preparing feedstock filament with altered mechanical characteristics for a prototype produced using FDM
	(c) Al powder	Prepare ABS and nylon composites that work well together to link different plastics.
	(d) Al/Al_2O_3	Characterizing wear of the 3D-printed component
LDE	SiC/Al_2O_3	LDPE feedstock filament preparation with improved mechanical qualities
Nylon	(a) Al-Al_2O_3	Creation of a substitute feedstock with mechanical characteristics equal to ABS filament
	(b) CF, GF, and KF fiber	Comparative analysis of several fiber-reinforced nylon matrix materials' mechanical performance
	(c) Al powder	Prepare ABS and nylon composites that work well together for use with friction welding.

results with an ultimate compressive strength of 12 MPa and a compressive modulus of 263 MPa [86]. Nylon's thermal conductivity was elevated by reinforcing it with iron (Fe) powder, surpassing that of virgin nylon matrices and other composites. High-density polyethylene (HDPE) was also combined with ultra-high molecular weight polyethylene (UHMWPE) to enhance wear resistance and mechanical strength for biomedical implant applications [87]. GF and maleic anhydride polyolefin (POE-g-MA) reinforcements, varying from 0 to 30 wt% in the PP matrix, were investigated as well. Addition of GF increased tensile modulus but reduced flexibility, while POE-g-MA improved flexibility but reduced the tensile strength of the fabricated parts. Studies on fiber orientation in FDM processing revealed the capability of FDM to align short fibers directionally, resulting in superior mechanical properties compared to conventional compression molding techniques.

1.2.6.2 Nanocomposites

Nanocomposites involve the incorporation of nanoparticles into a polymeric matrix, typically with particle sizes in the nanometer range (10^{-9} mm), while micro composites incorporate larger particles in the micrometer range (10^{-6} mm). Researchers have explored a variety of metallic, ceramic, and non-metallic reinforcements in both micro and nanocomposites, including clay, polymers, and wood. For example, a study investigated the incorporation of nano-sized CNTs ranging from 1 to 5 wt% into PEEK to enhance the tensile strength of FDM-printed objects. Although the addition of CNT nanoparticles reduced part porosity compared to pure PEEK, it did not yield significant improvements in mechanical properties [88]. Another experiment focused on assessing the biocompatibility and anisotropic behavior of PU/PLA reinforced with graphene oxide (GO), demonstrating substantial enhancements in mechanical and thermal properties upon the introduction of nanoscale GO [89]. PLA reinforced with multi-walled carbon nanotubes (MWCNTs) was also investigated for its high electrical conductivity, showing potential in sensor applications with optimized process parameters for enhancing the topography of printed prototypes [90].

Additionally, PU reinforced with MWCNTs was studied for strain sensor applications, revealing improvements in mechanical, electrical, and piezoresistive properties. Although there was a slight reduction in elastic modulus, the composite showed promising piezo resistivity suitable for FDM printing [91]. In another study, PLA was reinforced with 4 wt% of silica nano clay to enhance mechanical, dynamic, and morphological properties significantly, achieving a 115% increase in elastic modulus compared to pure PLA [92]. ABS reinforced with 4 wt% of graphene nanoparticles displayed increased elastic and storage modulus with higher heat stability based on FDM build orientation [93]. Comparison studies between polybutylene terephthalate (PBT) reinforced independently with CNT and graphene exhibited improved mechanical, morphological printability, and thermal stability in PBT/CNT composites [94]. Furthermore, researchers studied polyvinylidene fluoride (PVDF) matrix with barium titanates ($BaTiO_3$) to produce pyroelectric, piezoelectric, and dielectric capabilities, emphasizing three times greater piezoelectric constants in FDM-printed composites compared to solvent-cast equivalents [95]. Moreover, a multi-composite of MWCNTs/$BaTiO_3$/PVDF showed enhanced dielectric properties,

TABLE 1.4
Polymeric Nanocomposite Utilized for the Manufacture of Feedstock Filament for FDM Application

Polymer	Reinforcement	Studies Performed
PLA	MWCNT	Printing a working prototype for extremely conductive liquid sensors
PVDF	MWCNT/ $BaTiO_3$	Improve the multi-composite's dielectric characteristic for use in electronic sensor applications.
	$BaTiO_3$	Bring uniformity to the dispersion via FDM printing, and combine the BaTiO characteristics in the PVDF matrix.
ABS	Graphene	Dynamic, mechanical, and thermal aspects of a printed prototype using composite feedstock filament are examined.
PU	MWCNT	Thermoplastic-based strain sensors printed by FDM
PEEK	CNT	Evaluation of mechanical properties for the composite of PEEK and CNT

with optimized composite ratios achieving the desired dielectric and loss properties [96]. Table 1.4 summarizes various nanocomposites developed at the lab scale for preparing feedstock filaments suitable for FDM applications.

1.3 APPLICATION ON FUSION DEPOSITION MODELING

1.3.1 Shielding and Sensing Application

FDM's application capabilities have greatly increased with composite filaments. Adding conductive particles like graphene, CF, CNTs, and carbon black to FDM filaments allows to produce functional components at low voltages. The manufacture of conductive filament starts with the combination of matrix and reinforcement. Researchers frequently employ the solvent casting technique to correctly blend the matrix and reinforcing elements. Solvent casting involves:

- Using the appropriate solvent, incorporating the reinforcement into the polymer matrix.
- Stir the mixture to the appropriate length.
- Dry the mixture in a controlled environment to remove the solvent.

Hot-melt extrusion (HME) may turn solvent-casting composite film into filaments of the specified diameter [97]. Advanced HME systems combine matrix and reinforcement on their own, removing the need for solvent casting. 3D-printed sensors track resistance changes in response to external inputs.

The 3D-printed PVDF/CNT composite exhibits higher acetone sensitivity [98]. TPU/CNT nanocomposite-based multiaxial force sensor detects applied forces.

Sensor resistance fluctuates in response to external forces. The PLA/graphene temperature sensor can detect both air and water at temperatures up to 70°C. As the temperature increases, the resistance of the composite also rises. This 3D-printed PLA/graphene temperature sensor has a response time of 6 seconds and a recovery time of 14 seconds [99]. The 3D-printed smart hat uses conductive PLA filaments to measure temperature [100]. Composite filaments made from recycled ABS and reinforced with graphite, $ZnCl_2$, MnO_2, and NH_4Cl have been successfully used to generate electrical discharges (ESDs). The prepared ESD materials can operate at temperatures up to 95°C. Addition of acetone increased the ESD voltage from 0.04 to 0.8 V using ABS/CB conductive filaments to form a capacitive gap measurement sensor [101, 102]. The design process makes the capacitance value proportional to the liquid volume. The shielding performance of ABS varies depending on the ratio of carbon black, CNTs, and CNT/CB hybrid mixture. CNT outperforms other types of reinforcement in terms of shielding. Furthermore, shielding potential changes with printing direction [103]. Increasing reinforcement loading can improve shielding efficiency [104]. When printing conductive filament, it is important to avoid issues like nozzle blockage, brittleness, and warping. For optimal composite filament printing performance, ensure uniform reinforcement dispersion. The 3D-printed sensors are affected by air conditions. The force sensor's results vary depending on the material utilized. The study found that FDM pieces can function as shields and sensors, but further research is needed for improved results. Additional research could focus on creating a high-performance polymer composite capable of operating at high temperatures.

1.3.2 Drug Delivery Devices

The term "drug delivery" refers to the administration of a drug into the body to achieve a specific target. It uses the design management and drug delivery systems (DDS) shown in Figure 1.9. Diffusion occurs after the degradation, diffusion, and swelling stages. AM allows substances to be self-distributed and controlled. At the moment, 3D-printed SPRITAM medications have been authorized by the US Food and Drug Administration (FDA) [105]. The pharmaceutical business uses AM technologies such as pressure-assisted microsyringe printing, FDM, selective laser sintering, SLA, and inkjet 3D printing [106]. The indomethacin (IND) and hypromellose acetate succinate (HPMCAS)-loaded polyethylene glycol filament solution dissolves more quickly and has less bitterness [107]. Goyanes *et al.* [108] concluded that drug release depends on the nature and loading of the drug. Anti-inflammatory dissolvable capsules were prepared using PVA silk loaded with 4.14% budesonide standard drug [109]. To reduce the dosage frequency, hollow intragastric floating sustained release (FSR) tablets were developed using FDM technology [110]. 3D tablets formed using PVP/dipyridamole broken filaments dissolved in 30 minutes [110] and improved drug delivery. According to published research, FDM printing makes customization and personalization very similar to drug delivery. FDM-produced drug delivery device (DDD) provides good drug homogeneity and does not require post-processing. When designing a DDS, accurate control of the content is much more important than application. The drug and polymer must be mixed before and during extrusion to achieve a uniform consistency. Problems with FDM in DDD include insufficient filament, clogged nozzles, poor surface

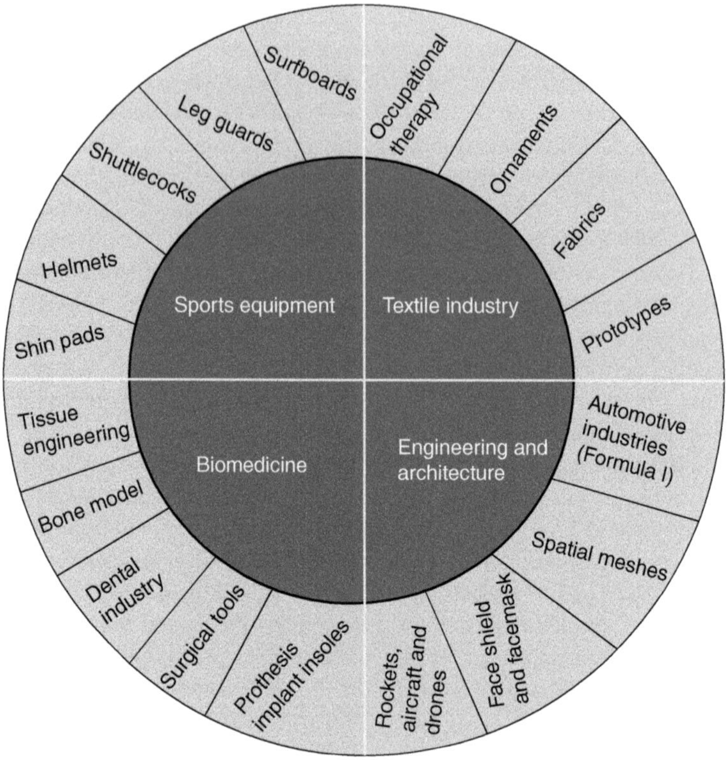

FIGURE 1.9 An overview of FDM method and its different applications (polymer materials, and their applications in the production of drug delivery systems).

smoothness, and minor defects. Prior knowledge of chemical products and printing is required for filament production and 3D printing.

1.3.3 Rapid Tooling

Rapid tooling (RT) is a speedier tool-making technique that can be either direct or indirect. It is an extension of RP. Direct RT methods use additively created tooling to prototypes and functional components can be made from a number of materials, whereas indirect procedures use normal processes and RP master designs to create tooling with the required strength. RT reduces labor time and cost compared to traditional methods; however, the materials that can be produced using RT are limited. Indirect RT technologies include silicone rubber molds, metal spray molds, and epoxy resin molds. Production costs for aluminum–magnesium alloy sheets can be reduced by up to 87% using epoxy punches and dies and indirect RT [112, 113]. Use indirect tooling to create wax filaments and FDM to create conformal cooling channels for wax. Injection molding inserts made from nylon/iron composite filament produced 70 LDPE and 40 ABS pieces [48]. The photolithography wax stamping technology used to create paper microdevices is costly. The term of a polydimethylsiloxane (PDMS) stamp for making paper microdevices was offered. The feasibility of using an FDM-printed mold to fabricate PDMS stamps was also examined. The minimal barrier and channel width for optimal flow in FDM molds were 1.1 and 1 mm, respectively, with failures occurring below these values. FDM-printed dies were tested for stamping capability using DC04 and S355MC materials. 3D printed dies successfully shaped 100 components of DC04 materials within tolerance limits. S355MC's high hardness makes it difficult to shape within tolerances and can distort the die after only a few parts [114]. Electrodes for electric discharge machining were made using FDM, with copper coated in a titanium alloy. Compared with copper electrodes, the material removal rate and tool wear rate of 3D printing electrodes are slightly lower [115]. Due to the high cost of making wax models, capital casting machines are only suitable for mass production. Accelerated investment uses rapid investment (RP) technology to create a sacrificial model. The use of RP in ICs can save time and money in the production of metal mold components [116]. Researchers have found problems such as cracking of the ceramic housing, failure of the structure, and burning of ash when applying RP models directly to ICs. The aluminum-filled epoxy mold used as a source for casting wax models has low thermal conductivity, allowing long-term use. Therefore, an aluminum-filled epoxy mold (fabricated using FDM) with conformal cooling channels is needed. The aluminum-filled epoxy molds were made with thin ABS cooling channels and then rinsed with acetone to remove adhesion. The proposed strategy improves heat transfer efficiency. FDM is not suitable for direct tooling due to its poor surface smoothness and lack of high-performance thermoplastic filaments. Although FDM can be used for indirect quick tooling, it is more time-consuming and costlier. FDM-based fast tooling allows for low-volume manufacturing of bespoke components. However, FDM-based fast tooling offers limited resolution options. FDM tooling is strengthened with reinforcements like iron and aluminum powder.

1.3.4 Aerospace and Automotive Applications

The aviation industry has adopted FDM due to flame retardants. Ultem 9085 has received approval from the US Federal Aviation Administration (FAA) as a flame-retardant polymer suitable for use in airplanes [117]. The study (Figure 1.10) concluded that lightweight FDM-printed parts, such as engine access doors, acoustic shrouds, and intake guide vanes, needed replacement. Stratasys investigated the feasibility of creating non-metallic gas turbines and reported on advancements in Environmental Control System (ECS) ducting for aerospace applications. To reduce noise and fuel consumption, FDM components, including inlet guide vanes, acoustic shrouds, and engine access doors, were designed and tested in an engine simulation environment. Research suggests that using ceramic additives and FDM-based polymers instead of metal in airplanes could decrease weight, lower design costs, reduce fuel consumption, and cut down CO_2 emissions. FDM is a well-established method for drone manufacturing. According to ref. [118], FDM technology is mainly used in the production of unmanned aerial vehicles (UAVs). This method can be used

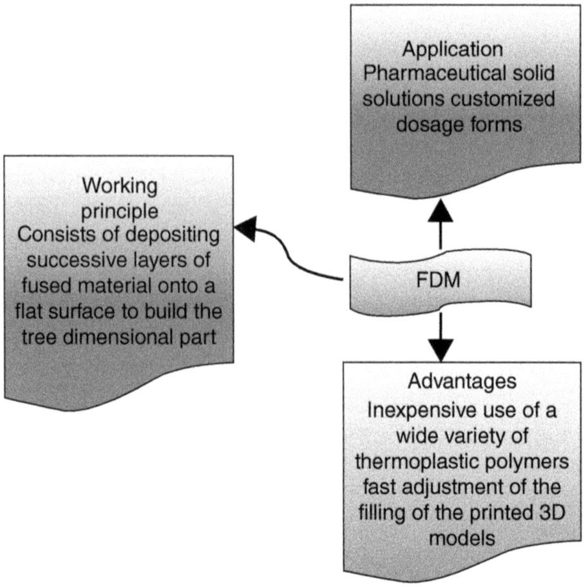

FIGURE 1.10 Application of FDM process.

to produce electronic components from thermoplastic polymers [119]. The goal of the automotive industry is to provide products with deep performance. FDM is one of the manufacturing processes that meets these requirements. FDM-produced fixtures and accessories improve ergonomics and productivity during vehicle assembly. Printed light assemblies and fixtures can reduce equipment effort and worker fatigue [120]. Although FDM holds great promise in the automotive industry, its adoption still faces many challenges. Technology limitations, lack of skills, lack of trust in technology, threats to intellectual property in digital products, etc. have an impact on the overall technology. These limitations can be addressed through research into the use of good technology, the use of additional equipment to determine the potential of the original product, and appropriate training in the delivery of the technology. The ability of FDM technology to create lighting patterns at the required performance levels has opened up many opportunities for the automotive industry.

1.3.5 Microfluidic Devices

Microfluidic devices leverage the unique physical and chemical properties of gases and liquids at the nanoscale. These devices are capable of separating liquids, preparing samples, and conducting tests with precision and control. PDMS has been widely used in the fabrication of microfluidic devices because it is transparent, non-toxic, porous, and cures at low temperatures through soft lithography techniques. However, the process of creating PDMS microfluidic devices can be time-consuming and may result in suboptimal outcomes. Additionally, PDMS devices can have a low profile due to their inherent weakness. Therefore, researchers have started using

AM techniques to create microfluidic devices. Therefore, create devices with robust connections, complex workflows, and integrated devices. Cell-on-chip FDM was used to create square cross-section (800 mm × 800 mm) microchannels using pure polyethylene terephthalate filaments [121]. A microfluidic system using PLA fused deposition modeling is employed to analyze glucose in liquid phosphate-buffered saline (PBS) at pH 7.4 through calorimetric data. Influenza hemagglutinin (virus) and methicillin-resistant *Staphylococcus aureus* (bacteria) have been detected using microfluidic chips created with fused deposition models [122–124]. Most of the materials used in FDM are not transparent. Although PMMA and PLA provide transparency, PDMS-based polymers have better performance for microfluidic devices. Good techniques can help reduce these effects but cannot eliminate the drawbacks. It is important to evaluate the weight of FDM printable polymers before printing microfluidic devices.

The Society of Automotive Engineers (SAE) has employed FDM to fabricate an intake manifold for a 600 cc formula automobile engine [125]. Produced using FDM technology, the built brace and bag hook are ready for use. Additionally, Lam *et al.* [126–128] developed cylindrical scaffolds using polymer materials to enhance mechanical and chemical characteristics. Biomedical is also the dominant field of FDM applications. FDM has emerged as a pivotal technology in the biomedical sector, particularly in the fabrication of damaged organs. Melchels *et al.* [129] utilized FDM to engineer personalized breast prostheses, highlighting the technology's potential in creating custom medical solutions. Espalin *et al.* [130] explored the compatibility of various polymeric materials within FDM processes for biomedical applications, such as craniofacial reconstruction and orthopedic insertions. Furthermore, FDM technology has facilitated advancements in dental repair and fabrication [131].

In addition to medical applications, Brooks *et al.* [132] successfully developed custom-made lamps using FDM, demonstrating the versatility of this technology. Korpela *et al.* [133] investigated the use of polymer and bioactive glass mixtures for scaffold fabrication via FDM, which is crucial for tissue engineering. The precision of surgical guides fabricated through the FDM process for dental applications was also scrutinized [134]. Soriano *et al.* [135] employed FDM to develop prosthetic hands incorporating shape-memory alloys (SMAs), while Okwuosa *et al.* [136] produced customized tablets to meet patient-specific demands using the same technology. Additionally, FDM has been applied in biomimetic projects, and its development now extends to non-polymeric materials for use in structural and robotic applications [137, 138].

The specific characteristics required for FDM products vary across different application domains. Engineering applications often demand tailored mechanical and surface properties, whereas the medical field necessitates attributes such as biodegradability, biocompatibility, durability, and strength. These properties can be meticulously controlled through the selection of appropriate manufacturing materials and the optimization of process conditions. Components were produced via FDM, showcasing the technology's capability to develop intricate assemblies [139]. Overall, FDM technology has established itself as a crucial manufacturing method in the contemporary industrial sector. Previous research underscores that the quality and performance of products fabricated using FDM are profoundly influenced by the settings of processing parameters and the properties of the materials employed. This necessitates the precise identification and optimization of these parameters before commencing the actual manufacturing process to ensure the production of high-quality parts.

1.4 CONCLUSION

FDM has become a crucial technology in AM, providing a flexible and effective approach to developing 3D things. The technology is defined by its layer-by-layer fabrication method, enabling the production of intricate geometries and tailored products across multiple sectors, including aerospace, automotive, and biomedical industries. Advancements in materials science and precision engineering have markedly improved FDM capabilities, allowing for the fabrication of functional components that closely mimic the mechanical qualities and performance of traditionally manufactured products.

The capacity to employ a diverse array of thermoplastic materials, along with the technology's affordability and user friendliness, establishes FDM as a favored option for both prototype and final applications. Additionally, current research endeavors focused on refining process parameters, including nozzle diameter, layer thickness, and raster orientation, which persist in enhancing the quality and efficiency of FDM manufactured components. This renders FDM appropriate for industrial applications and progressively attainable for educational purposes and small-scale manufacturing.

1.5 FUTURE SCOPE

The future of FDM technology is promising, with several avenues for further exploration and development:

- *Material Innovation:* Continued research into new thermoplastics and composite materials can expand the range of applications for FDM. Developing materials with enhanced mechanical properties or specific functionalities (e.g., biocompatibility for medical applications) will broaden its utility.
- *Process Optimization:* Advances in machine learning and artificial intelligence could lead to more sophisticated algorithms for optimizing printing parameters in real time, enhancing print quality while reducing waste.
- *Integration with Other Technologies:* Combining FDM with other AM techniques or hybrid systems could leverage the strengths of multiple processes, leading to improved product characteristics and capabilities.
- *Sustainability Initiatives:* The development of biodegradable or recycled filament materials can address environmental concerns associated with plastic waste in 3D printing. Exploring sustainable practices within FDM processes will be crucial as industries move toward greener manufacturing solutions.
- *Broader Applications:* As industries seek more customized solutions, FDM's role in producing specialized components ranging from automotive parts to

intricate biomedical devices will likely expand. The integration of smart technologies into FDM systems may also pave the way for producing smart parts that incorporate sensors or electronics.

FDM has already established itself as a significant player in AM, and ongoing advancements hold the potential to further enhance its capabilities, making it an even more integral part of modern manufacturing landscapes.

REFERENCES

1. B. Lippert and R. Lachmayer, "A design method for restriction oriented lightweight design by using selective laser melting," in *21st Int. Conf. Eng. Des. ICED 2017*, vol. 5, no. DS87-5, pp. 1–10, 2017, [Online]. Available: https://scopus.com/inward/record.uri?eid=2-s2.0-85029822002&partnerID=40&md5=67817167902e1eb7ec6fd2b8b0e359ed
2. K.-M. M. Tam and C. T. Mueller, "Additive manufacturing along principal stress lines," *3D Print. Addit. Manuf.*, vol. 4, no. 2, pp. 63–81, Jun. 2017, doi: 10.1089/3dp.2017.0001.
3. M. Vaezi, H. Seitz, and S. Yang, "A review on 3D micro-additive manufacturing technologies," *Int. J. Adv. Manuf. Technol.*, vol. 67, no. 5–8, pp. 1721–1754, 2013, doi: 10.1007/s00170-012-4605-2.
4. E. Atzeni and A. Salmi, "Economics of additive manufacturing for end-usable metal parts," *Int. J. Adv. Manuf. Technol.*, vol. 62, no. 9–12, pp. 1147–1155, 2012, doi: 10.1007/s00170-011-3878-1.
5. A. Bellini, S. Gu¨c¸eri, and M. Bertoldi, "Liquefier dynamics in fused deposition," *J. Manuf. Sci. Eng.*, vol. 126, no. 2, pp. 237–246sMay 2004, doi: 10.1115/1.1688377.
6. H. Bikas, P. Stavropoulos, and G. Chryssolouris, "Additive manufacturing methods and modelling approaches: a critical review," *Int. J. Adv. Manuf. Technol.*, vol. 83, no. 1–4, pp. 389–405, Mar. 2016, doi: 10.1007/s00170-015-7576-2.
7. L. Suárez and M. Domínguez, "Sustainability and environmental impact of fused deposition modelling (FDM) technologies," *Int. J. Adv. Manuf. Technol.*, vol. 106, no. 3–4, pp. 1267–1279, Jan. 2020, doi: 10.1007/s00170-019-04676-0.
8. K. Raney, E. Lani, and D. K. Kalla, "Experimental characterization of the tensile strength of ABS parts manufactured by fused deposition modeling process," *Mater. Today Proc.*, vol. 4, no. 8, pp. 7956–7961, 2017, doi: 10.1016/j.matpr.2017.07.132.
9. M. Srivastava and S. Rathee, "Optimisation of FDM process parameters by Taguchi method for imparting customised properties to components," *Virtual Phys. Prototyp.*, vol. 13, no. 3, pp. 203–210, Jul. 2018, doi: 10.1080/17452759.2018.1440722.
10. S. Rathee, M. Srivastava, S. Maheshwari, and A. N. Siddiquee, "Effect of varying spatial orientations on build time requirements for FDM process: a case study," *Def. Technol.*, vol. 13, no. 2, pp. 92–100, Apr. 2017, doi: 10.1016/j.dt.2016.11.006.
11. S. Ahn, M. Montero, D. Odell, S. Roundy, and P. K. Wright, "Anisotropic material properties of fused deposition modeling ABS," *Rapid Prototyp. J.*, vol. 8, no. 4, pp. 248–257, Oct. 2002, doi: 10.1108/13552540210441166.
12. A. K. Sood, V. Chaturvedi, S. Datta, and S. S. Mahapatra, "Optimization of process parameters in fused deposition modeling using weighted principal component analysis," *J. Adv. Manuf. Syst.*, vol. 10, no. 02, pp. 241–259, Dec. 2011, doi: 10.1142/S0219686711002181.

13. O. A. Mohamed, S. H. Masood, and J. L. Bhowmik, "Optimization of fused deposition modeling process parameters: a review of current research and future prospects," *Adv. Manuf.*, vol. 3, no. 1, pp. 42–53, Mar. 2015, doi: 10.1007/s40436-014-0097-7.
14. L. Xinhua, L. Shengpeng, L. Zhou, Z. Xianhua, C. Xiaohu, and W. Zhongbin, "An investigation on distortion of PLA thin-plate part in the FDM process," *Int. J. Adv. Manuf. Technol.*, vol. 79, no. 5–8, pp. 1117–1126, 2015, doi: 10.1007/s00170-015-6893-9.
15. F. Rayegani and G. C. Onwubolu, "Fused deposition modelling (FDM) process parameter prediction and optimization using group method for data handling (GMDH) and differential evolution (DE)," *Int. J. Adv. Manuf. Technol.*, vol. 73, no. 1–4, pp. 509–519, Jul. 2014, doi: 10.1007/s00170-014-5835-2.
16. S. Dev and R. Srivastava, "Experimental investigation and optimization of FDM process parameters for material and mechanical strength," *Mater. Today Proc.*, vol. 26, pp. 1995–1999, 2020, doi: 10.1016/j.matpr.2020.02.435.
17. M. Fernandez-Vicente, W. Calle, S. Ferrandiz, and A. Conejero, "Effect of infill parameters on tensile mechanical behavior in desktop 3D printing," *3D Print. Addit. Manuf.*, vol. 3, no. 3, pp. 183–192, Sep. 2016, doi: 10.1089/3dp.2015.0036.
18. B. M. Tymrak, M. Kreiger, and J. M. Pearce, "Mechanical properties of components fabricated with open-source 3-D printers under realistic environmental conditions," *Mater. Des.*, vol. 58, pp. 242–246, Jun. 2014, doi: 10.1016/j.matdes.2014.02.038.
19. B. Rankouhi, S. Javadpour, F. Delfanian, and T. Letcher, "Failure analysis and mechanical characterization of 3D printed ABS with respect to layer thickness and orientation," *J. Fail. Anal. Prev.*, vol. 16, no. 3, pp. 467–481, Jun. 2016, doi: 10.1007/s11668-016-0113-2.
20. A. K. Sood, R. K. Ohdar, and S. S. Mahapatra, "Experimental investigation and empirical modelling of FDM process for compressive strength improvement," *J. Adv. Res.*, vol. 3, no. 1, pp. 81–90, Jan. 2012, doi: 10.1016/j.jare.2011.05.001.
21. M. Gurr and R. Mülhaupt, "Rapid prototyping," in *Reference Module in Materials Science and Materials Engineering*, Elsevier, 2016. doi: 10.1016/B978-0-12-803581-8.01477-6.
22. B. Podsiadły, P. Matuszewski, A. Skalski, and M. Słoma, "Carbon nanotube-based composite filaments for 3D printing of structural and conductive elements," *Appl. Sci.*, vol. 11, no. 3, p. 1272, Jan. 2021, doi: 10.3390/app11031272.
23. D. Bourell, J. P. Kruth, M. Leu, G. Levy, D. Rosen, A. M. Beese, and A. Clare,., "Materials for additive manufacturing," *CIRP Ann.*, vol. 66, no. 2, pp. 659–681, 2017, doi: 10.1016/j.cirp.2017.05.009.
24. C. Abeykoon, P. Sri-Amphorn, and A. Fernando, "Optimization of fused deposition modeling parameters for improved PLA and ABS 3D printed structures," *Int. J. Light. Mater. Manuf.*, vol. 3, no. 3, pp. 284–297, Sep. 2020, doi: 10.1016/j.ijlmm.2020.03.003.
25. P. Wang, B. Zou, S. Ding, L. Li, and C. Huang, "Effects of FDM-3D printing parameters on mechanical properties and microstructure of CF/PEEK and GF/PEEK," *Chinese J. Aeronaut.*, vol. 34, no. 9, pp. 236–246, Sep. 2021, doi: 10.1016/j.cja.2020.05.040.
26. A. Gebisa and H. Lemu, "Investigating effects of fused-deposition modeling (FDM) processing parameters on flexural properties of ULTEM 9085 using designed experiment," *Materials (Basel)*, vol. 11, no. 4, p. 500, Mar. 2018, doi: 10.3390/ma11040500.
27. B. Akhoundi, A. H. Behravesh, and A. Bagheri Saed, "An innovative design approach in three-dimensional printing of continuous fiber–reinforced thermoplastic

composites via fused deposition modeling process: in-melt simultaneous impregnation," *Proc. Inst. Mech. Eng. Part B J. Eng. Manuf.*, vol. 234, no. 1–2, pp. 243–259, Jan. 2020, doi: 10.1177/0954405419843780.
28. P. Bettini, G. Alitta, G. Sala, and L. Di Landro, "Fused deposition tchnique for continuous fiber reinforced thermoplastic," *J. Mater. Eng. Perform.*, vol. 26, no. 2, pp. 843–848, Feb. 2017, doi: 10.1007/s11665-016-2459-8.
29. G. Dong, Y. Tang, D. Li, and Y. F. Zhao, "Mechanical properties of continuous Kevlar fiber reinforced composites fabricated by fused deposition modeling process," *Procedia Manuf.*, vol. 26, pp. 774–781, 2018, doi: 10.1016/j.promfg.2018.07.090.
30. N. W. Y. Omar, N. A. Shuaib, M. H. J. A. Hadi, and A. I. Azmi, "Mechanical properties of carbon and glass fibre reinforced composites produced by additive manufacturing: a short review," *IOP Conf. Ser. Mater. Sci. Eng.*, vol. 670, no. 1, p. 012020, Nov. 2019, doi: 10.1088/1757-899X/670/1/012020.
31. J. Shang, X. Tian, M. Luo, W. Zhu, D. Li, Y. Qin, and Z. Shan, "Controllable inter-line bonding performance and fracture patterns of continuous fiber reinforced composites by sinusoidal-path 3D printing," *Compos. Sci. Technol.*, vol. 192, p. 108096, May 2020, doi: 10.1016/j.compscitech.2020.108096.
32. G. Sodeifian, S. Ghaseminejad, and A. A. Yousefi, "Preparation of polypropylene/short glass fiber composite as fused deposition modeling (FDM) filament," *Results Phys.*, vol. 12, pp. 205–222, Mar. 2019, doi: 10.1016/j.rinp.2018.11.065.
33. W. Zhang, W Zhang, C. Cotton, J. Sun, D. Heider, B. Gu, B. Sun, and T. Chou, "Interfacial bonding strength of short carbon fiber/acrylonitrile-butadiene-styrene composites fabricated by fused deposition modeling," *Compos. Part B Eng.*, vol. 137, pp. 51–59, Mar. 2018, doi: 10.1016/j.compositesb.2017.11.018.
34. Z. Weng, J. Wang, T. Senthil, and L. Wu, "Mechanical and thermal properties of ABS/montmorillonite nanocomposites for fused deposition modeling 3D printing," *Mater. Des.*, vol. 102, pp. 276–283, Jul. 2016, doi: 10.1016/j.matdes.2016.04.045.
35. L. Yang, Y. Chen, M. Wang, S. Shi, and J. Jing, "Fused deposition modeling 3D printing of novel poly(vinyl alcohol)/graphene nanocomposite with enhanced mechanical and electromagnetic interference shielding properties," *Ind. Eng. Chem. Res.*, vol. 59, no. 16, pp. 8066–8077, Apr. 2020, doi: 10.1021/acs.iecr.0c00074.
36. V. Francis and P. K. Jain, "Achieving improved dielectric, mechanical, and thermal properties of additive manufactured parts via filament modification using OMMT-based nanocomposite," *Prog. Addit. Manuf.*, vol. 2, no. 3, pp. 109–115, Sep. 2017, doi: 10.1007/s40964-017-0031-1.
37. M. R. Khosravani, "Composite materials manufacturing processes," *Appl. Mech. Mater.*, vol. 110–116, pp. 1361–1367, Oct. 2011, doi: 10.4028/www.scientific.net/AMM.110-116.1361.
38. B. Podsiadły, A. Skalski, B. Wałpuski, and M. Słoma, "Heterophase materials for fused filament fabrication of structural electronics," *J. Mater. Sci. Mater. Electron.*, vol. 30, no. 2, pp. 1236–1245, Jan. 2019, doi: 10.1007/s10854-018-0391-4.
39. J. Gonçalves, P. Lima, B. Krause, P. Pötschke,U. Lafont, J. R. Gomes, C. S. Abreu, M. C. Paiva, and J. A. Covas, "Electrically conductive polyetheretherketone nanocomposite filaments: from production to fused deposition modeling," *Polymers (Basel)*, vol. 10, no. 8, p. 925, Aug. 2018, doi: 10.3390/polym10080925.
40. P. Lamberti, G. Spinelli, P. P. Kuzhir, L. Guadagno, C. Naddeo, V. Romano, R. Kotsilkova, P. Angelova, and V. Georgiev, "Evaluation of thermal and electrical conductivity of carbon-based PLA nanocomposites for 3D printing," 2018, p. 020158. doi: 10.1063/1.5046020.

41. H. I. Medellin-Castillo and J. Zaragoza-Siqueiros, "Design and manufacturing strategies for fused deposition modelling in additive manufacturing: a review," *Chinese J. Mech. Eng. (English Ed.)*, vol. 32, no. 1, 2019, doi: 10.1186/s10033-019-0368-0.
42. M. Lalegani Dezaki and M. K. A. Mohd Ariffin, "The effects of combined infill patterns on mechanical properties in FDM process," *Polymers (Basel)*, vol. 12, no. 12, p. 2792, Nov. 2020, doi: 10.3390/polym12122792.
43. A. Dey and N. Yodo, "A systematic survey of FDM process parameter optimization and their influence on part characteristics," *J. Manuf. Mater. Process.*, vol. 3, no. 3, p. 64, Jul. 2019, doi: 10.3390/jmmp3030064.
44. O. A. Mohamed, S. H. Masood, and J. L. Bhowmik, "Optimization of fused deposition modeling process parameters for dimensional accuracy using I-optimality criterion," *Measurement*, vol. 81, pp. 174–196, Mar. 2016, doi: 10.1016/j.measurement.2015.12.011.
45. L. Di Angelo, P. Di Stefano, and A. Marzola, "Surface quality prediction in FDM additive manufacturing," *Int. J. Adv. Manuf. Technol.*, vol. 93, no. 9–12, pp. 3655–3662, Dec. 2017, doi: 10.1007/s00170-017-0763-6.
46. B. N. Panda, K. Shankhwar, A. Garg, and Z. Jian, "Performance evaluation of warping characteristic of fused deposition modelling process," *Int. J. Adv. Manuf. Technol.*, vol. 88, no. 5–8, pp. 1799–1811, Feb. 2017, doi: 10.1007/s00170-016-8914-8.
47. M. Domingo-Espin, J. M. Puigoriol-Forcada, A.-A. Garcia-Granada, J. Llumà, S. Borros, and G. Reyes, "Mechanical property characterization and simulation of fused deposition modeling Polycarbonate parts," *Mater. Des.*, vol. 83, pp. 670–677, Oct. 2015, doi: 10.1016/j.matdes.2015.06.074.
48. S. Masood and W. Song, "Development of new metal/polymer materials for rapid tooling using Fused deposition modelling," *Mater. Des.*, vol. 25, no. 7, pp. 587–594, Oct. 2004, doi: 10.1016/j.matdes.2004.02.009.
49. R. C. Pennington, N. L. Hoekstra, and J. L. Newcomer, "Significant factors in the dimensional accuracy of fused deposition modelling," *Proc. Inst. Mech. Eng. Part E J. Process Mech. Eng.*, vol. 219, no. 1, pp. 89–92, Feb. 2005, doi: 10.1243/095440805X6964.
50. B. H. Lee, J. Abdullah, and Z. A. Khan, "Optimization of rapid prototyping parameters for production of flexible ABS object," *J. Mater. Process. Technol.*, vol. 169, no. 1, pp. 54–61, Oct. 2005, doi: 10.1016/j.jmatprotec.2005.02.259.
51. T.-M. Wang, J.-T. Xi, and Y. Jin, "A model research for prototype warp deformation in the FDM process," *Int. J. Adv. Manuf. Technol.*, vol. 33, no. 11–12, pp. 1087–1096, Aug. 2007, doi: 10.1007/s00170-006-0556-9.
52. Y. Zhang and K. Chou, "A parametric study of part distortions in fused deposition modelling using three-dimensional finite element analysis," *Proc. Inst. Mech. Eng. Part B J. Eng. Manuf.*, vol. 222, no. 8, pp. 959–968, Aug. 2008, doi: 10.1243/09544054JEM990.
53. M. Saravanan, "Investigation of thermal stress development by layer manufacturing on fused deposition modelling parts," *Int. J. Enterp. Netw. Manag.*, vol. 4, no. 2, p. 183, 2010, doi: 10.1504/IJENM.2010.037934.
54. A. Rosochowski and A. Matuszak, "Rapid tooling: the state of the art," *J. Mater. Process. Technol.*, vol. 106, no. 1–3, pp. 191–198, Oct. 2000, doi: 10.1016/S0924-0136(00)00613-0.
55. C. K. Chua, C. Feng, C. W. Lee, and G. Q. Ang, "Rapid investment casting: direct and indirect approaches via model maker II," *Int. J. Adv. Manuf. Technol.*, vol. 25, no. 1–2, pp. 26–32, Jan. 2005, doi: 10.1007/s00170-004-1865-5.

56. S. Pattnaik, D. B. Karunakar, and P. K. Jha, "Developments in investment casting process—a review," *J. Mater. Process. Technol.*, vol. 212, no. 11, pp. 2332–2348, Nov. 2012, doi: 10.1016/j.jmatprotec.2012.06.003.
57. P. M. Dickens, R. Stangroom, M. Greul, B. Holmer, K. K. B. Hon, R. Hovtun, R. Neumann, S. Noeken, and D. Wimpenny, "Conversion of RP models to investment castings," *Rapid Prototyp. J.*, vol. 1, no. 4, pp. 4–11, Dec. 1995, doi: 10.1108/13552549510104401.
58. A. V. Agapovichev, A. V. Balaykin, V. G. Smelov, and A. V. Agapovichev, "Application of additive technologies in the production of aircraft engine parts," *Mod. Appl. Sci.*, vol. 9, no. 4, Jan. 2015, doi: 10.5539/mas.v9n4p151.
59. C. K. Chua, K. H. Hong, and S. L. Ho, "Rapid tooling technology. Part 1. A comparative study," *Int. J. Adv. Manuf. Technol.*, vol. 15, no. 8, pp. 604–608, Jul. 1999, doi: 10.1007/s001700050108.
60. A. Bernard, J. Delplace, N. Perry, and S. Gabriel, "Integration of CAD and rapid manufacturing for sand casting optimisation," *Rapid Prototyp. J.*, vol. 9, no. 5, pp. 327–333, Dec. 2003, doi: 10.1108/13552540310502220.
61. D. Sahebrao Ingole, A. Madhusudan Kuthe, S. B. Thakare, and A. S. Talankar, "Rapid prototyping—a technology transfer approach for development of rapid tooling," *Rapid Prototyp. J.*, vol. 15, no. 4, pp. 280–290, Jul. 2009, doi: 10.1108/13552540910979794.
62. L. M. Galantucci, F. Lavecchia, and G. Percoco, "Experimental study aiming to enhance the surface finish of fused deposition modeled parts," *CIRP Ann.*, vol. 58, no. 1, pp. 189–192, 2009, doi: 10.1016/j.cirp.2009.03.071.
63. R. Singh, "Three dimensional printing for casting applications: a state of art review and future perspectives," *Adv. Mater. Res.*, vol. 83–86, pp. 342–349, Dec. 2009, doi: 10.4028/www.scientific.net/AMR.83-86.342.
64. C. Gouldsen and P. Blake. *Investment Casting Using FDM/ABS Rapid Prototype Patterns*, Rapid ToolworX and Stratasys Inc., 1998.
65. A. D. Valino, J. R. C. Dizon, A. H. Espera, Q. Chen, J. Messman, and R. C. Advincula, "Advances in 3D printing of thermoplastic polymer composites and nanocomposites," *Prog. Polym. Sci.*, vol. 98, p. 101162, Nov. 2019, doi: 10.1016/j.progpolymsci.2019.101162.
66. S. Ahn, C. S. Lee, and W. Jeong, "Development of translucent FDM parts by post-processing," *Rapid Prototyp. J.*, vol. 10, no. 4, pp. 218–224, Sep. 2004, doi: 10.1108/13552540410551333.
67. M. Dawoud, I. Taha, and S. J. Ebeid, "Mechanical behaviour of ABS: an experimental study using FDM and injection moulding techniques," *J. Manuf. Process.*, vol. 21, pp. 39–45, Jan. 2016, doi: 10.1016/j.jmapro.2015.11.002.
68. A. Equbal, A. K. Sood, V. Toppo, R. K. Ohdar, and S. S. Mahapatra, "Prediction and analysis of sliding wear performance of fused deposition modelling-processed ABS plastic parts," *Proc. Inst. Mech. Eng. Part J J. Eng. Tribol.*, vol. 224, no. 12, pp. 1261–1271, Dec. 2010, doi: 10.1243/13506501JET835.
69. A. Equbal and A. K. Sood, "Investigations on metallization in FDM build ABS part using electroless deposition method," *J. Manuf. Process.*, vol. 19, pp. 22–31, Aug. 2015, doi: 10.1016/j.jmapro.2015.03.002.
70. H. Jami, S. H. Masood, and W. Q. Song, "Dynamic response of FDM made ABS parts in different part orientations," *Adv. Mater. Res.*, vol. 748, pp. 291–294, Aug. 2013, doi: 10.4028/www.scientific.net/AMR.748.291.

71. A. Kantaros and D. Karalekas, "Fiber Bragg grating based investigation of residual strains in ABS parts fabricated by fused deposition modeling process," *Mater. Des.*, vol. 50, pp. 44–50, Sep. 2013, doi: 10.1016/j.matdes.2013.02.067.
72. R. Singh and G. Singh, "Investigations for statistically controlled investment casting solution of FDM-based ABS replicas," *Rapid Prototyp. J.*, vol. 20, no. 3, pp. 215–220, Apr. 2014, doi: 10.1108/RPJ-03-2013-0036.
73. D. Drummer, S. Cifuentes-Cuéllar, and D. Rietzel, "Suitability of PLA/TCP for fused deposition modeling," *Rapid Prototyp. J.*, vol. 18, no. 6, pp. 500–507, Sep. 2012, doi: 10.1108/13552541211272045.
74. M. F. Afrose, S. H. Masood, P. Iovenitti, M. Nikzad, and I. Sbarski, "Effects of part build orientations on fatigue behaviour of FDM-processed PLA material," *Prog. Addit. Manuf.*, vol. 1, no. 1–2, pp. 21–28, Jun. 2016, doi: 10.1007/s40964-015-0002-3.
75. Y. Jia, H. He, X. Peng, S. Meng, J. Chen, and Y. Geng, "Preparation of a new filament based on polyamide-6 for three-dimensional printing," *Polym. Eng. Sci.*, vol. 57, no. 12, pp. 1322–1328, Dec. 2017, doi: 10.1002/pen.24515.
76. P. Cruz, E. D. Shoemake, P. Adam, and J. Leachman, "Tensile strengths of polyamide based 3D printed polymers in liquid nitrogen," *IOP Conf. Ser. Mater. Sci. Eng.*, vol. 102, p. 012020, Dec. 2015, doi: 10.1088/1757-899X/102/1/012020.
77. M. Domingo-Espin, S. Borros, N. Agullo, A.-A. Garcia-Granada, and G. Reyes, "Influence of building parameters on the dynamic mechanical properties of polycarbonate fused deposition modeling parts," *3D Print. Addit. Manuf.*, vol. 1, no. 2, pp. 70–77, Jun. 2014, doi: 10.1089/3dp.2013.0007.
78. I. Gajdoš, Ľ. Kaščák, E. Spišák, and J. Slota, "Flexural properties of FDM prototypes made with honeycomb and sparse structure," *Key Eng. Mater.*, vol. 635, pp. 169–173, Dec. 2014, doi: 10.4028/www.scientific.net/KEM.635.169.
79. N. Hill and M. Haghi, "Deposition direction-dependent failure criteria for fused deposition modeling polycarbonate," *Rapid Prototyp. J.*, vol. 20, no. 3, pp. 221–227, Apr. 2014, doi: 10.1108/RPJ-04-2013-0039.
80. W. C. Smith and R. W. Dean, "Structural characteristics of fused deposition modeling polycarbonate material," *Polym. Test.*, vol. 32, no. 8, pp. 1306–1312, Dec. 2013, doi: 10.1016/j.polymertesting.2013.07.014.
81. O. S. Carneiro, A. F. Silva, and R. Gomes, "Fused deposition modeling with polypropylene," *Mater. Des.*, vol. 83, pp. 768–776, Oct. 2015, doi: 10.1016/j.matdes.2015.06.053.
82. G. Scotti, S. M. E. Nilsson, M. Haapala, P. P. G. B. Gennäs, J. Y. Kauhaluomaa, and T. Kotiaho, "A miniaturised 3D printed polypropylene reactor for online reaction analysis by mass spectrometry," *React. Chem. Eng.*, vol. 2, no. 3, pp. 299–303, 2017, doi: 10.1039/C7RE00015D.
83. P. Bedi, R. Singh, and I. Ahuja, "Effect of SiC/Al_2O_3 particle size reinforcement in recycled LDPE matrix on mechanical properties of FDM feed stock filament," *Virtual Phys. Prototyp.*, vol. 13, no. 4, pp. 246–254, Oct. 2018, doi: 10.1080/17452759.2018.1496605.
84. R. W. Gray, D. G. Baird, and J. Helge Bøhn, "Effects of processing conditions on short TLCP fiber reinforced FDM parts," *Rapid Prototyp. J.*, vol. 4, no. 1, pp. 14–25, Mar. 1998, doi: 10.1108/13552549810197514.
85. S. Kalita, J. Finley, S. Bose, H. Hosick, and A. Bandyopadhyay, "Development of porous polymer-ceramic composites as bone grafts," *MRS Proc.*, vol. 726, p. Q5.8, Feb. 2002, doi: 10.1557/PROC-726-Q5.8.
86. R. Kumar, R. Singh, and I. P. S. Ahuja, "Investigations of mechanical, thermal and morphological properties of FDM fabricated parts for friction welding

applications," *Measurement*, vol. 120, pp. 11–20, May 2018, doi: 10.1016/j.measurement.2018.02.006.

87. S. H. Masood and W. Q. Song, "Thermal characteristics of a new metal/polymer material for FDM rapid prototyping process," *Assem. Autom.*, vol. 25, no. 4, pp. 309–315, Dec. 2005, doi: 10.1108/01445150510626451.
88. S. Berretta, R. Davies, Y. T. Shyng, Y. Wang, and O. Ghita, "Fused deposition modelling of high temperature polymers: exploring CNT PEEK composites," *Polym. Test.*, vol. 63, pp. 251–262, Oct. 2017, doi: 10.1016/j.polymertesting.2017.08.024.
89. Q. Chen, J. D. Mangadlao, J. Wallat, A. De Leon, J. K. Pokorski, and R. C. Advincula, "3D printing biocompatible polyurethane/poly(lactic acid)/graphene oxide nanocomposites: anisotropic properties," ACS Appl. Mater. Interfaces, vol. 9, no. 4, pp. 4015–4023, Feb. 2017, doi: 10.1021/acsami.6b11793.
90. K. Chizari, M. A. Daoud, A. R. Ravindran, and D. Therriault, "3D printing of highly conductive nanocomposites for the functional optimization of liquid sensors," *Small*, vol. 12, no. 44, pp. 6076–6082, Nov. 2016, doi: 10.1002/smll.201601695.
91. J. F. Christ, N. Aliheidari, A. Ameli, and P. Pötschke, "3D printed highly elastic strain sensors of multiwalled carbon nanotube/thermoplastic polyurethane nanocomposites," *Mater. Des.*, vol. 131, pp. 394–401, Oct. 2017, doi: 10.1016/j.matdes.2017.06.011.
92. B. Coppola, N. Cappetti, L. Di Maio, P. Scarfato, and L. Incarnato, "Layered silicate reinforced polylactic acid filaments for 3D printing of polymer nanocomposites," in *2017 IEEE 3rd International Forum on Research and Technologies for Society and Industry (RTSI)*, IEEE, Sep. 2017, pp. 1–4. doi: 10.1109/RTSI.2017.8065892.
93. S. Dul, L. Fambri, and A. Pegoretti, "Fused deposition modelling with ABS–graphene nanocomposites," *Compos. Part A Appl. Sci. Manuf.*, vol. 85, pp. 181–191, Jun. 2016, doi: 10.1016/j.compositesa.2016.03.013.
94. K. Gnanasekaran, T. Heijmans, S. van Bennekom, H. Woldhuis, S. Wijnia, G. de With, and H. Friedrich, "3D printing of CNT- and graphene-based conductive polymer nanocomposites by fused deposition modeling," *Appl. Mater. Today*, vol. 9, pp. 21–28, Dec. 2017, doi: 10.1016/j.apmt.2017.04.003.
95. H. Kim, T. Fernando, M. Li, Y. Lin, and T.-L. B. Tseng, "Fabrication and characterization of 3D printed $BaTiO_3$/PVDF nanocomposites," *J. Compos. Mater.*, vol. 52, no. 2, pp. 197–206, Jan. 2018, doi: 10.1177/0021998317704709.
96. H. Kim, J. Johnson, L. A. Chavez, C. A. Garcia Rosales, T.-L. B. Tseng, and Y. Lin, "Enhanced dielectric properties of three phase dielectric MWCNTs/$BaTiO_3$/PVDF nanocomposites for energy storage using fused deposition modeling 3D printing," *Ceram. Int.*, vol. 44, no. 8, pp. 9037–9044, Jun. 2018, doi: 10.1016/j.ceramint.2018.02.107.
97. M. Anoop , P. Senthil, and T. Sathies, "A review on advancements in applications of fused deposition modelling process," *Rapid Prototyp. J.*, vol. 26, no. 4, pp. 669–687, Jan. 2020, doi: 10.1108/RPJ-08-2018-0199.
98. Z. C. Kennedy, J. F. Christ, K. A. Evans, B. W. Arey, L. E. Sweet, M. G. Warner, R. L. Eriksonb, and C. A. Barrett, "3D-printed poly(vinylidene fluoride)/carbon nanotube composites as a tunable, low-cost chemical vapour sensing platform," *Nanoscale*, vol. 9, no. 17, pp. 5458–5466, 2017, doi: 10.1039/C7NR00617A.
99. M. Sajid, J. Z. Gul, S. W. Kim, H. B. Kim, K. H. Na, and K. H. Choi, "Development of 3D-printed embedded temperature sensor for both terrestrial and aquatic environmental monitoring robots," *3D Print. Addit. Manuf.*, vol. 5, no. 2, pp. 160–169, Jun. 2018, doi: 10.1089/3dp.2017.0092.

100. S. L. Marasso, M. Cocuzza, V. Bertana, F. Perrucci, A. Tommasi, S. Ferrero, L. Scaltrito, and C. F. Pirri, "PLA conductive filament for 3D printed smart sensing applications," *Rapid Prototyp. J.*, vol. 24, no. 4, pp. 739–743, May 2018, doi: 10.1108/RPJ-09-2016-0150.
101. R. Singh, H. Singh, I. Farina, F. Colangelo, and F. Fraternali, "On the additive manufacturing of an energy storage device from recycled material," *Compos. Part B Eng.*, vol. 156, pp. 259–265, Jan. 2019, doi: 10.1016/j.compositesb.2018.08.080.
102. N. Jayanth and P. Senthil, "Application of 3D printed ABS based conductive carbon black composite sensor in void fraction measurement," *Compos. Part B Eng.*, vol. 159, pp. 224–230, Feb. 2019, doi: 10.1016/j.compositesb.2018.09.097.
103. D. P. Schmitz, L.G. Ecco, S. Dul, E. C. L. Pereira, B. G. Soares, G. M. O. Barra, and A. Pegoretti, "Electromagnetic interference shielding effectiveness of ABS carbon-based composites manufactured via fused deposition modelling," *Mater. Today Commun.*, vol. 15, pp. 70–80, Jun. 2018, doi: 10.1016/j.mtcomm.2018.02.034.
104. K. Prashantha and F. Roger, "Multifunctional properties of 3D printed poly(lactic acid)/graphene nanocomposites by fused deposition modeling," *J. Macromol. Sci. Part A*, vol. 54, no. 1, pp. 24–29, Jan. 2017, doi: 10.1080/10601325.2017.1250311.
105. L. K. Prasad and H. Smyth, "3D Printing technologies for drug delivery: a review," *Drug Dev. Ind. Pharm.*, vol. 42, no. 7, pp. 1019–1031, Jul. 2016, doi: 10.3109/03639045.2015.1120743.
106. M. R. P. Araújo, L. L. Sa-Barreto, T. Gratieri, G. M. Gelfuso, and M. Cunha-Filho, "The digital pharmacies era: how 3D printing technology using fused deposition modeling can become a reality," *Pharmaceutics*, vol. 11, no. 3, p. 128, Mar. 2019, doi: 10.3390/pharmaceutics11030128.
107. K. Rycerz, K. A. Stepien, M. Czapiewska, B T. Arafat, R. Habashy, A. Isreb , M. Peak, and M. A. Alhnan, "Embedded 3D printing of novel bespoke soft dosage form concept for pediatrics," *Pharmaceutics*, vol. 11, no. 12, p. 630, Nov. 2019, doi: 10.3390/pharmaceutics11120630.
108. A. Goyanes, M. Kobayashi, R. Martínez-Pacheco, S. Gaisford, and A. W. Basit, "Fused-filament 3D printing of drug products: microstructure analysis and drug release characteristics of PVA-based caplets," *Int. J. Pharm.*, vol. 514, no. 1, pp. 290–295, Nov. 2016, doi: 10.1016/j.ijpharm.2016.06.021.
109. A. Goyanes, A. B. M. Buanz, G. B. Hatton, S. Gaisford, and A. W. Basit, "3D printing of modified-release aminosalicylate (4-ASA and 5-ASA) tablets," *Eur. J. Pharm. Biopharm.*, vol. 89, pp. 157–162, Jan. 2015, doi: 10.1016/j.ejpb.2014.12.003.
110. X. Chai, H. Chai, X. Wang, J. Yang, J. Li, Y. Zhao, W. Cai, T. Tao, X. Xiang, "Fused deposition modeling (FDM) 3D printed tablets for intragastric floating delivery of domperidone. *Sci. Rep.*, vol. 7, no. 1, p. 2829, Jun 6, 2017, doi: 10.1038/s41598-017-03097-x, PMID: 28588251, PMCID: PMC5460192.
111. S. Lamichhane, S. Bashyal, T. Keum, G. Noh, S. J. Eun, R. Bastola, C. Jaewoong, S. D. Hwan, and L. Sangkil, "Complex formulations, simple techniques: can 3D printing technology be the Midas touch in pharmaceutical industry?," *Asian J. Pharm. Sci.*, vol. 14, no. 5, pp. 465–479, Sep. 2019, doi: 10.1016/j.ajps.2018.11.008.
112. C.-C. Kuo and M.-R. Li, "A cost-effective method for rapid manufacturing sheet metal forming dies," *Int. J. Adv. Manuf. Technol.*, vol. 85, no. 9–12, pp. 2651–2656, Aug. 2016, doi: 10.1007/s00170-015-8139-2.
113. S. Feng, A. M. Kamat, and Y. Pei, "Design and fabrication of conformal cooling channels in molds: review and progress updates," *Int. J. Heat Mass Transf.*, vol. 171, p. 121082, Jun. 2021, doi: 10.1016/j.ijheatmasstransfer.2021.121082.

114. I. Durgun, "Sheet metal forming using FDM rapid prototype tool," *Rapid Prototyp. J.*, vol. 21, no. 4, pp. 412–422, Jun. 2015, doi: 10.1108/RPJ-01-2014-0003.
115. A. Equbal, M. Israr Equbal, I. A. Badruddin, and A. Algahtani, "A critical insight into the use of FDM for production of EDM electrode," *Alexandria Eng. J.*, vol. 61, no. 5, pp. 4057–4066, May 2022, doi: 10.1016/j.aej.2021.09.033.
116. C. M. Cheah, L. H. Tan, C. Feng, C. W. Lee, and C. K. Chua, "Rapid investment casting: direct and indirect approaches via fused deposition modelling," *Int. J. Adv. Manuf. Technol.*, vol. 23, no. 1–2, pp. 93–101, Jan. 2004, doi: 10.1007/s00170-003-1694-y.
117. M. S. Thompson, "Current status and future roles of additives in 3D printing—a perspective," *J. Vinyl Addit. Technol.*, vol. 28, no. 1, pp. 3–16, Feb. 2022, doi: 10.1002/vnl.21887.
118. H. Klippstein, A. Diaz De Cerio Sanchez, H. Hassanin, Y. Zweiri, and L. Seneviratne, "Fused deposition modeling for unmanned aerial vehicles (UAVs): a review," *Adv. Eng. Mater.*, vol. 20, no. 2, Feb. 2018, doi: 10.1002/adem.201700552.
119. J. M. Gardner, G. Sauti, J. W. Kim, R. J. Cano, R. A. Wincheski, C. J. Stelter, B. W. Grimsley, D. C. Working, and E. J. Siochi, "3-D printing of multifunctional carbon nanotube yarn reinforced components," *Addit. Manuf.*, vol. 12, pp. 38–44, Oct. 2016, doi: 10.1016/j.addma.2016.06.008.
120. R. V. Pazhamannil, V. N. Jishnu Namboodiri, P. Govindan, and A. Edacherian, "Property enhancement approaches of fused filament fabrication technology: a review," *Polym. Eng. Sci.*, vol. 62, no. 5, pp. 1356–1376, May 2022, doi: 10.1002/pen.25948.
121. G. W. Bishop, J. E. Satterwhite, S. Bhakta, K. Kadimisetty, K. M. Gillette, E. Chen, and J. F. Rusling, "3D-printed fluidic devices for nanoparticle preparation and flow-injection amperometry using integrated Prussian blue nanoparticle-modified electrodes," *Anal. Chem.*, vol. 87, no. 10, pp. 5437–5443, May 2015, doi: 10.1021/acs.analchem.5b00903.
122. J. E. Sosa-Hernández, S. Hernández, J. Eduardo, A. M. V. Rodríguez, K. D. Romero-Castillo, M. A. Aguilar-Aguila-Isaías, I. E. García-Reyes, A. Hernández-Antonio, I. Ahmed, A. Sharma, R. Parra-Saldívar, and H. M. N. Iqbal., "Organs-on-a-chip module: a review from the development and applications perspective," *Micromachines*, vol. 9, no. 10, p. 536, Oct. 2018, doi: 10.3390/mi9100536.
123. L. Krejcova, L. Nejdl, M. A. Rodrigo, M. Zurek, M. Matousek, D. Hynek, O. Zitka, P. Kopel, V. Adam, and R. Kizek, "3D printed chip for electrochemical detection of influenza virus labeled with CdS quantum dots," *Biosens. Bioelectron.*, vol. 54, pp. 421–427, Apr. 2014, doi: 10.1016/j.bios.2013.10.031.
124. R. Ilardo and C. B. Williams, "Design and manufacture of a formula SAE intake system using fused deposition modeling and fiber-reinforced composite materials," *Rapid Prototyp. J.*, vol. 16, no. 3, pp. 174–179, Apr. 2010, doi: 10.1108/13552541011034834.
125. E. Asadollahi-Yazdi, J. Gardan, and P. Lafon, "Toward integrated design of additive manufacturing through a process development model and multi-objective optimization," *Int. J. Adv. Manuf. Technol.*, vol. 96, no. 9–12, pp. 4145–4164, Jun. 2018, doi: 10.1007/s00170-018-1880-6.
126. C. X. F. Lam, X. M. Mo, S. H. Teoh, and D. W. Hutmacher, "Scaffold development using 3D printing with a starch-based polymer," *Mater. Sci. Eng. C*, vol. 20, no. 1–2, pp. 49–56, May 2002, doi: 10.1016/S0928-4931(02)00012-7.
127. N. Guo and M. C. Leu, "Additive manufacturing: technology, applications and research needs," *Front. Mech. Eng.*, vol. 8, no. 3, pp. 215–243, Sep. 2013, doi: 10.1007/s11465-013-0248-8.

128. N. Boyard, O. Christmann, M. Rivette, O. Kerbrat, and S. Richir, "Support optimization for additive manufacturing: application to FDM," *Rapid Prototyp. J.*, vol. 24, no. 1, pp. 69–79, Jan. 2018, doi: 10.1108/RPJ-04-2016-0055.
129. F. P. W. Melchels, M. A. N. Domingos, T. J. Klein, J. Malda, P. J. Bartolo, and D. W. Hutmacher, "Additive manufacturing of tissues and organs," *Prog. Polym. Sci.*, vol. 37, no. 8, pp. 1079–1104, Aug. 2012, doi: 10.1016/j.progpolymsci.2011.11.007.
130. D. Espalin, K. Arcaute, D. Rodriguez, F. Medina, M. Posner, and R. Wicker, "Fused deposition modeling of patient-specific polymethylmethacrylate implants," *Rapid Prototyp. J.*, vol. 16, no. 3, pp. 164–173, Apr. 2010, doi: 10.1108/13552541011034825.
131. R. van Noort, "The future of dental devices is digital," *Dent. Mater.*, vol. 28, no. 1, pp. 3–12, Jan. 2012, doi: 10.1016/j.dental.2011.10.014.
132. Brooks, B. J., Arif, K.M., Dirven, S. et al. Robot-assisted 3D printing of biopolymer thin shells. *Int. J. Adv. Manuf. Technol.* vol. 89, pp. 957–968 (2017). https://doi.org/10.1007/s00170-016-9134-y
133. Korpela J, Kokkari A, Korhonen H, Malin M, Närhi T, Seppälä J. 2013. Biodegradable and bioactive porous scaffold structures prepared using fused deposition modeling. *J. Biomed. Mater. Res. Part B*, vol. 101B, pp. 610–619 (2013).
134. M. Juneja, N. Thakur, D. Kumar, A. Gupta, B. Bajwa, and P. Jindal, "Accuracy in dental surgical guide fabrication using different 3-D printing techniques," *Addit. Manuf.*, vol. 22, pp. 243–255, Aug. 2018, doi: 10.1016/j.addma.2018.05.012.
135. Soriano-Heras, E., Blaya-Haro, F., Molino, C. *et al.* Rapid prototyping prosthetic hand acting by a low-cost shape-memory-alloy actuator. *J Artif Organs* **21**, 238–246 (2018). https://doi.org/10.1007/s10047-017-1014-1
136. Okwuosa TC, Pereira BC, Arafat B, Cieszynska M, Isreb A, Alhnan MA. Fabricating a shell-core delayed release tablet using dual FDM 3D printing for Patient-centred therapy. *Pharm. Res.*, vol. 34, no. 2, pp. 427–437, Feb 2017. doi: 10.1007/s11095-016-2073-3. Epub 2016 Dec 9. PMID: 27943014.
137. N. Xu, Y. Xiaojian, W. Daixu, Z. Jian, C. Yuyun, X. Guohua, and H. Dannong, "3D artificial bones for bone repair prepared by computed tomography-guided fused deposition modeling for bone repair," *ACS Appl. Mater. Interfaces*, vol. 6, no. 17, pp. 14952–14963, Sep. 2014, doi: 10.1021/am502716t.
138. J. Podroužek, M. Marcon, K. Ninčević, and R. Wan-Wendner, "Bio-inspired 3D infill patterns for additive manufacturing and structural applications," *Materials (Basel).*, vol. 12, no. 3, p. 499, Feb. 2019, doi: 10.3390/ma12030499.
139. V. K. Tiwary, P. Arunkumar, and V. R. Malik, "An overview on joining/welding as post-processing technique to circumvent the build volume limitation of an FDM-3D printer," *Rapid Prototyp. J.*, vol. 27, no. 4, pp. 808–821, Jun. 2021, doi: 10.1108/RPJ-10-2020-0265.

2 Design for Manufacturability with Fused Deposition Modeling

Albert E. Patterson and Yong Hoon Lee

2.1 INTRODUCTION

Fused deposition modeling (FDM), sometimes also called fused filament fabrication or simply material extrusion additive manufacturing (AM), is a key AM process. It is typically used to process glassy thermoplastic polymers and polymer matrix composites. In recent years, it has emerged as a useful end-user manufacturing process, particularly for small-batch manufacturing. It is a scanning-type AM process that selectively extrudes a bead or "road" of molten thermoplastic or thermoplastic composite material to scribe out the geometry of each layer in the part. Polymerization between layers has been the subject of much research but is thought to be the result of a fast polymer healing reaction [1–5]. The position of the extruder follows a calculated toolpath, similar to a computer numerical control (CNC) milling machine [3, 4, 6–9]. The basic system components are demonstrated in Figure 2.1. There are several common configurations of the various hardware components, based primarily on the motion of the various axes as demonstrated in Figure 2.2. In addition, the regular three-axis design categories, versions built around delta robots and robotic arms also exist. It is assumed that the essential design principles of parts to be printed will be the same regardless of the printer configuration, as the essential consideration is the relative position of the extruder and the surface (build plate or part) onto which the material is being extruded.

The final bulk material processed via FDM contains numerous small defects and inclusions, making the parts anisotropic [10]. However, these defects are usually patterned and laid out in a specific way that is designable. This allows FDM to be used to produce dense designed or graded materials by optimizing the distribution of these defects [11–14]. Therefore, the effective design of FDM-processed parts requires careful and deep consideration of the printing and material layout parameters. The most common ones discussed in the literature are packing density, element layout, layer thickness, nozzle size, build plate orientation, print speed,

DOI: 10.1201/9781003686941-2

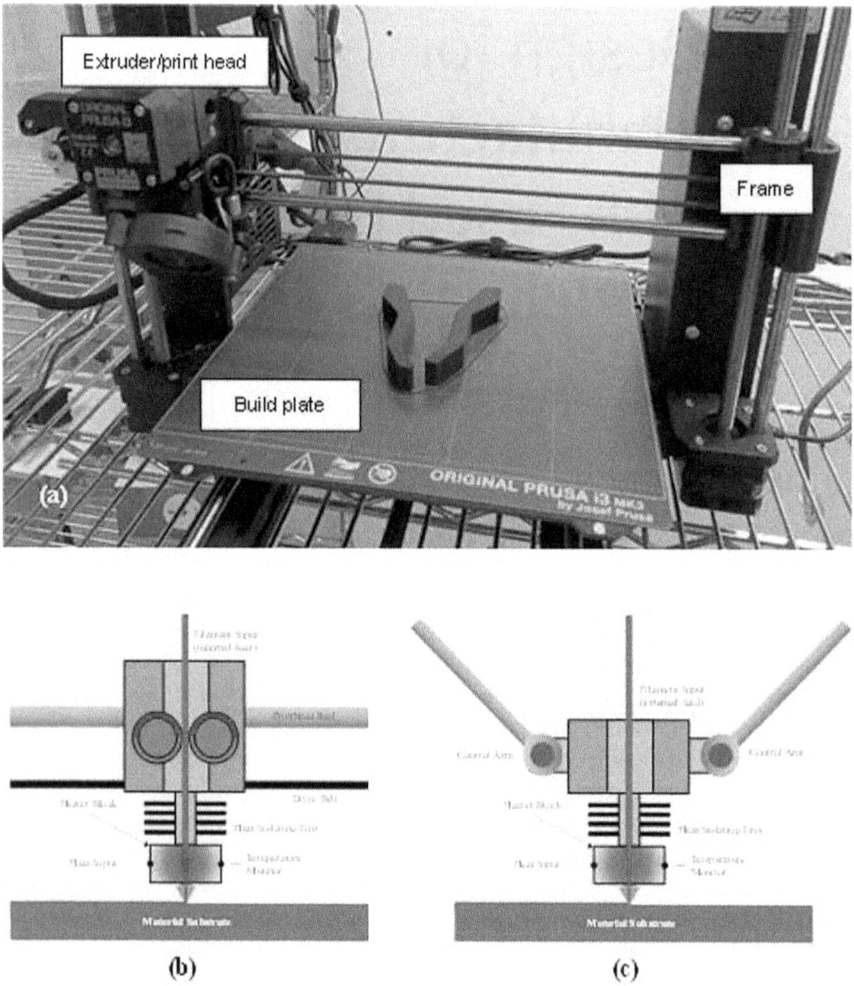

FIGURE 2.1 Usual hardware configuration for the FDM/FFF process with typical print head details for (a) gantry-based and (b) delta- or robotic arm-based machines.

and processing temperature. Some examples of constraints for these are shown in Table 2.1. Material selection is not considered as a major parameter here, as the best parameters can differ significantly depending on material choice. The use of semi-crystalline or amorphous thermoplastics also affects parameter choice. Semi-crystalline materials such as polylactide (PLA) and polyamide (nylon) can provide interesting advantages as engineering materials and print nicely using FDM. Amorphous thermoplastics such as acrylonitrile butadiene styrene (ABS) and polycarbonate (PC) can offer more control over the material structure due to more predictable shrinkage of the elements but are more difficult to print and more subject to residual stresses.

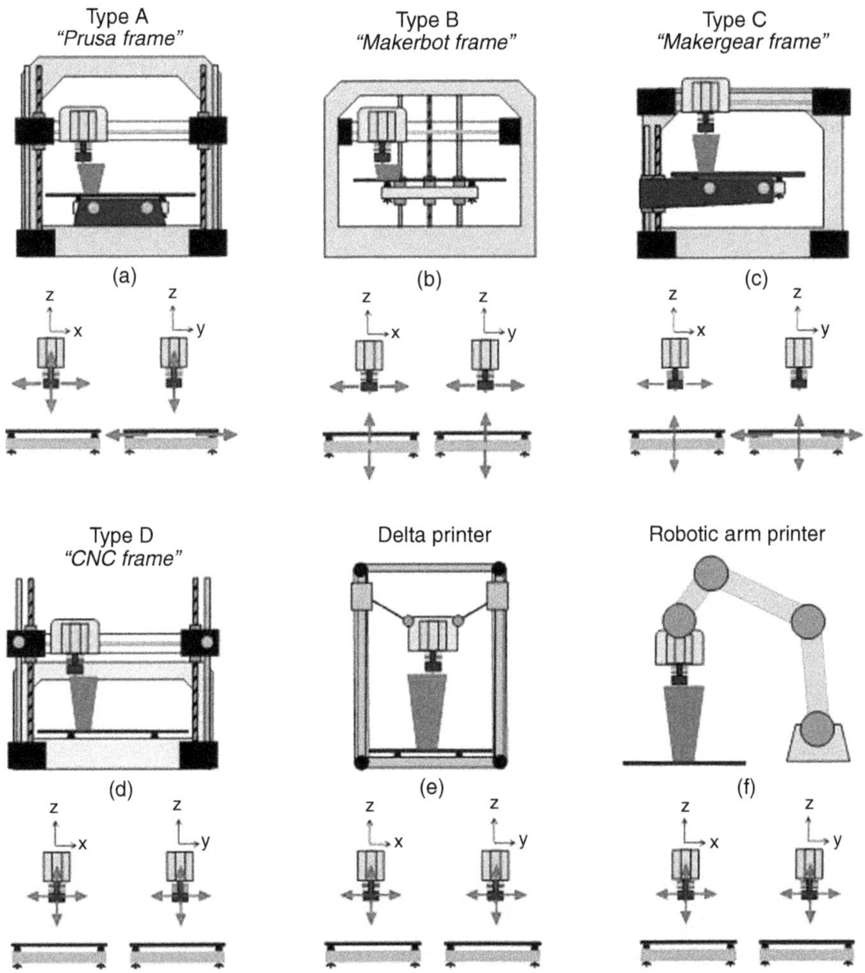

FIGURE 2.2 Common FDM hardware configurations.

Source: Adapted from ref. [9], reproduced under a CC-BY license.

TABLE 2.1
Example Constraints for Common Parameters

Parameter	Example Simple Constraints
Nominal density	Maximum actual density (< 100%) or the minimum stable density
Raster angle	Allowable extremes for the raster angle in cases where a raster layout is used
Extrusion speed	Allowable extremes for the print speed. This will be based on polymer rheology, hardware stability/vibration constraints, and limits on stepper motor performance

(continued)

TABLE 2.1 (Continued)
Example Constraints for Common Parameters

Parameter	Example Simple Constraints
Extrusion temperature	Extrema of temperatures for desired polymer rheology, operating temperatures for proper hardware function
Surrounding environment	Enclosed or open environmental, as well as temperature and humidity during processing
Layer thickness	Layer thickness for desired polymer rheology, minimum and maximum thickness relative to print time, required thickness for surface quality, surface integrity considerations
Build orientation	Build orientation relative to the number of degrees of freedom (DoF) for the fabrication hardware (most printers operate with three DoF)
Extruder nozzle size	Maximum nozzle size needs to be less than the filament size while the minimum nozzle size dependent on feedstock melting behavior and temperature
Number of shells	Minimum number of shells is determined by the minimal stable layer boundary while the maximum number of shells is limited by the smallest layer dimension divided by double the nozzle diameter
Compensation for diameter	At a minimum no compensation for variances in diameter, while maximum extra material that will not disturb surface finish or create nodules/strings

Source: Adapted from ref. [9], reproduced under a CC-BY license.

2.2 DEFINING MANUFACTURABILITY CONSTRAINTS

2.2.1 Design Problem Formulation

As previously discussed, the process parameters are the primary drivers of the final bulk properties and density, assuming a properly tuned and maintained machine. They are an excellent source of design or decision variables for the construction of useful objective functions and constraints [15–17]. Proper formulation of a design problem, regardless of the form or design method, requires at least one design objective, constraints to bound the objective or set of objectives, and a well-defined design space [18]. There are some design problems, particularly those based on convex functions, which do not require constraints, but these are rare in design for manufacturing (DFM) problems and so are not of great concern for this chapter. Although the objective function always takes the form of some function or relationship between the design or decision variables, the constraints can take on two major forms. The first and most common form is a bound on the variable, while the second is a function of design or decision variables like the objective function. The bound constraints are always inequality constraints, while the function constraints may take the form of either equality or inequality constraints [14, 19–21]. For DFM problems, inequality constraints are by far the most common and the most straightforward to work with. Like all design problems, the constraints need to be as well-defined and accurate as possible while also making sure the problem is solvable within available mental, experimental, or computational resources [20, 22–27].

2.2.2 Relationship of Design and Manufacturing

When collecting and mapping manufacturing process information and knowledge into useful DFM constraints, it is vital to consider not only the limitations of the process in question (as is typically done in traditional DFM [25–27]) but also the advantages and best practices that have been developed by regular process users. To take advantage of these best practices, it is very helpful for the designers to have first-hand knowledge of the process mechanics or at least have a competent technical point of contact who can provide insight during design decisions. Fortunately, FDM is a very accessible process that is open for many people to learn successfully. While rigorous and careful design problem formulation is more straightforward in simple cases (e.g., optimizing the thickness of a wall or overhanging feature), it becomes more important for complex design problems. Many of the advanced and automated design methods becoming more widely used, partially because of the widespread adoption of AM, depend heavily on careful problem formulation; otherwise, the design outputs may be non-manufacturable. Examples include generative design, topology optimization, and design guided by large language models.

To ensure the most rigorous possible problem formulation, practical and technical knowledge of the manufacturing process must be mapped into the design problem [28–36]. The purpose of this is to shrink the design space to the point where only manufacturable design candidates are produced. In many cases, manufacturability must be evaluated before essentials such as the cost, performance, and reliability of the part or system being designed. This is particularly important for algorithm-generated design methods, as it is usually impossible to introduce small changes to the design or implement a re-manufacturing plan for the final designs. In these cases, a design that is mismatched with the manufacturing process (and therefore non-manufacturable) must be completely scrapped and re-generated, completely wasting the invested time and resources. This is becoming a somewhat less concerning issue as computational resources become cheaper and more readily available but will always present some kind of cost in terms of time and resources for the stakeholders in the design process.

The first step in the mapping is to define the process in question in terms of its parameters. For FDM, this is relatively straightforward due to the well-defined parameters and their clear and definable influence on the outcome of the printing. Manufacturing processes are generally classified as *subtractive*, *formative*, and *AM*. Subtractive manufacturing involves the removal of materials (generally by cutting), formative manufacturing works by forming the material (generally, the volume of material is the same throughout the process), and AM adds material in layers and fuses the layers together. Each of the manufacturing types can be further subdivided into process families and individual process mechanics. For this discussion, the definition of a *manufacturing process* is *one or more machines or tools working to accomplish a pre-defined objective,* such as adding or removing material in a particular pattern or plan. It is assumed that each step in the process will add value to the workpiece by improving its utility or value as a consumer product [37].

2.2.3 Mapping Manufacturing Process to Design Constraints

Once the manufacturing process or subset of feasible processes to be used has been identified during the design process, a list of important process characteristics can be identified or defined. These characteristics should then be divided into *advantages*, *disadvantages*, and *best practices*. Each of the advantages opens up the available design space and provides a reduction in design risk related to cost and schedule. In contrast, the disadvantages reduce the design space and can add risk and uncertainty to the design decisions. The best practices could be considered as soft constraints which serve as advice and not major constraints; often, these will be used to make decisions when many feasible design candidates are offered. These *manufacturing considerations* can cover a very large domain and are not specific enough to be used directly in design, as shown graphically in Figure 2.3 [34–36].

Once the manufacturing considerations are defined or observed, they can easily be mapped to *manufacturing constraints*, which are physical limitations on what the process can actually do. An illustrative example is the heat generated by milling processes; the manufacturing considerations here are that the contract between the tool and the workpiece causes a significant amount of force and friction to be applied to the part being manufactured. The resulting manufacturability constraint would be the minimum thickness of a machined part needed to both withstand the applied force and maintain the desired tolerance and to dissipate the heat from friction without

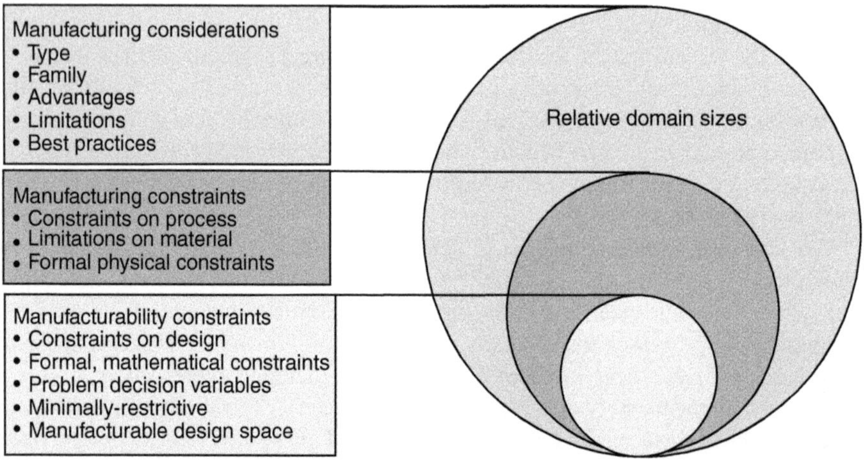

FIGURE 2.3 Relative domain sizes for manufacturing considerations, manufacturing constraints (on the process), and manufacturability constraints (on the product or system being made).

destroying the chemical structure of the material. As shown in Figure 2.3, the manufacturability constraint domain is far smaller than the manufacturing consideration domain as it has been formalized for a particular process or small set of processes. It is also assumed that the material or at least the class of materials to be used is known at the stage where the manufacturing constraints are generated. Once an effective set of manufacturing constraints has been generated, it is relatively straightforward to map them to design constraints that are specific to the part or system being made. These final constraints are the *manufacturability constraints*. These can be added directly to the design problem as they will use the same design variables as the main objective function and the non-manufacturability constraints such as those generated by performance, reliability, cost, and sustainability requirements. As shown in Figure 2.3, this final domain is the smallest of the three, as the data generated or collected is specific to the design under consideration. As with any design problem, the final set of constraints (both from the manufacturability and other sources) will need to be reduced and cleaned up before the generation of designs or design candidates [18, 34–36].

2.3 MANUFACTURABILITY CONSTRAINTS FOR FDM

2.3.1 FDM Manufacturing Considerations

On examining the FDM process, it becomes clear that four different levels of technical analysis are necessary to consider for the design of products to be made using it. First, the entire system, followed by the extruder, the design of the individual elements, and the layout of the elements. This is shown graphically in Figure 2.4. Based on this conceptual model of the system, eight key manufacturing considerations emerge:

- FDM is a scanning-type AM process. Each layer is built from a selective layout of elements of uniform size and properties. It is generally assumed that all the elements have identical cross sections and the only difference between them is the length and layout pattern (Figures 2.4a, 2.4d, and 2.4e) [5, 10–12, 14, 21, 38–40].
- During printing, the material is extruded through an extruder die and deposited in a molten state to form the elements (Figures 2.4b, 2.4c, 2.4d, and 2.4e) [10, 21, 38, 40–42].
- The fusion between layers and elements in the same layer is accomplished via polymer healing (Figure 2.4b, 2.4c, and 2.4d) [40, 43–49].
- The bulk material is subject to repeated thermal cycling above and below the transition temperature for significant periods of time (Figure 2.4b) [43–45, 47, 50, 51].
- It has been determined and widely claimed that the elements are isotropic or transversely isotropic on the local level. Given this and the fact that the elements are solid and uniform in cross section, they can be conceptualized as beams to be laid out in truss forms during printing (Figure 2.4b, 2.4c, and 2.4d) [10, 40, 52–57].

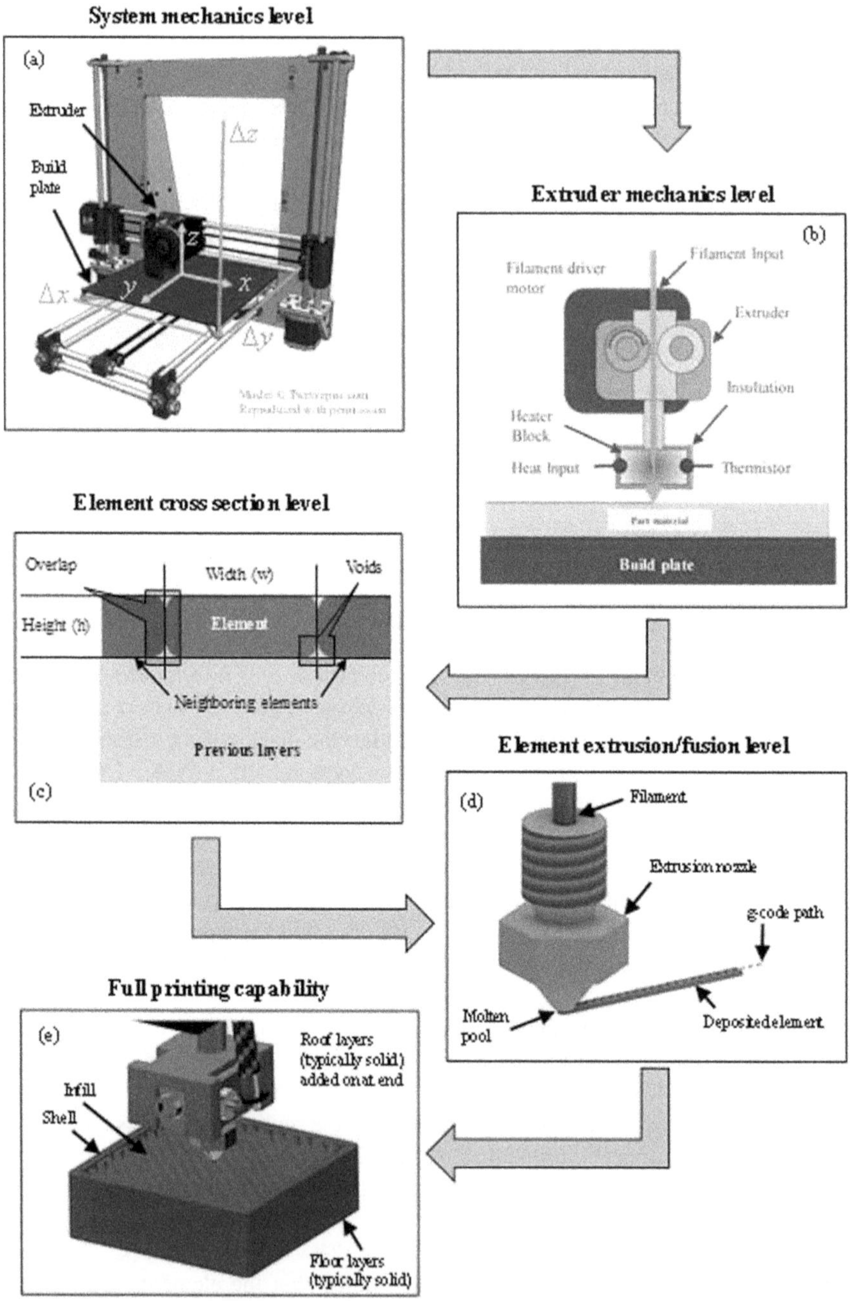

FIGURE 2.4 (a) FDM system level, (b) FDM extruder level, (c) FDM element level, (d) FDM deposition and fusion level, and (e) FDM full process.

- The process input is a series of toolpath commands in the form of g-code (Figure 2.4a and 2.4d) [10, 21, 40, 44, 58–61].
- The motion of the X–Y axes constructs each layer by scanning out the geometry in the form of uniform material elements (Figure 2.4a).
- The motion of the Z axis adds layers to the part being built (Figure 2.4a).

It is also vital to consider the design scale at which different design decisions are made. Note that all of these decisions are being made about the product to be printed using FDM, not the design of the process or printing hardware itself. For FDM, these analysis scales, in order of size, are:

- *Macroscale Product:* Final useful product.
- *Macroscale Feature:* A major feature or are of the product being designed that may need its own design process separate from the rest of the part or system.
- *Mesoscale Layer:* As FDM parts are always anisotropic, they will all have a mesoscale. This design domain is the design of a layer or a significant collection of elements.
- *Mesoscale Element:* Design of the element or bead cross section, angle, placement, and similar. Based on the manufacturing considerations, this would be similar to the design of a beam with a uniform cross section.
- *Microscale:* The microscopic features of the input feedstock, mixed composite materials, and similar.
- *Sub-Microscale:* Material type, including the design, composition, and level of crystallinity of the polymer chains, given that FDM processes thermoplastic polymers.

The structured nature of the printed bulk material necessitates this multi-scale analysis and decision-making [10–12, 38–40]. Figure 2.5 shows the relationship and relevance of each of the manufacturing considerations at least of the levels of analysis for FDM.

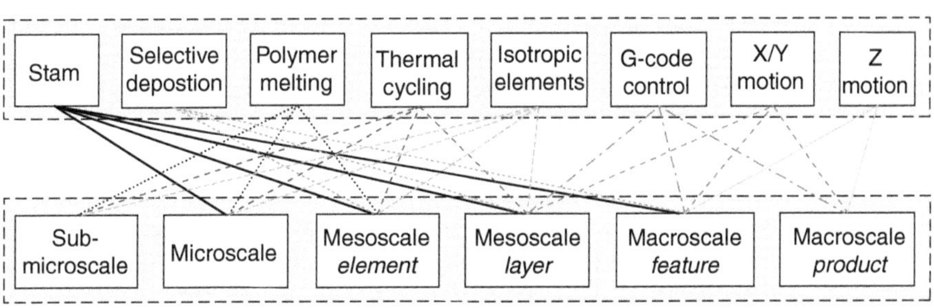

FIGURE 2.5 Effective mapping from manufacturing considerations to manufacturing constraints for FDM.

2.3.2 FDM Manufacturing Constraints

From the eight manufacturing considerations discussed in the previous section, the manufacturing constraints can be derived at each of the main levels of analysis (Figure 2.5).

2.3.2.1 Sub-Microstructure Level

At the sub-micro level, the basic material properties of the polymers being printed drive the manufacturing constraints. Most importantly, the crystallinity, the homogeneity of the local material units (polymer chains for FDM), and the rheological properties of the polymer must be accounted for. Fortunately for the designer and process user, these properties are generally fixed and only need to be observed, defined, or characterized, and are generally not designable using FDM. However, it is vital that they are known and as accurate as possible, as incorrect technical data can lead to poor design decisions and parameter settings at higher levels of analysis.

2.3.2.2 Microstructure Level

In contrast to the sub-microstructure level, where basic material properties drive the manufacturing constraints, the constraints at the microstructure level are significantly influenced by the printing parameters used. Important considerations include the head effects on the polymer formulation, thermal cycling, amount of embedded residual stresses, amount of shrinkage, heating and cooling rate, and similar factors. As with the manufacturing constraints at the sub-microstructure level, the appropriate parameters and the manufacturing constraints generally must be determined or tuned experimentally and are not easily designable.

2.3.2.3 Mesostructure Level

For the purposes of this analysis, *mesostructure* is the "middle" level of analysis between the micro- and macrostructures. All anisotropic materials have a defined (and sometimes designable) structure at this level. The presence of one or more levels of mesostructure in all AM materials offers both benefits and challenges. Most importantly, the structural and fatigue properties will be inferior to those of molded or machined versions of the same material. However, the mesostructure also enables sophisticated design of the materials to control the properties to a significant degree. Of all the design levels, the mesoscale is the most challenging but most designable level. Given that FDM is a scanning-type AM process, the elements are approximately uniform in cross section and can be laid out in the form of beams and trusses. There can be more than one level of mesostructure with this material configuration, which allows the design of the element layouts within the layers as one of the design levels. The other is the design of the layer outlines themselves, which ties well to the macroscale design level.

2.3.2.4 Macrostructure Level

At the macroscale or "useful" level, it is assumed that a full part or at least a significant designable feature is being made. It is best to consider the impact of feature design and whole part design to be separate problems, which allows multiple levels of design freedom as seen with the mesostructure design. While parts are going to generally be designed fully, the individual features may be standardized and defined using a feature catalog or part family algorithm, as seen in the examples presented in refs. [62–67].

2.3.3 FDM Manufacturability Constraints

From the derived manufacturing constraints (constraints on process capabilities), the design constraints on printed products can be derived. Driving from manufacturing constraints to manufacturability constraints for FDM is fairly straightforward, as these are essentially the limits on design that enable the manufacturing constraints to exist effectively. The resulting manufacturability constraints can then be used directly in the design problem and to make design decisions. Some may not apply in all design cases and some may be inactive or redundant, so it is essential that the design stakeholders carefully consider how they are defined and used in the design process. It is very likely that only a relatively small subset of the available manufacturability constraints will be used in a realistic design, with the remaining constraints converted into fixed parameters or included in the Factor of Safety calculations. As shown in several studies on this question [9, 34–36, 68, 69], the manufacturability constraints for FDM fall into five basic categories.

2.3.3.1 Build Parameters

These are restrictions on the build parameters which directly affect the designed product and generally appear in the form of boundary constraints. In the form of manufacturability constraints, these can also include considerations such as the maximum printing speed to keep vibration and dimensional error within acceptable tolerances. For FDM, the major manufacturability constraints are as follows:

- Acceptable environmental parameters, such as surrounding ambient temperature, air pressure, and humidity.
- Extremes for extruder temperature. As many thermoplastic materials have a wide range of temperatures over which they can be printed with FDM, the ranges must be specified based on the design of the part being made.
- Ambient temperature inside of a protective encloser, if used.
- Minimum and maximum print speed, which is strongly coupled to temperature and the thermal and rheological properties of the material being processed.
- Minimum and maximum build plate temperature.
- Defect and inclusion tolerance.
- Degree of expected homogeneity in the printed structure.
- Realistic minimum and maximum element packing density. This can be related to material properties as well, as the shrinkage and level of crystallinity can affect this. For amorphous thermoplastics, the maximum possible packing density is around 99%, while it is about 94% for semi-crystalline materials [68, 70–73].

- Minimum and maximum extruder nozzle size, assuming a standard round nozzle is used. Using a different nozzle shape could add additional manufacturability constraints locally.
- Nozzle material, typically brass, stainless steel, or carbon steel. This can have an effect on the printing quality and consistency, particularly for thin and delicate features.
- Minimum and maximum deposited element width. Some function of the nozzle width, usually somewhere between 70% and 130% of the nozzle width, partially depending on the rheology of the polymer used. This can be used as a tool to improve surface finish on a printing part or partially fill gaps between printed elements to increase effective density.
- Element layout restrictions.
- Minimum raster angle (if raster pattern is used).
- Minimum stand-alone element length. This is an important constraint for making thin-walled parts or designing material structures to steer and control cracks.
- Minimum and maximum corner/turn radius.
- Minimum contact length with the outer layer shell.
- Minimum contact length or area with the material in the previous layer.
- Minimum contact length of the shell elements to the previous layer.
- Minimum and maximum overlap or air gap between neighbor elements within a layer.
- Shell thickness. This could be in the form of a number of shell elements or total thickness as a geometric distance.
- Minimum stable number of layers. For most parts, this is 5–10 layers, but it could be less or significantly more depending on the material used. This is one of the main constraints putting a lower bound on the part size that can be printed using FDM.
- Minimum floor or roof thickness for parts that are not printed at 100% density.
- Ability to be evaluated during inspection, verification, validation, and certification activities.

2.3.3.2 Geometric Constraints

These may be driven by heuristic design rules [63, 64], geometric relationships, calculated geometry, and similar. The most common form will be dimensional constraints on part features, but they can also appear as equality constraints in some cases. The essential manufacturability constraints will be:

- Support material placement and removal.
- Support material does not significantly damage the features or surface integrity of the printed part.
- Allowed variability in the layer height. This constraint refers to intentional and designed layer height variability, as natural process variability and dimensional uncertainty will be considered under the allowable dimensional error constraint.

- Allowable printed element width/height (*W/H*) ratio. Generally, the larger the *W/H* ratio, the better the printing quality and the smaller the voids in the material. However, this dramatically affects the printing time and therefore the cost of the final part. Most experts recommend a *W/H* ratio of at least 3/2.
- Shell-infill ratio for each layer.
- Maximum overhang distance (element or feature).
- Maximum bridge distance (element of feature).
- Minimum corner and sharp feature radius to avoid stress concentrations during printing.
- Overhang feature geometry.
- Bridge feature geometry.
- Hole feature geometry.
- Printed threads.
- Feature geometry for walls and thin areas in the printed part.
- Minimum overall feature size to dissipate heat input without damaging part features.
- Surface preparation and access for repair/post-processing.
- Constraints for post-processing.
- Measurable and consistent for verification, validation, and certification activities.

2.3.3.3 Material Property Restrictions

These are restrictions on the material used relating to the behavior of the process and the desired fabricated part or system. Usually, these constraints are related to shrinkage, residual stresses, warpage, dimensional stability, polymer solidification speed, and similar for FDM. In many cases, particularly if a specific material is chosen for the design, these constraints can be simplified to fixed constant values. The important manufacturability constraints for this related to the produced part or system are as follows:

- Acceptable material purity and quality. In general, the higher the quality and purity, the better the printing quality. However, setting a lower acceptable bound on this decision variable can serve as a trade-off in terms of cost or supply chain simplification. It can also be used as a component in the Factor of Safety calculations for the final design.
- Additives and colors in the material feedstock.
- Reinforcing or filler materials used for thermoplastic polymer matrix composites, which are chopped carbon fiber, metal or ceramic powder, and similar.
- Level of crystallinity after printing for semi-crystalline materials.
- Maximum allowable shrinkage, specified as either a percentage or dimension. This constraint is vital for design cases where warping, residual stresses, and cracking of the part present a risk during use (or even printing, in extreme cases).

- Allowable uncertainty and variance in the basic material properties of the feedstock.
- Measurability during verification, validation, and certification activities.

2.3.3.4 Allowable Dimensional Error

- Allowable dimensional error in the feedstock dimensions, particularly in cases where a filament is used.
- Dimensional tolerance (height and width) on shell elements.
- Dimensional tolerance (height and width) on infill elements.
- Dimensional tolerance on layer thickness.
- Dimensional tolerance on element placement and nodes.
- Dimensional tolerance on contact error with previous layer.
- Dimensional tolerance on the contact error between the shell and infill elements.
- Dimensional tolerance on the major part features, particularly if they need to interface with other parts.
- Dimensional tolerance on overall part dimensions.
- Allowable surface roughness.
- Measurability of these parameters during inspection for verification, validation, and certification activities.

2.3.3.5 Confounding Factors

Constraints on factors which are not directly related to the process or cannot be easily controlled. Examples include vibration in the surrounding environment, temperature, and humidity changes over time (assuming basic knowledge of the ambient conditions to be related to build parameters), and similar. In some cases, they are directly related to or even proxies for other constraints and sometimes are simply required to be considered by the stakeholders to explain error and inconsistency in the final printing quality.

2.4 REALISTIC CONSTRAINT DETERMINATION

In practice, many of these constraints will need to be determined by the design stakeholders or held as constants during the design process. In practice, there are eight different approaches to determining the values and bounds, as well as which constraints are vital. Depending on the product or system design and the material used, there may be hybrid methods used or data collected using several different methods used to make the design decision. These eight approaches are:

1. *Stakeholder Decision*: Generally based on expert intuition or previous experience on other similar projects. For this approach, there may be little or no direct data collection or experimentation.
2. *Observation*: Some of the parameters and constraints may be directly observable, allowing easy and quick collection of data. Observation may be used in coordination with stakeholder decision.

3. *Partial Observation*: Similar to observation but involving the use of incomplete or uncertain information. The best example of this would be using commercial data sheets from material suppliers instead of characterizing the material in each case.
4. *Experiment*: As has been well-established in the literature up to this point, most of the geometric, material, and dimensional constraints can be determined experimentally. However, this can be resource- and time-consuming in many cases.
5. *Simulation*: Many of the geometric constraints, particularly those related to stress and thermal effects, can be estimated using simulation models.
6. *Hardware Examination*: Limitations on printing hardware can generate a significant number of the FDM manufacturability constraints and may be directly measured from the hardware itself. Information on these constraints may also be available from industry handbooks such as *Machinery's Handbook* [74] or the *Design for Manufacturability Handbook* [75].
7. *Calculations*: Certain constraints can be calculated easily with existing models and data, such as ratios and relationships between elements. Design around stress or strain patterns and material behavior within the linear-elastic realm may also be able to be easily calculated using existing models.
8. *Geometry Catalog*: Design feature catalogs and experimental testing methods exist and can be consulted during design.

To be effective for the design in question, the constraints must be specified in the same terms and variables as the rest of the design problem. In the case of mathematical optimization or decision-making, the constraints may be formulated as mathematical equality and inequality constraints and used directly in the optimization model. For other types of design problems, the constraints may take the form of heuristics, design rules, requirement definitions, and success criteria for the design effort. With this in mind, the final list of manufacturability constraints needs to be refined before being the design candidates are generated officially. Numerous handbooks, such as the *NASA Systems Engineering Handbook* [76], and work from Patterson *et al.* [9, 34–36] and others recommend a six-step process to prepare the final constraint set:

1. *Collection*: Collect all relevant manufacturability constraints, noting their type and form. Reformulate or redefine as needed.
2. *Refinement*: Determine which constraints should be simplified or require additional technical information or development.
3. *Completion*: For constraints requiring additional technical information or development, perform the tests or collect data from the published literature.
4. *Condensation*: Check for inactive, redundant, and duplicate constraints and remove them from the design problem.
5. *Application*: Apply to the design and generate design candidates.
6. *Sensitivity analysis*: Given that imperfect information and assumptions were used, complete a sensitivity analysis to ensure that the final design candidates are reasonable and are likely to result in successful manufactured products or systems.

2.5 SENSITIVITY ANALYSIS

For FDM design problems, a number of assumptions and simplifications typically must be made. Table 2.2 shows an example comprehensive sensitivity analysis adapted from Patterson *et al.* [9] for FDM.

2.6 CASE STUDIES

To best demonstrate the DFM for FDM parts at different scales and types, three simple case studies were taken from the published literature and described in terms of their design approaches and use of manufacturability constraints. It should be noted that none of the described case studies use all of the generated manufacturability constraints but make effective use of the ones that were expected to actually affect the final design.

TABLE 2.2
Sensitivity Analysis for FDM

Sensitivity	Risk	Probability	Mitigation
High-quality feedstock	Larger tolerances and lower manufacturing repeatability	Low to moderate	Careful selection of feedstock
Consistent environmental conditions	Loss of control of some material properties	Moderate	Proper conditioning, storage, and preparation of the feedstock materials
Effective control of acceleration and velocity	Overshooting, dimensional error, bad surface properties, inconsistent material properties	Low	Proper parameter settings and periodic evaluation of performance
Amorphous materials will not see a fundamental change in properties during printing	Less predictability of material properties and structure	Very low	None
Semi-crystalline materials will not see a fundamental change in properties during printing	Material properties can change and become unpredictable	Moderate to high	Careful selection of printing parameters
Shrinkage during cooling is predictable and consistent during printing and does not deviate significantly from OEM specifications	Lower effective density and lower repeatability	Low	None

TABLE 2.2 (Continued)
Sensitivity Analysis for FDM

Sensitivity	Risk	Probability	Mitigation
The layer height throughout the structure does not change significantly unless specifically designed to do so	Cracking and poor layer fusion	Low	Proper machine maintenance and periodic inspections
A single element is sufficient for stable printing	Inconsistent material structure, increased chance of localized material failure (cracks, voids)	Very low	None
Entire layers and individual elements behave similarly for bridges and overhangs	Bridges and overhang failure without extra support material, wasted material, and printing time	Low	None
Dimensional error does not have a significant influence from the part size and analysis scale	Less geometric control of the element placement at some scales	Moderate	Print with internal overlap, carefully maintain the mechanical hardware, and reduce print speed and jerk.
Roof and floor thicknesses have a significant effect for parts not printed fully dense	Wasted material and longer print time	Low to moderate	None
Minimum length scale is a function of the nozzle diameter to allow for two shells and some infill, with the exception of single-line prints	Reduced design freedom, wasted material, and increased print time	Low to moderate	None

Source: Adapted from ref. [9] and reproduced under a CC-BY license.

2.6.1 Case Study 1: Stress-Driven Layout to Toughen Material

This case study was taken from ref. [77] and reproduced here under a CC-BY license. The objective of this case study was to calculate a simple element layout using the predicted Von Mises stress field around the crack tip for a standard ASTM D5045 compact tension specimen. The size of the stress field was determined by the expected material properties of the three materials used (ABS, PLA, and PC), so they are expected to vary in size. The essential manufacturability constraints, particularly the geometric constraints on the mesoscale elements, were essential for ensuring that the designs were actually manufacturable. Given the Von Mises yield criterion,

$$2\sigma_{yield}^2 = (\sigma_1 - \sigma_2)^2 + (\sigma_2 - \sigma_3)^2 + (\sigma_3 - \sigma_1)^2 \tag{2.1}$$

Assuming the parts are thin enough so that $\sigma_3 = 0$ and assuming a Mode I crack opening,

$$2\sigma_{yield}^2 = (\sigma_1 - \sigma_2)^2 + \sigma_2^2 + \sigma_1^2 = \frac{K_I^2}{2\pi r}\cos^2\frac{\theta}{2}(6\sin^2\frac{\theta}{2} + 2) \tag{2.2}$$

Solving for r,

$$r = \frac{K_I^2}{4\pi\sigma_{yield}^2}\left(1 + \cos\theta + \frac{3}{2}\sin^2\theta\right) \tag{2.3}$$

Calculating the stress intensity factor for a compact tension geometry:

$$K_I = \frac{P}{B}\sqrt{\frac{\pi}{W}}(16.7\left(\frac{a}{W}\right)^{\frac{1}{2}} - 104.7\left(\frac{a}{W}\right)^{\frac{3}{2}} + 369.9\left(\frac{a}{W}\right)^{\frac{5}{2}}$$

$$-573.8\left(\frac{a}{W}\right)^{\frac{7}{2}} + 360.5\left(\frac{a}{W}\right)^{\frac{9}{2}} \tag{2.4}$$

where P is the applied force, B is the sample thickness, W is the size factor, and a is the crack length in the sample. Assuming P = 150 N, a/W = 0.5, B = 5 mm, and typical FDM material parameters, the Von Mises stress fields for each material can be calculated. These are shown in Figure 2.6. With the designed layouts with all the manufacturability constraints applied, the material layouts can be seen in Figure 2.7. Note that it is common to cut the notches in these samples after printing, so the material that will be removed is still shown in the figures. The successfully printed versions are shown in Figure 2.8.

2.6.2 Case Study 2: 2.5-Dimensional Designs with Stable Features

The second case study is reproduced from ref. [9] under a CC-BY license and explores three simple problems (Figure 2.9) that could be represented as two-dimensional (2D) or 2.5D (2D + thickness) designs. For Design 1a, w = 0.4 mm was used with a layer

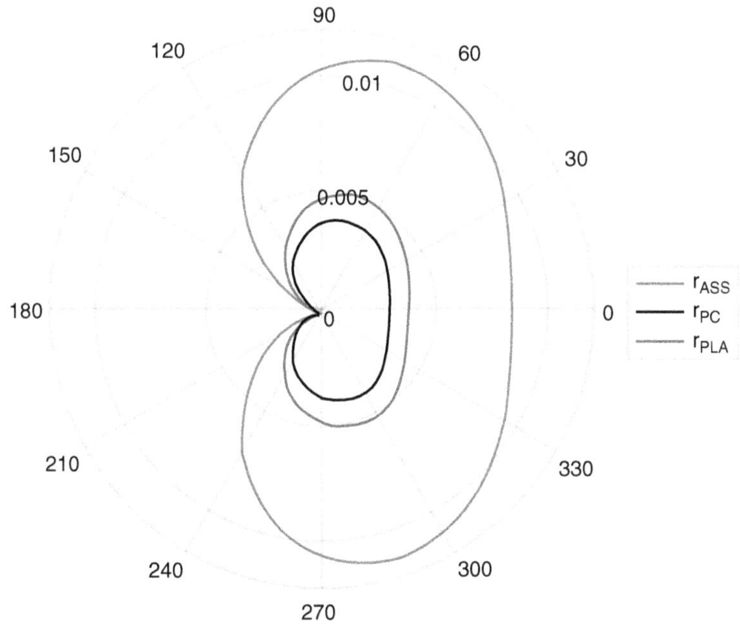

FIGURE 2.6 Von Mises stress fields for Case Study 1.

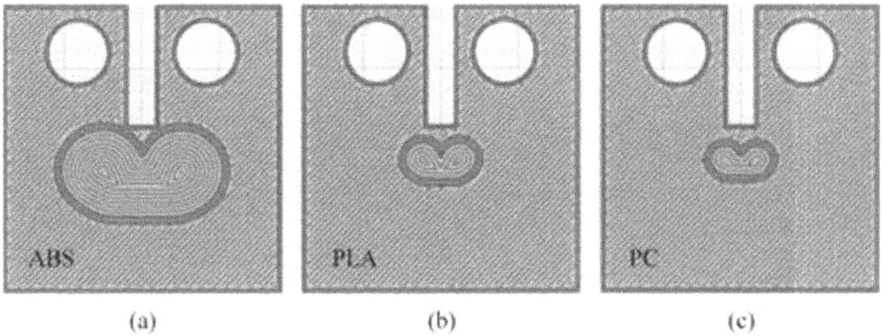

FIGURE 2.7 Case study layouts (before notching).

thickness of h = 0.2 mm (W/H = 2), with each of the thicker features being two elements wide and the thinner being one element wide. For Design 1b, a nozzle of w = 1.0 mm (W/H = 5) was used, with resulting changes in the design. In the case of Design 2, the spring coil was manually designed and laid out, with the rest of the part being automatically generated after imposition of the manufacturability constraints. Design 3 was a simple truss problem, calculated using a truss design algorithm. All designs were easily manufacturable on the first attempt, showing the value and power of using manufacturability constraints.

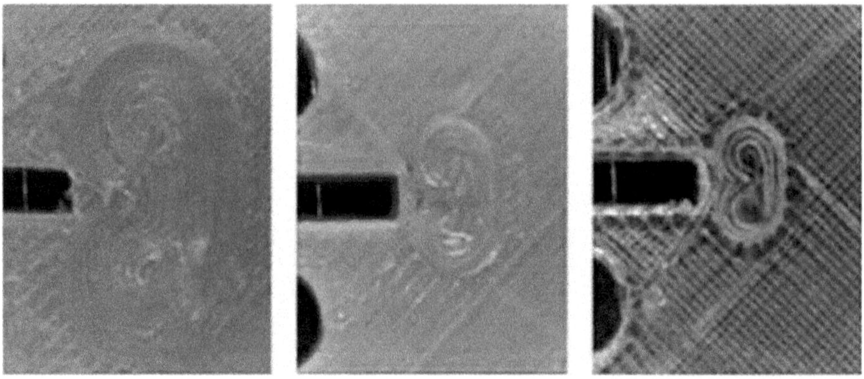

FIGURE 2.8 Printed stress fields for each material (samples not notched), which match Figure 2.6 very well.

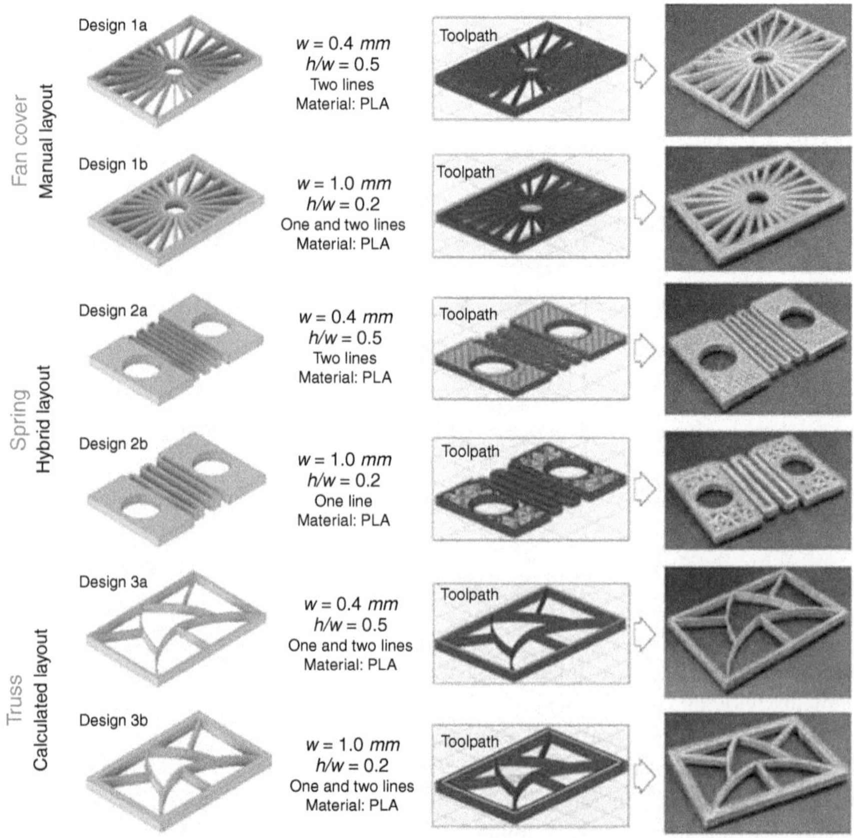

FIGURE 2.9 Results of Case Study 2.(Reproduced from ref. [9] under the terms of a CC-BY license.)

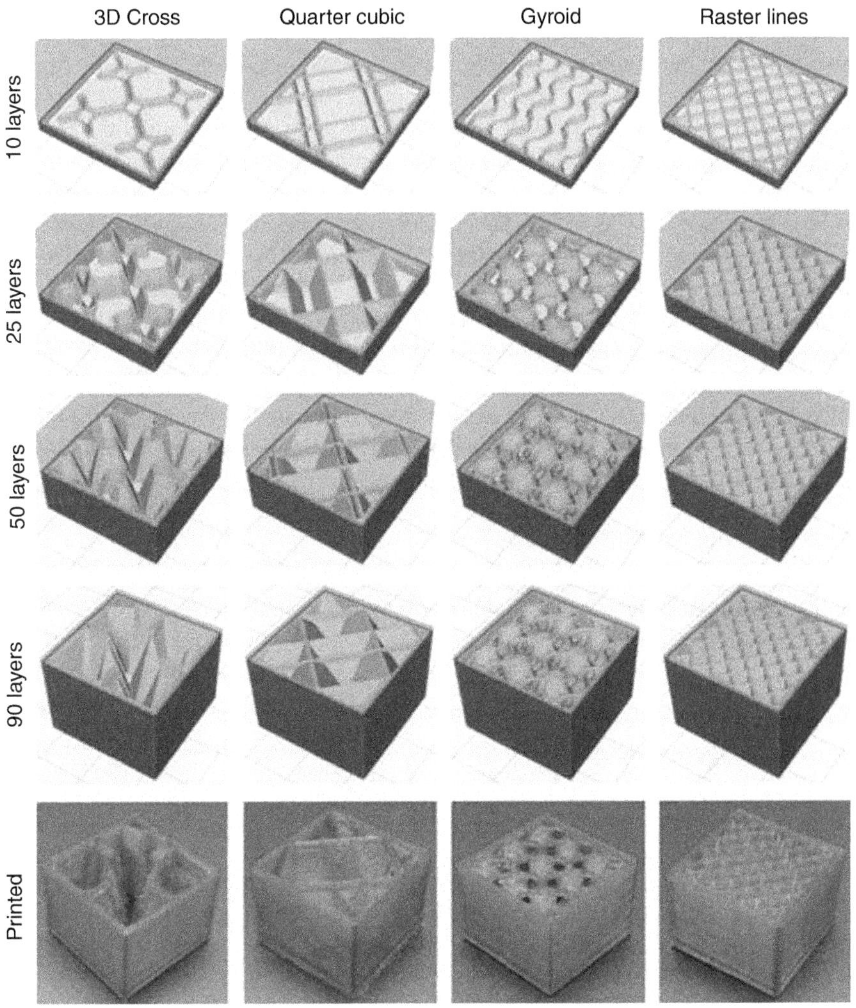

FIGURE 2.10 Results of Case Study 3.(Reproduced from ref. [9] under the terms of a CC-BY license.)

2.6.3 Case Study 3: Designed Infill Geometry in Defined Space

The third case study was reproduced from ref. [9] under a CC-BY license and dealt with complex 3D designs based on computational tools within a 20-mm cubic shell. The designs and printed outcome are shown in Figure 2.10. All the appropriate manufacturability constraints were applied in the design problem before generating the design candidates. In addition to the standard layout commands in Ultimaker Cura and application of manufacturability constraints, the geometry was only allowed to be one layer thick, the deposition nozzle used was 0.5 mm, with a 0.505 mm deposition width to account for shrinkage, and the *W/H* ratio used was 5/2.

2.7 CONCLUSIONS AND CLOSING REMARKS

In this chapter, the manufacturability considerations and design constraints related to the FDM process were explored and outlined. This comprehensive analysis encompassed various aspects of manufacturing and manufacturability, focusing on the different levels of the FDM process, feature and design dimensional scales, conditions, materials, and process characteristics. Case studies demonstrated how DFM principles for FDM are applied at different scales and design problem types, and how these principles influence the final designs. This chapter emphasized the necessity of integrating process-specific DFM from the initial stages of the design process. Early incorporation of these design constraints ensures that the final product is feasible within the selected manufacturing process. Aligning design problem formulation with manufacturing and manufacturability constraints minimizes potential design iterations and optimizes product performance by leveraging the unique characteristics of the process. Specifically, FDM produces anisotropic materials due to inherent small defects, printing directions, and layered construction. By accounting for these characteristics, designers can optimize the structural integrity and behavior of the final product. This work will not only improve the knowledge and tools available to designers considering FDM for the final manufacturing process but will also enable more widespread use of FDM as a standard manufacturing process for end-user production.

REFERENCES

1. O. A. Mohamed, S. H. Masood, and J. L. Bhowmik, "Experimental investigations of process parameters influence on rheological behavior and dynamic mechanical properties of FDM manufactured parts," *Materials and Manufacturing Processes*, vol. 31, pp. 1983–1994, Dec. 2015.
2. O. A. Mohamed, S. H. Masood, and J. L. Bhowmik, "Optimization of fused deposition modeling process parameters for dimensional accuracy using I-optimality criterion," *Measurement*, vol. 81, pp. 174–196, Mar. 2016.
3. B. N. Turner and S. A. Gold, "A review of melt extrusion additive manufacturing processes: II. Materials, dimensional accuracy, and surface roughness," *Rapid Prototyping Journal*, vol. 21, pp. 250–261, Apr. 2015.
4. B. N. Turner, R. Strong, and S. A. Gold, "A review of melt extrusion additive manufacturing processes: I. process design and modeling," *Rapid Prototyping Journal*, vol. 20, pp. 192–204, Apr. 2014.
5. S. Messimer, T. Pereira, A. Patterson, M. Lubna, and F. Drozda, "Full-density fused deposition modeling dimensional error as a function of raster angle and build orientation: large dataset for eleven materials," *Journal of Manufacturing and Materials Processing*, vol. 3, no. 1, p. 6, 2019.
6. S. Hongyao, Y. Xiaoxiang, and F. Jianzhong, "Research on the flexible support platform for fused deposition modeling," *The International Journal of Advanced Manufacturing Technology*, vol. 97, pp. 3205–3221, May 2018.
7. C. Kim, D. Espalin, A. Cuaron, M. A. Perez, E. MacDonald, and R. B. Wicker, "A study to detect a material deposition status in fused deposition modeling technology," in *2015 IEEE International Conference on Advanced Intelligent Mechatronics (AIM)*, IEEE, Jul. 2015.

8. Y. Wang, J. Huang, Y. Wang, S. Feng, T. Peng, H. Yang, and J. Zou, "A CNN-based adaptive surface monitoring system for fused deposition modeling," *IEEE/ASME Transactions on Mechatronics*, vol. 25, pp. 2287–2296, Oct. 2020.
9. A. E. Patterson, C. Chadha, and I. M. Jasiuk, "Identification and mapping of manufacturability constraints for extrusion-based additive manufacturing," *Journal of Manufacturing and Materials Processing*, vol. 5, p. 33, Apr. 2021.
10. S.-H. Ahn, M. Montero, D. Odell, S. Roundy, and P. K. Wright, "Anisotropic material properties of fused deposition modeling ABS," *Rapid Prototyping Journal*, vol. 8, no. 4, pp. 248–257, 2002.
11. J. Gardan, A. Makke, and N. Recho, "A method to improve the fracture toughness using 3D printing by extrusion deposition," *Procedia Structural Integrity*, vol. 2, pp. 144–151, 2016.
12. J. Gardan, "Method for characterization and enhancement of 3D printing by binder jetting applied to the textures quality," *Assembly Automation*, vol. 37, no. 2, pp. 162–169, 2017.
13. A. Alafaghani, A. Qattawi, B. Alrawi, and A. Guzman, "Experimental optimization of fused deposition modelling processing parameters: a design-for-manufacturing approach," *Procedia Manufacturing*, vol. 10, pp. 791–803, 2017.
14. R. Rezaie, M. Badrossamay, A. Ghaie, and H. Moosavi, "Topology optimization for fused deposition modeling process," *Procedia CIRP*, vol. 6, pp. 521–526, 2013.
15. C. Abeykoon, P. Sri-Amphorn, and A. Fernando, "Optimization of fused deposition modeling parameters for improved PLA and ABS 3D printed structures," *International Journal of Lightweight Materials and Manufacture*, vol. 3, pp. 284–297, Sep. 2020.
16. S. N. Cerda-Avila, H. I. Medellin-Castillo, and D. F. de Lange, "Analysis and numerical simulation of the structural performance of fused deposition modeling samples with variable infill values," *Journal of Engineering Materials and Technology*, vol. 141, Dec. 2018.
17. A. Dey, D. Hoffman, and N. Yodo, "Optimizing multiple process parameters in fused deposition modeling with particle swarm optimization," *International Journal on Interactive Design and Manufacturing (IJIDeM)*, vol. 14, pp. 393–405, Oct. 2019.
18. P. Y. Papalambros and D. J. Wilde, *Principles of Optimal Design: Modeling and Computation* (2nd Edition). Cambridge University Press, 2000.
19. N. Morris, A. Butscher, and F. Iorio, "A subtractive manufacturing constraint for level set topology optimization," *Structural and Multidisciplinary Optimization*, vol. 61, pp. 1573–1588, Feb. 2020.
20. S. L. Vatanabe, T. N. Lippi, C. R. de Lima, G. H. Paulino, and E. C. Silva, "Topology optimization with manufacturing constraints: a unified projection-based approach," *Advances in Engineering Software*, vol. 100, pp. 97–112, 2016.
21. J. V. Carstensen, "Topology optimization with nozzle size restrictions for material extrusion-type additive manufacturing," *Structural and Multidisciplinary Optimization*, vol. 62, pp. 2481–2497, Jun. 2020.
22. A. Sutradhar, J. Park, P. Haghighi, J. Kresslein, D. Detwiler, and J. J. Shah, "Incorporating manufacturing constraints in topology optimization methods: a survey," in *Volume 1: 37th Computers and Information in Engineering Conference*, ASME, 2017.
23. Y. H. Lee, J. K. Schuh, R. H. Ewoldt, and J. T. Allison, "Enhancing full-film lubrication performance via arbitrary surface texture design," *Journal of Mechanical Design*, vol. 139, no. 5, p. 053401, 2017.

24. E. Lutters, F. J. van Houten, A. Bernard, E. Mermoz, and C. S. Schutte, "Tools and techniques for product design," *CIRP Annals*, vol. 63, no. 2, pp. 607–630, 2014.
25. J. G. Bralla, *Design for Manufacturability Handbook* (2nd Edition). McGraw-Hill Education, 1998.
26. W. ElMaraghy, H. ElMaraghy, T. Tomiyama, and L. Monostori, "Complexity in engineering design and manufacturing," *CIRP Annals*, vol. 61, no. 2, pp. 793–814, 2012.
27. B. S. Blanchard and W. J. Fabrycky, *Systems Engineering and Analysis* (4th Edition). Prentice Hall, 2005.
28. A. E. Patterson and J. T. Allison, "Manufacturability constraint formulation for design under hybrid additive-subtractive manufacturing," in *ASME IDETC: Volume 4: 23rd Design for Manufacturing and the Life Cycle Conference*, ASME, 2018.
29. A. E. Patterson and J. T. Allison, "Generation and mapping of minimally-restrictive manufacturability constraints for mechanical design problems," in *24th Design for Manufacturing and the Life Cycle Conference*, American Society of Mechanical Engineers, Aug. 2019.
30. A. E. Patterson, Y. H. Lee, and J. T. Allison, "Overview of the development and enforcement of process-driven manufacturability constraints in product design," in *24th Design for Manufacturing and the Life Cycle Conference*, American Society of Mechanical Engineers, Aug. 2019.
31. O. Borgue, J. Muller, A. Leicht, M. Panarotto, and O. Isaksson, "Constraint replacement-based design for additive manufacturing of satellite components: ensuring design manufacturability through tailored test artefacts," *Aerospace*, vol. 6, p. 124, Nov. 2019.
32. O. Borgue, F. Valjak, M. Panarotto, and O. Isaksson, "Supporting additive manufacturing technology development through constraint modeling in early conceptual design: a satellite propulsion case study," *Proceedings of the Design Society: DESIGN Conference*, vol. 1, pp. 817–826, May 2020.
33. J. R. Muller, O. Borgue, M. Panarotto, and O. Isaksson, "Mapping the design space in function and geometry models supporting redesign for additive manufacturing," *Journal of Design Research*, vol. 18, no. 1/2, p. 37, 2020.
34. A. E. Patterson, Y. H. Lee, and J. T. Allison, "Generation and enforcement of process-driven manufacturability constraints: a survey of methods and perspectives for product design," *Journal of Mechanical Design*, vol. 143, May 2021.
35. A. E. Patterson and J. T. Allison, "Mapping and enforcement of minimally restrictive manufacturability constraints in mechanical design," *ASME Open Journal of Engineering*, vol. 1, Jan. 2022.
36. A. E. Patterson, C. Chadha, and I. M. Jasiuk, "Manufacturing process-driven structured materials (MPDSMs): design and fabrication for extrusion-based additive manufacturing," *Rapid Prototyping Journal*, vol. 28, pp. 716–731, Oct. 2021.
37. J. T. Black and R. A. Kohser, *DeGarmo's Materials and Processes in Manufacturing* (11th Edition). Wiley, 2011.
38. K. Alvarez, R. F. Lagos, and M. Aizpun, "Investigating the influence of infill percentage on the mechanical properties of fused deposition modelled ABS parts," *Ingeniria e Investigacion*, vol. 36, p. 110, Dec. 2016.
39. J. Gardan, A. Makke, and N. Recho, "Improving the fracture toughness of 3D printed thermoplastic polymers by fused deposition modeling," *International Journal of Fracture*, vol. 210, no. 1–2, pp. 1–15, 2017.

40. A. Cattenone, S. Morganti, G. Alaimo, and F. Auricchio, "Finite element analysis of additive manufacturing based on fused deposition modeling: distortions prediction and comparison with experimental data," *Journal of Manufacturing Science and Engineering*, vol. 141, Nov. 2018.
41. A. Bellini, S. Guceri, and M. Bertoldi, "Liquefier dynamics in fused deposition," *Journal of Manufacturing Science and Engineering*, vol. 126, pp. 237–246, May 2004.
42. N. Shadvar, E. Foroozmehr, M. Badrossamay, I. Amouhadi, and A. S. Dindarloo, "Computational analysis of the extrusion process of fused deposition modeling of acrylonitrile-butadiene-styrene," *International Journal of Material Forming*, vol. 14, pp. 121–131, Dec. 2019.
43. V. Srinivas, C. S. van Hooy-Corstjens, and J. A. Harings, "Correlating molecular and crystallization dynamics to macroscopic fusion and thermodynamic stability in fused deposition modeling: a model study on polylactides," *Polymer*, vol. 142, pp. 348–355, Apr. 2018.
44. T. J. Coogan and D. O. Kazmer, "In-line rheological monitoring of fused deposition modeling," *Journal of Rheology*, vol. 63, pp. 141–155, Jan. 2019.
45. J. Zhang, X. Z. Wang, W. W. Yu, and Y. H. Deng, "Numerical investigation of the influence of process conditions on the temperature variation in fused deposition modeling," *Materials & Design*, vol. 130, pp. 59–68, Sep. 2017.
46. Y. Zhou, T. Nyberg, G. Xiong, and D. Liu, "Temperature analysis in the fused deposition modeling process," in *2016 3rd International Conference on Information Science and Control Engineering (ICISCE)*, IEEE, Jul. 2016.
47. Y. Zhang and V. Shapiro, "Linear-time thermal simulation of as-manufactured fused deposition modeling components," *Journal of Manufacturing Science and Engineering*, vol. 140, Apr. 2018.
48. C. Bellehumeur, L. Li, Q. Sun, and P. Gu, "Modeling of bond formation between polymer filaments in the fused deposition modeling process," *Journal of Manufacturing Processes*, vol. 6, pp. 170–178, Jan. 2004.
49. S. Costa, F. Duarte, and J. Covas, "Estimation of filament temperature and adhesion development in fused deposition techniques," *Journal of Materials Processing Technology*, vol. 245, pp. 167–179, Jul. 2017.
50. Y. Kim, D. Alcantara, and T. I. Zohdi, "Thermal state estimation of fused deposition modeling in additive manufacturing processes using Kalman filters," *International Journal for Numerical Methods in Engineering*, vol. 123, no. 13, pp. 3021–3033, Aug. 2020.
51. S. L. Messimer, A. E. Patterson, N. Muna, A. P. Deshpande, and T. R. Pereira, "Characterization and processing behavior of heated aluminum-polycarbonate composite build plates for the FDM additive manufacturing process," *Journal of Manufacturing and Materials Processing*, vol. 2, p. 12, Feb. 2018.
52. J. F. Rodriguez, J. P. Thomas, and J. E. Renaud, "Mechanical behavior of acrylonitrile butadiene styrene fused deposition materials modeling," *Rapid Prototyping Journal*, vol. 9, pp. 219–230, Oct. 2003.
53. M. Shofner, F. Rodriguez-Macias, R. Vaidyanathan, and E. Barrera, "Single wall nanotube and vapor grown carbon fiber reinforced polymers processed by extrusion freeform fabrication," *Composites Part A: Applied Science and Manufacturing*, vol. 34, pp. 1207–1217, Dec. 2003.

54. G. Alaimo, S. Marconi, L. Costato, and F. Auricchio, "Influence of meso-structure and chemical composition on FDM 3D-printed parts," *Composites Part B: Engineering*, vol. 113, pp. 371–380, Mar. 2017.
55. N. P. Levenhagen and M. D. Dadmun, "Interlayer diffusion of surface segregating additives to improve the isotropy of fused deposition modeling products," *Polymer*, vol. 152, pp. 35–41, s. 2018.
56. S. Lin, L. Xia, G. Ma, S. Zhou, and Y. M. Xie, "A maze-like path generation scheme for fused deposition modeling," *The International Journal of Advanced Manufacturing Technology*, vol. 104, pp. 1509–1519, Jul. 2019.
57. A. Bellini and S. Guceri, "Mechanical characterization of parts fabricated using fused deposition modeling," *Rapid Prototyping Journal*, vol. 9, pp. 252–264, Oct. 2003.
58. E. G. Gordeev, A. S. Galushko, and V. P. Ananikov, "Improvement of quality of 3D printed objects by elimination of microscopic structural defects in fused deposition modeling," *PLoS One*, vol. 13, p. e0198370, Jun. 2018.
59. H. H. Nadiyapara and S. Pande, "A review of variable slicing in fused deposition modeling," *Journal of The Institution of Engineers (India): Series C*, vol. 98, pp. 387–393, Jun. 2016.
60. E. C. Balta, D. M. Tilbury, and K. Barton, "Control-oriented modeling and layer-to-layer stability for fused deposition modeling: a kernel basis approach," in *2019 American Control Conference (ACC)*, IEEE, Jul. 2019.
61. D. Aksoy, E. C. Balta, D. M. Tilbury, and K. Barton, "A control-oriented model for bead cross-sectional geometry in fused deposition modeling," in *2020 American Control Conference (ACC)*, IEEE, Jul. 2020.
62. P. Pradel, Z. Zhu, R. Bibb, and J. Moultrie, "A framework for mapping design for additive manufacturing knowledge for industrial and product design," *Journal of Engineering Design*, vol. 29, no. 6, pp. 291–326, 2018.
63. G. A. Adam and D. Zimmer, "Design for additive manufacturing – element transitions and aggregated structures," *CIRP Journal of Manufacturing Science and Technology*, vol. 7, no. 1, pp. 20–28, 2014.
64. G. A. O. Adam and D. Zimmer, "On design for additive manufacturing: evaluating geometrical limitations," *Rapid Prototyping Journal*, vol. 21, no. 6, pp. 662–670, 2015.
65. S. B. Maidin, I. Campbell, and E. Pei, "Development of a design feature database to support design for additive manufacturing," *Assembly Automation*, vol. 32, no. 3, pp. 235–244, 2012.
66. M. K. Thompson, G. Moroni, T. Vaneker, G. Fadel, R. I. Campbell, I. Gibson, A. Bernard, J. Schulz, P. Graf, B. Ahuja, and F. Martina, "Design for additive manufacturing: trends, opportunities, considerations, and constraints," *CIRP Annals*, vol. 65, no. 2, pp. 737–760, 2016.
67. S. Liu, Q. Li, W. Chen, L. Tong, and G. Cheng, "An identification method for enclosed voids restriction in manufacturability design for additive manufacturing structures," *Frontiers of Mechanical Engineering*, vol. 10, no. 2, pp. 126–137, 2015.
68. K. Rajan, M. Samykano, K. Kadirgama, W. S. W. Harun, and M. M. Rahman, "Fused deposition modeling: process, materials, parameters, properties, and applications," *The International Journal of Advanced Manufacturing Technology*, vol. 120, pp. 1531–1570, Feb. 2022.
69. R. J. Urbanic and R. Hedrick, "Fused deposition modeling design rules for building large, complex components," *Computer-Aided Design and Applications*, vol. 13, no. 3, pp. 348–368, 2016.

70. A. E. Patterson, T. R. Pereira, J. T. Allison, and S. L. Messimer, "IZOD impact properties of full-density fused deposition modeling polymer materials with respect to raster angle and print orientation," *Proceedings of the Institution of Mechanical Engineers, Part C: Journal of Mechanical Engineering Science*, p. 095440621984038, Apr. 2019.
71. M. Galetto, E. Verna, and G. Genta, "Effect of process parameters on parts quality and process efficiency of fused deposition modeling," *Computers & Industrial Engineering*, vol. 156, p. 107238, Jun. 2021.
72. C. M. Vicente, T. S. Martins, M. Leite, A. Ribeiro, and L. Reis, "Influence of fused deposition modeling parameters on the mechanical properties of ABS parts," *Polymers for Advanced Technologies*, vol. 31, pp. 501–507, Nov. 2019.
73. H. Rezayat, W. Zhou, A. Siriruk, D. Penumadu, and S. S. Babu, "Structure–mechanical property relationship in fused deposition modelling," *Materials Science and Technology*, vol. 31, pp. 895–903, Jun. 2015.
74. E. Oberg, F. Jones, H. Horton, H. Ryffel, and C. McCauley, *Machinery's Handbook*, 32nd edn. Industrial Press, Inc., 2024.
75. J. Bralla, *Design for Manufacturability Handbook*, 2nd edn. McGraw-Hill, 1998.
76. NASA, *NASA Systems Engineering Handbook: NASA/Sp-2016-6105 Rev2 - Full Color Version*. 12th Media Services, 2017.
77. A. E. Patterson, *Meso-Scale FDM Material Layout Design Strategies under Manufacturability Constraints and Fracture Conditions*. Doctoral Dissertation, University of Illinois at Urbana–Champaign, 2021.

3 Additive Manufacturing of Metallic Parts Using Fused Deposition Modeling

A Quest for New Materials

R. Durga Prasad Reddy, Mohit Kumar, Aswani Kumar Singh, Ramandeep Singh, and Varun Sharma

3.1 INTRODUCTION

Additive manufacturing (AM) or three-dimensional (3D) printing technology has emerged as one of the most revolutionary tools in the field of manufacturing. This technology creates parts with minimal material volumes, easy design change, and high design complexity. This is achieved due to layer-by-layer building approach of this technology by acquiring data from computerized computer-aided design (CAD) models through Standard Tessellation Language (STL) files. Owing to its ingenious approach, this technology has found its applications in a wide range of industries like construction, aerospace, automobile, biomedicine, pharmacy, food, fashion, etc. [1–3]. Initially introduced as an approach for creating prototypes, this technology emerged as a promising manufacturing methodology. As a result, it has been found to be slowly replacing the traditional methods. Over the years, various AM techniques have been introduced based on material, deposition, curing method, etc. Based on the ISO/ASTM 52900:2015 standards, the AM techniques are classified into seven types, as shown in Figure 3.1.

Of the different techniques introduced, fused deposition modeling (FDM) method was found to be one of the most preferred and used printing methodologies due to its simple architecture, ease of handling, and low-cost operation [4]. However, this technique has material limitations, and it was found to process thermoplastics, wax, and plastic-based composites in most of the studies. This has hampered the capability and utility of FDM technique for wider manufacturing sectors.

Metals and alloys are among the most widely used materials in AM to obtain industry-ready parts. Owing to its versatility, metal AM has emerged as a juggernaut

DOI: 10.1201/9781003686941-3

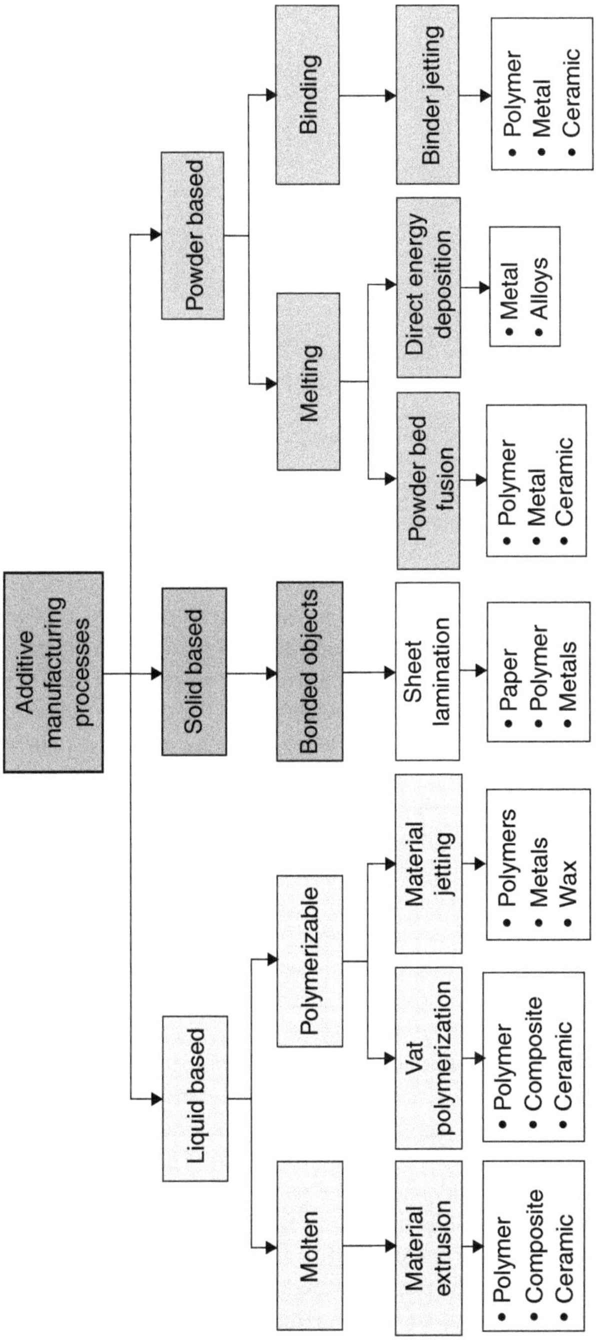

FIGURE 3.1 Additive manufacturing techniques.

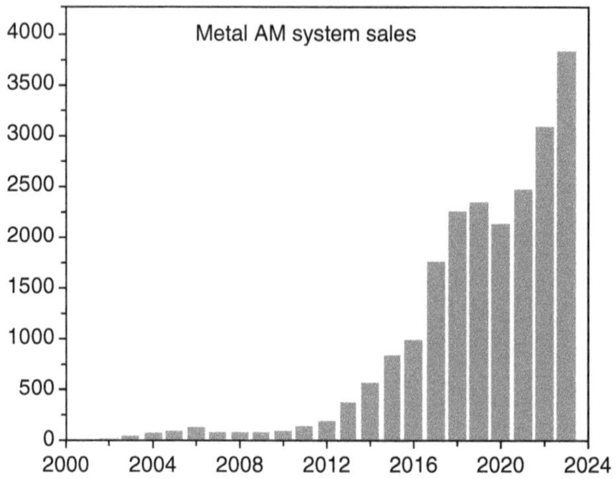

FIGURE 3.2 Sales of metal additive manufacturing systems as per Wohler's Report, 2024.

in niche industries where customization and accuracy go hand in hand. Figure 3.2 shows the increasing usage of AM techniques for metal printing. It can be observed that the demand for these systems has skyrocketed in the last 4 years. Metal AM has different methodologies and challenges, as opposed to polymers and resins. As per the studies done over last two decades, powder bed fusion techniques like selective laser melting (SLM), electron beam additive manufacturing (EBAM), and direct metal laser sintering (DMLS) have been widely used for printing metallic parts [5, 6]. However, these techniques are very expensive and complex to use. There are various other requirements for handling and operating these techniques, like an inert atmosphere and arduous post-processing methods that can adversely affect the production line and environment. Additionally, these techniques can be hazardous to the operator in case of emergency due to exposure to laser, electron beam, etc.

Extrusion-based metal AM can be seen as an alternative to produce parts at lower costs, simpler usage without compromising dimensional accuracy and clandestine designs. Fabricating parts by selectively depositing materials through nozzle/orifice, this mechanism can also process materials like metals, ceramics, cermets, and reinforced materials. There is good feasibility to attempt various combinations of materials with sufficient percentage variation and homogeneity.

Metal FDM is one of the extrusion-based AM techniques which can be explored to produce metallic parts. In this method, metallic filaments are heated into a semi-solid state and deposited onto the printing platform. These filaments are produced using metal powders and binders. So, the parts obtained can be comparable to the metal injection molding, an injection-based fabrication process for producing small metallic parts. However, the binder used in the case of metal FDM will be higher to increase the printability of the fabricated filament. Moreover, it has also been seen that by producing the green part for metal injection molded parts its cost is decreased for small batch production [7]. The process printing parts using metal FDM initiates with the preparation of metallic filament by using binders with the metallic

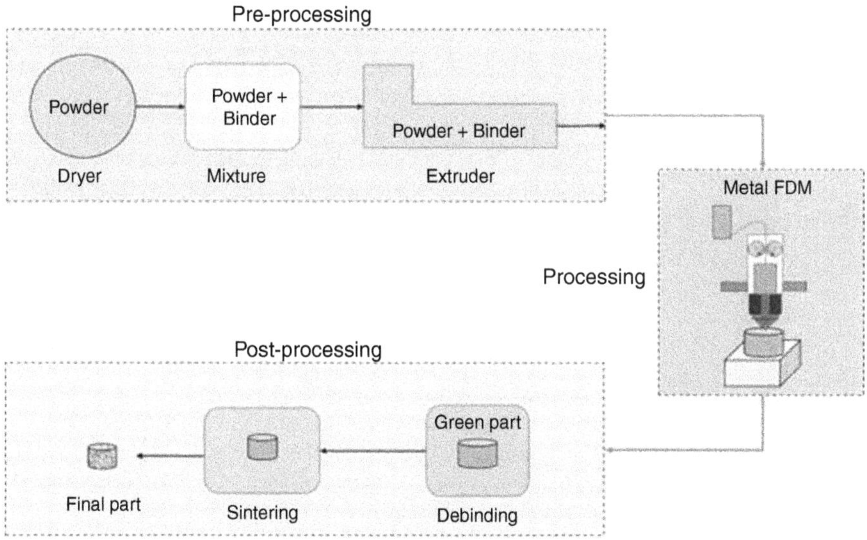

FIGURE 3.3 Metal part fabrication using metal FDM process.

powders. An optimized ratio of these materials is extremely important to ensure sufficient printability of the filaments obtained. These filaments are then fed to the heated nozzle using a roller or screw-based mechanism to get the material deposited on the printing platform. Finally, the printed part undergoes post-processing in the form of debinding, sintering, etc., as shown in Figure 3.3. A comparison of metal FDM with other widely used metal AM processes is presented in Table 3.1.

The aim of this chapter is to identify the potency of metal FDM technique and to deal with the steps encountered in the fabrication of metallic parts using metal FDM in detail. These steps are studied elaborately, and it is found out how the final product is obtained with low cost, ease of usage, safe handling, and improved accessibility, as compared to existing perilous techniques. Based on the feeding methods, deposition, trademarks, and the process of preparation, this methodology has been represented using different names like bound metal deposition (BMD) [8], shaping, debinding, and sintering (SDS) [9], MF3 (metal fused filament fabrication) [10], etc. Then, the microstructural and mechanical properties are also discussed. Later, the advantages and challenges involved in this process are also illustrated along with future scope of this printing method.

3.2 FILAMENT PREPARATION

Metals, in powder form, are generally preferred for producing metallic filaments. The steps involved in the process of metal FDM are illustrated in Figure 3.4. The process of filament preparation involves mixing metal powders with binders at a critical temperature lying between binder degradation temperature and its melting point. The rheological properties of the mixture play a crucial role in assessing the feasibility of

TABLE 3.1
Comparison of Different AM Processes Used for Metal Part Fabrication [11]

AM Process	Resolution	Materials Used	Application
Selective laser melting	10–50	Metals, alloys, composites [stainless steel (316L, 17-4 PH), tool steel (H13, A2, D2), titanium alloys (Ti-6Al-4V, Ti-6Al-2Sn-4Zr-2Mo), aluminum alloys (AlSi10Mg, Al6061), nickel-based superalloys (Inconel 718, Inconel 625, Nimonic 80A), cobalt-chromium alloys (CoCrMo, CoCr), copper and copper alloys (copper, CuCrZr)]	Wide range of metallic parts, High accuracy, no supports needed
Direct energy deposition	50–250	Metals, alloys, composites [titanium, aluminum, nickel, copper, steel, Inconel, Ti-6Al-4V, maraging steel, stainless steel (316L, 17-4 PH), cobalt-chrome (CoCr), Hastelloy, aluminum alloys (AlSi10Mg, Al6061), metal matrix composites (MMCs), titanium carbide (TiC) reinforced composites, tungsten carbide (WC) reinforced composites, aluminum–silicon carbide (Al–SiC) composites]	High flexibility, wide range of metals and alloys
Electron beam melting	50–100	Metals, alloys, composites [titanium, Ti-6Al-4V, Inconel 718, Inconel 625, CoCr (cobalt–chromium), TiAl (titanium aluminide), NiTi (nickel-titanium), titanium carbide (TiC) reinforced composites]	Fast production, large components can be printed
Extrusion-based metal AM	50–100	Metals, alloys, ceramics, composites [copper, Ti-6Al-4V, Inconel 718 and 625, AlSi10Mg, tungsten, aluminum oxide, zirconia, silicon carbide, titanium carbide reinforced composites, aluminum–silicon carbide composites]	Low cost, suitable for small components
Binder Jetting	20-100	Metals, alloys, composites, ceramics [stainless steel 316L, stainless steel 17-4 PH, Inconel 625 and 718, copper, titanium, cobalt–chrome (CoCr), silicon carbide (SiC), aluminum oxide, zirconia, silica, tungsten carbide (WC)–cobalt]	Low cost, mass production

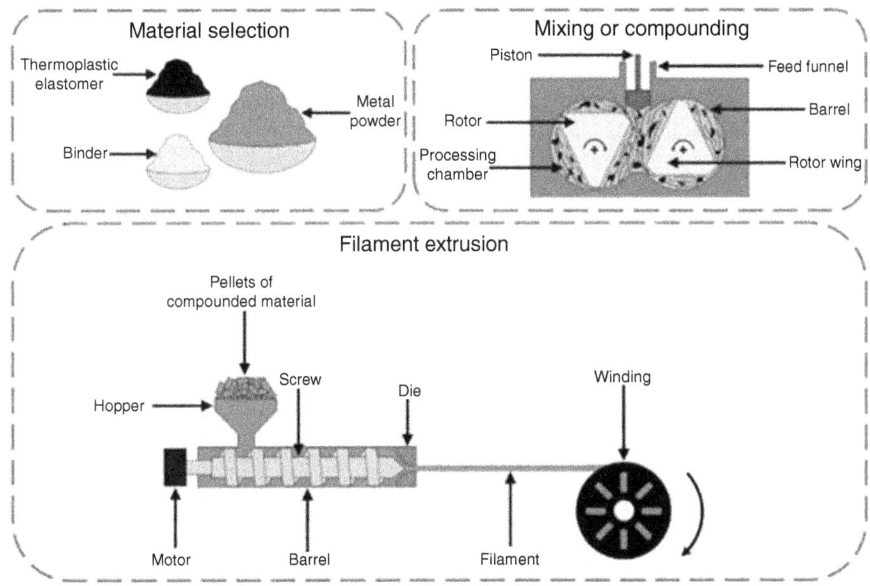

FIGURE 3.4 Steps involved in metal FDM. (Adapted from Bankapalli et al. [12] with permission from Elsevier).

extrusion. These rheological properties depend on the material properties. Initially, the choice of metal powder takes into account the aspects discussed below [12].

(A) Percentage of Powder Loading

Using more metal particles in composites by proportion than polymer additives is recommended as it may have an adverse impact due to sintering on metallic components for low distortion, good final density, and uniform shrinkage. An increase in metal percentage was found to result in high-temperature sensitivity, thereby enabling extrusion at a wider temperature range. Nevertheless, it can also be noted that the filament becomes brittle as the powder concentration increases. There will be an increase in the melt viscosity during filament extrusion, which further increases the force needed for extruding the filament.

(B) Powder Size

The powder size can also impact the rheology of the powder–binder mixture. As per Singh et al., the size of fine particles can be in the range of 0–30 μm, while coarser particles are > 30 μm in size [10]. The powder size also affects the properties of the feedstock in terms of metal loading and viscosity. Finer particles increase the mixing torque of extrusion owing to higher surface area of contact than that of coarser powder.

(c) Binder and Additives

Careful consideration is crucial to obtain the best possible binding system. An additive's degradation action can affect the complete binder system's performance. In most of the studies, thermoplastic elastomers or polymers act as the main binder. These enable adequate bonding between layers. Then, there is a backbone binder that helps maintain the shape of the part after solvent debinding. There is another additive called surfactant that helps in reducing the torque required for extruding the filament of metal and additive mixtures.

3.2.1 MATERIAL SELECTION

This is a crucial stage in the filament fabrication process. Depending on the requirement, selecting a metallic powder of the right metal is a straightforward step in the material selection process. The metallic powder is the staple ingredient for the filament preparation. However, selecting the proper powder size can be regarded as an important aspect of material selection, as this influences the rheological behavior during filament extrusion and the subsequent step of 3D printing. As discussed in the previous section, a higher powder percentage will also increase viscosity, thereby impeding the AM process and filament extrusion. As a result, it would be essential to optimize the filaments' metallic powder content.

Besides, a binder plays a significant role in this process. Binder helps to bind the metal particles together in the filament and green part. It provides necessary structural integrity so that the filaments can be extruded smoothly and the desired part be obtained in the required shape. The addition of a binder acts as a saving grace from the risky metal AM printers that involve hazardous environments and high costs. Moreover, the use of binder enables us to print metallic parts by employing common commercial FDM printers at very low temperatures compared to other metal printers. The composition of metal powder and other additives used in filament production must be taken into account in addition to particle size and percentage for smooth printing process and easy handling.

3.2.2 MIXING

Mixing or compounding is another important step in the metal FDM process after the ingredients have been chosen in accordance with the specifications. To create a composite using metal powder, binders, and additives that have a consistent composition, the selected components should be combined or compounded. One technique to improve the flowability of the polymer/metal composite melt inside the extruder barrel is to better mix and homogenize the feedstock. There are two methods for mixing metal and binder systems, as shown in Figure 3.5a using single-screw extruders, and Figure 3.5b using twin-screw extruders.

In the case of small-scale production, single extruder type printers are very useful. In this method, the mixture particles are introduced into the barrel containing a single screw that processes and extrudes material in the form of filament. This filament can be pelletized and introduced again into the extruder for better homogeneity. Moreover,

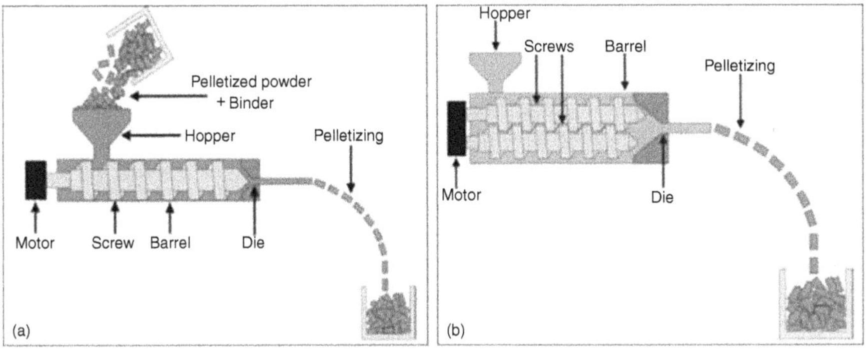

FIGURE 3.5 Mixing and pelletizing in (a) single-screw extruder, and (b) twin-screw extruder. (Adapted from Bankapalli et al. [12] with permission from Elsevier.)

this method takes more iterations to attain sufficient homogeneity as compared to the twin-extruder type. In the twin extruder type, the mixture of metal, binder and additive materials is combined using two rollers and extruded. The extruded material is pelletized prior to moving on to the next stage of the process. It may be necessary to perform the procedure multiple times to attain a consistent composition. Different types of metals and alloys, like iron, aluminum, titanium, steel, Inconel, etc., can be processed using appropriate binders and additives [13]. This type of extruder can be utilized effectively for mass-production applications. However, the feedstock required is very high, and this method is expensive to procure and maintain.

3.2.3 EXTRUSION

This is the final step, where final filaments are obtained from the extruder and stored in a spool using a winding setup, as shown in Figure 3.4. The material selection and mixing determine the feedstock's rheology, which can result in the easy extrusion of the material as filament. Two types of extrusion are possible.

(A) Piston or Pneumatic Type

In this type, the feedstock material is loaded into the cylinder, which is maintained at an elevated temperature higher than the melting point and pressure is applied. This pressure pushes the hot and semi-solid material through the nozzle, and filaments are obtained. However, this method cannot fabricate long filaments continuously. Small fragments of filaments can be obtained using this type of extrusion process.

(B) Screw Type

This type of extruder consists of a screw inside a hollow barrel that is fitted with an electric motor shaft. The cylindrical hollow barrel has a nozzle at the end and is connected to a heater, which heats both the nozzle and cylinder to a very high temperature above the melting point of the feedstock. Through the hopper, the feedstock is fed into the cylinder, and the screw's rotating action transfers the feedstock into the device. The crushing of heated feedstock between the cylinder wall and screw melts it. The screw rotation gathers the molten feedstock from the barrel surface and pumps it toward the end of the extruder at the nozzle's end. Finally, the filaments are extruded after the feedstock gets gradually melted by the time it reaches the screw's end for extrusion. Screw spinning imparts a continuous supply of feedstock through the hopper, resulting in continuous filament extrusion until either the rotation ceases or the feeding through the hopper is halted. This type of extrusion is widely followed due to its simple principle and ease of handling along with impressive output.

3.3 PRINTING OF GREEN COMPONENT

First, a green component is printed by depositing filaments in a layer-by-layer fashion. There are two approaches for printing using metal FDM:

(a) *Continuous Printing:* In this type, the printing is carried out by loading granules into the printer. This method was utilized to develop the first metal FDM printer, named The EFeSTO printer, which was extruder-based metal injection molding (MIM). The deposition plate of this machine is movable while the extruding system remains stationary. This printing method is suitable for research laboratory applications and cost-effective [14].
(b) *Filament-Based Printing:* Printing using this method is done by loading material feedstock into the printing in the form of filaments using rollers. This methodology is suitable for commercial applications and is utilized by established firms of 3D printing like BASF, Markforged, etc.

It is to be noted that the green part undergoes significant shrinkage during the post-processing operations, which is carried out after the printing step. This phenomenon should be considered while designing the part. Thus, support is provided during the designing stage to preserve the shape, form, and stability of the printed part [15].

Various parameters are involved at different steps, and optimizing these parameters can be crucial to obtaining the best possible metallic part with impressive mechanical and microstructural properties. These parameters can be optimized using traditional techniques like response surface methodology or nature-based optimization algorithms such as genetic algorithms [16–18]. Nowadays, it has become feasible to use futurist concepts like machine learning to optimize these parameters. Moreover, such models are generative and can build their own metal-binder database for new types of materials [19, 20].

3.4 POST-PROCESSING

Originally developed for polymers, FDM has evolved to accommodate metal filaments, making it possible to produce complex, custom-designed parts with significant material savings and reduced lead times.

TABLE 3.2
Parameters Related to Different Steps in Metal FDM Printing [5]

Steps	Parameters
Designing of part	Design, shape, dimensions, allowances
Material preparation	Metal powder ratio in mixture, pressure and temperature of the extruder, nozzle temperature and diameter, and mixture viscosity.
Printing	Infill pattern, infill percentage, layer thickness, platform temperature, build orientation, print speed, raster angle, filament melting, and glass transition temperature
Post-processing	Type of equipment, sintering cycle time and temperature, furnace temperature and pressure.

However, FDM-produced metal components often come with certain challenges, such as rough surface finishes, internal porosity, and lower mechanical strength compared to conventionally manufactured metal parts. To address these issues, different post-processing techniques are applied, including heat treatments to improve strength, chemical treatments to reduce porosity, and mechanical finishing methods to enhance surface quality. These post-processing steps are essential for refining the properties of FDM metal components, ensuring they meet the performance and reliability standards required for industrial applications.

The components produced using FDM exhibit rough surfaces because the deposited layers appear as visible lines (refer Figure 3.6) [21, 22]. The waviness on the surface of the SS316L part can be seen before and after sintering. Consequently, it is more practical and energy-efficient to treat the surfaces of green parts compared to sintered metal or ceramic surfaces, which are particularly difficult to polish or alter using high-energy lasers. The surface quality or finish of the fabricated samples can be improved through one of five methods: grinding, sandblasting, laser polishing or structuring, milling, and thermal post-treatment.

3.4.1 Grinding, Sandblasting and Laser Polishing

It is more effective to improve the surface finish by grinding the part while it is in the green state rather than after sintering, using standard grinding paper. The green part demonstrated sufficient stability for manual grinding. The sintered surface of a hand-ground sample is compared to an as-printed sample after sintering [refer Figure 3.7a and 3.7b)]. Although grinding eliminates surface irregularities, the printing lines remain visible. Extending the grinding duration may further enhance the surface finish; however, for more complex geometries than a flat surface, grinding becomes increasingly challenging and may negatively affect the parts' dimensional accuracy. The green parts were sandblasted in an enclosed chamber using a standard sandblasting unit with HGH SM 2000 quartz blasting media. The surface of the samples was treated for 25 seconds, followed by hand grinding. This treatment resulted in a stable surface structure after sintering, though faint traces of the original printing pattern remained visible, as shown in Figure 3.7c. However, sandblasting

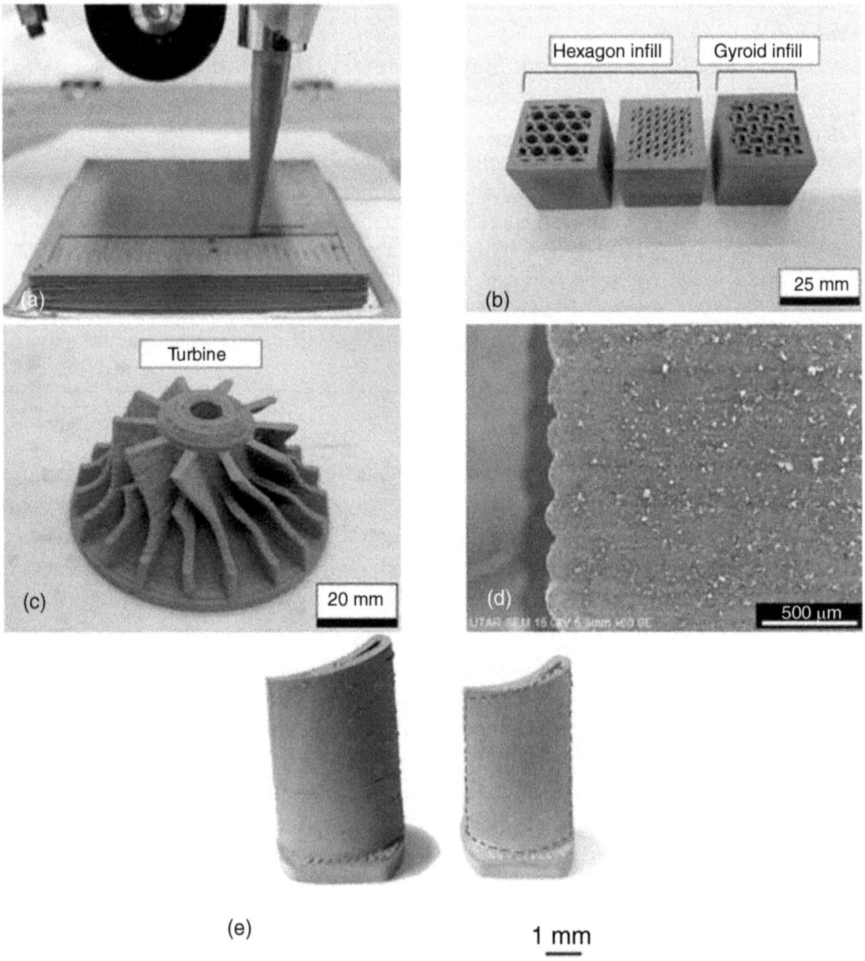

FIGURE 3.6 (a) FDM 3D printing process, (b) 3D-printed cuboids with different infill patterns, (c) 3D-printed turbine, (d) cross section of a 3D-printed green sample showing the layers fused properly, and (e) 316L stainless steel blade.(Adapted from Thompson et al. [21] and Ang et al. [22] with permission from Elsevier.)

is not effective for addressing printing surface defects, as it can further weaken the already compromised printed structure, as illustrated in Figure 3.7d.

The laser-polished sintered samples are illustrated in Figure 3.7e. By applying the correct laser polishing parameters, it is possible to completely remove the printing lines, resulting in a consistent surface texture. This improvement is clearly visible in Figure 3.7f, where a portion of the original printed surface (upper right corner) has been laser-polished. However, when applied to green parts, laser polishing has limited effectiveness for correcting surfaces with severe printing defects, as shown in Figure 3.7g. Laser technology can also be used to create specific surface patterns

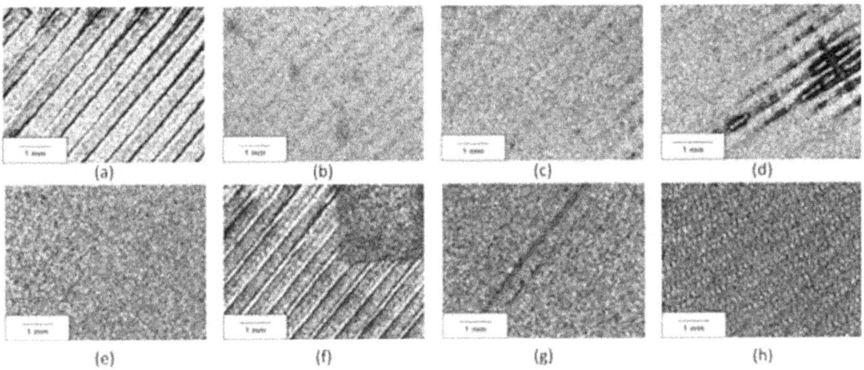

FIGURE 3.7 Post-processing using (a, b) grinding, (c, d) sandblasting, and (e–h) laser polishing.(Adapted from Bankapalli et al. [12] with permission from Elsevier.)

on green parts, as shown in Figure 3.7h, where a patterned structure is formed on the sintered surface. This capability is useful for various applications, such as creating biomaterials or crafting components for jewelery.

3.4.2 Milling

The surface quality of sintered parts can be improved through a process called green-state milling, where parts are milled while still in their green state. This involves using a milling machine to achieve the desired dimensions and remove any surface undulations created during 3D printing. Parenti et al. [23] introduced a hybrid technique that allows for milling during the 3D printing process, as illustrated in the schematic in Figure 3.8. Their study showed that this method significantly reduced the surface roughness of the printed workpiece from an initial R_a 70–120 μm to R_a 1.4–2.8 μm on the top and side surfaces of the green parts.

Although some upper edge damage occurred due to air-gap formation, green state milling was shown to be a viable technique for enhancing surface finish and creating precise features that can endure both debinding and sintering. The sharp edges are achieved through milling, with a smaller radius than the as-deposited surfaces. Scanning electron microscopy (SEM) analysis of the milled material revealed no surface gaps and sharp corners. However, milling can sometimes cause damage to green parts, as the cutting forces can weaken the parts and their deposited strands (refer Figure 3.9).

Damage can occur in two primary ways:

(a) Surface damage during facing operations: Damage to the upper surface, often resulting in single-layer separation, mainly around the outer edges, leading to localized, shallow flaws larger than the milling tool's diameter.
(b) Cracking and shell detachment during contouring: When a shell or perimeter deposition approach is used, cracks may form, and shells may detach due to poor adhesion between the outer and inner shells, which causes complete fractures under cutting pressure.

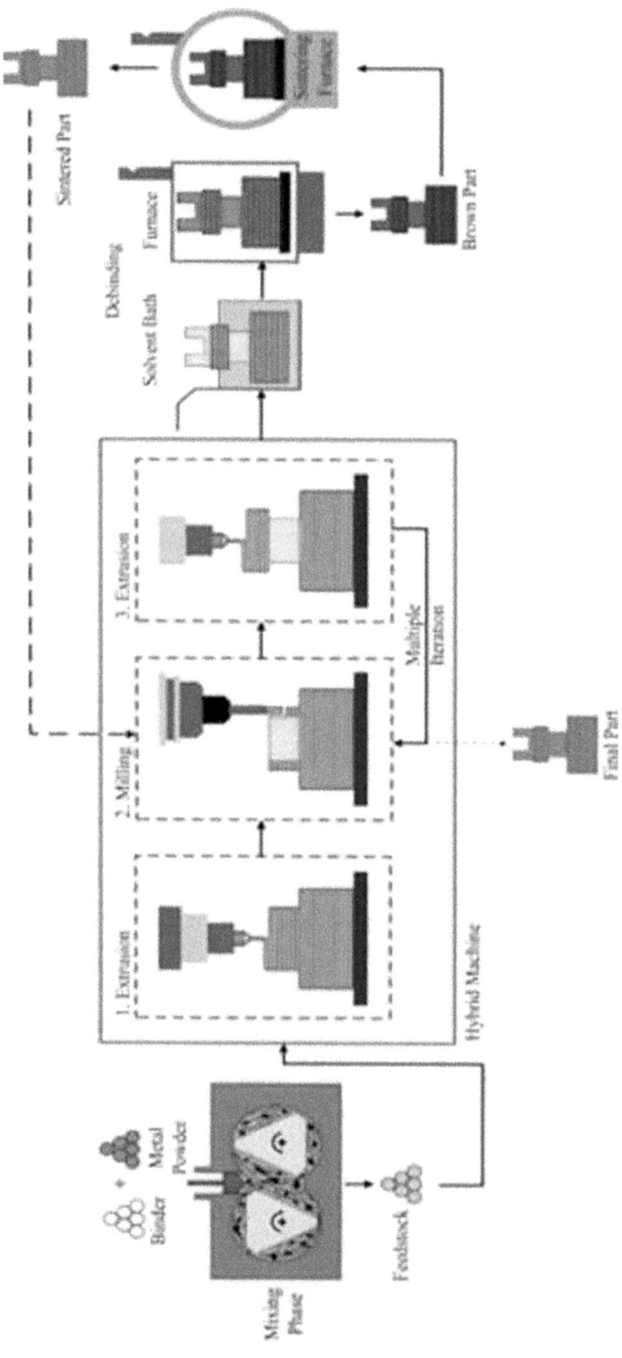

FIGURE 3.8 Hybrid manufacturing involving extrusion freeform fabrication and milling. (Adapted from Parenti et al. [23] with permission from Elsevier.)

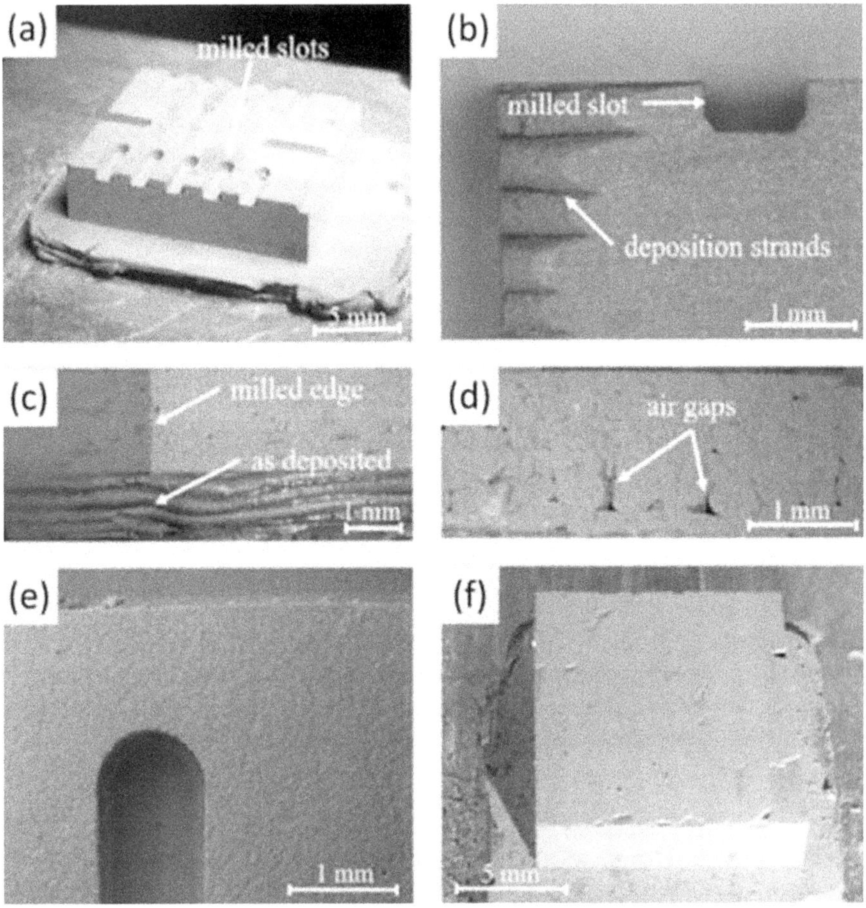

FIGURE 3.9 (a, b) Micro-milled parts at green state, (c) edge quality on the green part, (d) air gaps on the green part, (e) milled slot on the brown part, and (f) surface defects in green part. (Adapted from Parenti et al. [23] with permission from Elsevier.)

3.4.3 Thermal Post-Treatment

Improving bronze's thermal conductivity is key for better heat transfer and reduced energy use in systems like heating and cooling. FDM printing and thermal post-treatment have been used to efficiently produce sintered bronze (Cu89/Sn11). By optimizing printing parameters and sintering temperatures, the study aimed to maximize thermal conductivity while minimizing energy and material use. Figure 3.10a depicts the fabricated samples and Figure 3.10b shows porosity in the parts with different flow rates, and nozzle diameter of 1 mm. Figure 3.10c shows the sintering curve for the post-processing [24].

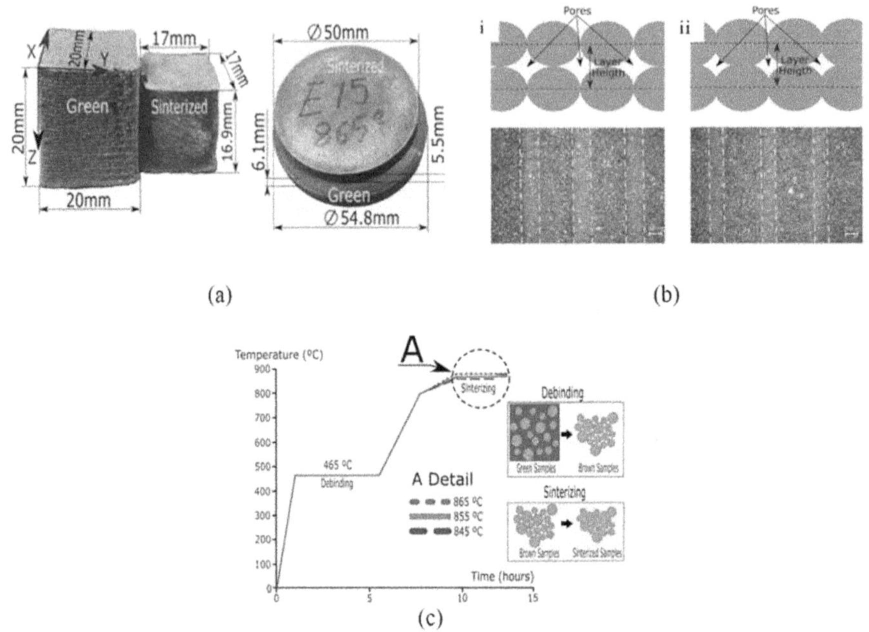

FIGURE 3.10 (a) FDM-fabricated samples, (b) porosity in the parts with different flow rate (1) flow rate of 100%, (2) flow rate of 110%, and (3) sintering curve for the post-processing. (Adapted from Lostado-Lorza et al. [24] with permission from Elsevier.)

Testing of 27 samples according to ASTM E1530:2019 revealed that the best results were achieved with a sintering temperature of 845°C, a 1.0 mm nozzle diameter, and a 110% flow rate, which resulted in the highest thermal conductivity (87.61 W/m·K) and density (7.52 g/cm^3). The lowest values were observed at 865°C sintering temperature, a 0.6 mm nozzle, and a 100% flow rate.

3.5 ADVANTAGES OF METAL FDM

- Metal FDM printing is capable of creating complex shapes with customized designs and dimensions.
- This type of printer is relatively easier and simpler to operate as it does not involve extra exotic attachments like laser, electro gun, etc.
- Metal FDM is comparatively faster than other widely used metal printing methodologies like SLM, EBM, etc.
- The metal FDM technique is cost-effective in terms of installation, operation, maintenance, etc.
- This printer can be installed in any industrial environment without requiring special arrangements.

3.6 INDUSTRIAL APPLICATIONS

The capabilities of this technique to create parts with material freedom and low cost have increased its usage in various fields of industries. According to Grand View Research, the metal 3D printing in the whole world was valued at USD 7.73 bn in 2023, and it is expected to see a significant 24.6% compound annual growth rate (CAGR) in the span of 2024–2030. With the growing importance of metal 3D printing in the industrial world, the prominence of metal FDM can rise manifold owing to its advancements and acceptance over the years. The green parts made using this technique have been used for MIM applications. Various types of metallic and polymeric tools have been printed for injection molding. Similarly, Strano et al. [25] reported that this method fabricated different types of blanking, shearing tools, and deep drawing punches. This technology has also been used to create custom-made critical parts like porous metal parts, EDM electrodes, metallic components, energy applications like turbine blades, and biomedical applications like implants. Some parts have also been fabricated for thermal applications like foam-structured heat exchanges, internal cooling channels, etc. [11].

This methodology is also being applied in niche areas like aerospace and aircraft production in the USA. Moreover, this technique is expected to emerge phenomenally due to the development of commercial printers by leading 3D printing companies like Ultimaker, Markforged, BASF, etc. Some 3D printing-based industries like Sparx 3D, Nieka Systems have opted to replace L-PBF with metal FDM printers because of high-quality output at low costs and easy handling. Several materials have been attempted successfully to manufacture various types of metallic parts. There is still a lot of scope for pushing this technology into the mainstream metal printing industry.

3.7 CHALLENGES AND FUTURE RESEARCH DIRECTIONS

The metal FDM has offered an additional route for fabrication of metallic parts with less cost and admirable properties. However, it is still in the nascent stage and requires a lot of improvements to scale up its use in an industrial environment in tandem with powder-bed fusion methods, as given below:

- To achieve a broad range of capabilities in terms of material combinations and applications.
- Ensuring better distribution in the mixture is a major challenge. A homogeneous mixture is essential for the fabrication of the best possible part.
- Developing low-cost binders and additives will be instrumental in making this method economically competitive.
- The percentage of metal in the mixture should be improved as much as possible without compromising the quality and printability of the filaments.

- More research would be required to enhance the mechanical properties of printed parts. A good control over the metallurgical characteristics can help produce parts with best possible properties as per requirement.
- As this process involves material mixture, hot extrusion, and shrinkage, ensuring accuracy and repeatability in the printed structures will be critical for better acceptability in industries.
- The acceptance of AM parts in terms of guaranteed dimensional accuracy and material properties can also be improved by introducing technical standards by established organizations.
- Optimizing the parameters at different processing steps is required to obtain the required part with necessary characteristics. Various futuristic techniques by using AI/ML models can be employed for this purpose.
- Numerical simulation of the metal FDM printed part has not been carried out sufficiently. This helps in predicting the printed part characteristics beforehand, thereby saving a lot of time and money.

3.8 CONCLUSIONS

Extrusion-based metal AM offers a new route to produce metallic parts in an economical and easy way without compromising quality and performance. This chapter dealt with the aspects of one of the extrusion-based methods called metal FDM, where parts were printed by loading feedstock in the form of filament. The steps in the process and the parameters considered have been discussed. Various factors that influence the steps of this process are presented with necessary illustrations. Moreover, a comparative analysis has also been done between existing metal AM techniques and metal FDM. The comparison showed the difference in terms of printing properties. It has also been discussed how metal FDM can be safer and more advantageous for small-scale production. Different types of printing approaches and post-processing methods have been explained with elaborate illustrations and case studies. Additionally, the advantages and applications in various industries have been explained. Finally, it has been concluded that the utilization and research on this technique are still in nascent stage and various directions of research have been discussed.

This technique has considerable potential in the realm of metal AM, and its importance is set to rise in the upcoming years with a growing acceptance of metal additive parts in the real world. Some of the future directions are presented in this work, along with the challenges involved in accomplishing them.

REFERENCES

1. Rouf S, Malik A, Singh N, Raina A, Naveed N, Siddiqui M I H, Haq M I U (2022) Additive manufacturing technologies: industrial and medical applications. *Sustainable Operations and Computers,* 3:258–74.
2. Kumar M, Sharma V (2021) Additive manufacturing techniques for the fabrication of tissue engineering scaffolds: a review. *Rapid Prototyping Journal,* 27(6):1230–72.

3. Reddy R D P, Sharma V (2024) Mechanistic modelling of drug dissolution for additively manufactured polymeric tablets: a novel layer-by-layer approach. *European Polymer Journal,* 204:112712.
4. Reddy R D P, Sahu A K, Mahapatra S S (2022) Investigation on performance of a copper coated hollow rapid electrode during electrical discharge machining. *Sādhanā*, 47(3):173.
5. Ramazani H, Kami A (2022) Metal FDM, a new extrusion-based additive manufacturing technology for manufacturing of metallic parts: a review. *Progress in Additive Manufacturing,* 7(4):609–26.
6. Singh E, Singh A K, Reddy R D P, Sharma V (2023) Drilling strategies for additively manufactured and casted SS316L stainless steel. *Journal of Materials Engineering and Performance,* 33:14096–106.
7. Korotchenko A Y, Khilkov D E, Tverskoy M V, Khilkova A A (2020) Use of additive technologies for metal injection molding. *Engineering Solid Mechanics,* 8(2):143–50.
8. Watson A, Belding J, Ellis B D (2020) Characterization of 17-4 PH processed via bound metal deposition (BMD), *TMS 2020 149th Annual Meeting & Exhibition Supplemental Proceedings*, pp. 205–16.
9. Kukla C, Gonzalez-Gutierrez J, Hampel S, Burkhardt C, Holzer C (2017) The SDS process: a viable way for the production of metal parts. *ICIT&APT 2017 10th International Conference on Industrial Tools and Advanced Processing Technologies.* TECOS – Slovenian Tool and Die Development Center, Vol 10.
10. Singh P, Balla V K, Tofangchi A, Atre S V, Kate K H (2020) Printability studies of Ti–6Al–4V by metal fused filament fabrication (MF3). *International Journal of Refractory Metals and Hard Materials,* 91:105249.
11. Rane K, Strano M (2019) A comprehensive review of extrusion-based additive manufacturing processes for rapid production of metallic and ceramic parts. *Advances in Manufacturing,* 7(2):155–73.
12. Bankapalli N K, Gupta V, Saxena P, Bajpai A, Lahoda C, Polte J (2023) Filament fabrication and subsequent additive manufacturing, debinding, and sintering for extrusion-based metal additive manufacturing and their applications: a review. *Composites Part B: Engineering,* 264:110915.
13. Bek M, Gonzalez-Gutierrez J, Kukla C, Pušnik Črešnar K, Maroh B, Slemenik Perše L (2020) Rheological behaviour of highly filled materials for injection moulding and additive manufacturing: effect of particle material and loading. *Applied Sciences,* 10(22):7993.
14. Giberti H, Sbaglia L, Silvestri M (2017) Mechatronic design for an extrusion-based additive manufacturing machine. *Machines,* 5(4):29.
15. Rosnitschek T, Hueter F, Alber-Laukant B (2020) FEM-based modelling of elastic properties and anisotropic sinter shrinkage of metal EAM. *International Journal of Simulation Modelling,* 19(2):197–208.
16. Singh P K, Singh A K, Sangle S (2023) Multi-objective parametric optimization of electrochemical machining of Inconel-625 superalloy. *Proceedings of the Institution of Mechanical Engineers, Part E: Journal of Process Mechanical Engineering,* 239(2):927–37.
17. Tiwari T, Nag A, Pramanik A, Dixit A R (2022) A comparative study of spark assisted bending process using teaching–learning based optimization, desirability approach and genetic algorithm *Applied Soft Computing,* 130:109712.

18. Singh A K, Sharma V (2022) Multi-objective optimization of grinding and vibration parameters of ultrasonic-assisted grinding with ultrasonically atomized novel green cutting fluid of Nimonic 80A *Journal of the Brazilian Society of Mechanical Sciences and Engineering,* 44(9):1–20.
19. Raj R, Mahato S, Moharana A P, Dixit A R (2024) Predicting dimensional accuracy in 3D printed polydimethylsiloxane-carbon nanotubes composites via machine learning. *Polymer Composites,* 45(4):2965–80.
20. Raj R, Dixit A R (2024) Characterizing polydimethylsiloxane–carbon nanotubes composites manufactured using material extrusion additive technology: A rheological and mechanical analysis. *Materials Letters,* 360:136019.
21. Thompson Y, Gonzalez-Gutierrez J, Kukla C, Felfer P (2019) Fused filament fabrication, debinding and sintering as a low cost additive manufacturing method of 316L stainless steel. *Additive Manufacturing,* 30:100861.
22. Ang X, Tey J Y, Yeo W H (2024) 3D printing of low carbon steel using novel slurry feedstock formulation via material extrusion method. *Applied Materials Today*, 38:102174.
23. Parenti P, Cataldo S, Grigis A, Covelli M, Annoni M (2019) Implementation of hybrid additive manufacturing based on extrusion of feedstock and milling. *Procedia Manufacturing,* 34:738–46.
24. Lostado-Lorza R, Corral-Bobadilla M, Sabando-Fraile C, Somovilla-Gómez F (2024) Enhancing thermal conductivity of sinterized bronze (Cu89/Sn11) by 3D printing and thermal post-treatment: energy efficiency and environmental sustainability. *Energy*, 299:131435.
25. Strano M, Rane K, Farid M A, Mussi V, Zaragoza V, Monno M (2021) Extrusion-based additive manufacturing of forming and molding tools. *International Journal of Advanced Manufacturing Technology,* 117:2059–71.

4 Study of Effect of Process Parameters on Mechanical Properties of Build Parts Using Fused Deposition Modeling

Medical Application

Neha Choudhary, Mohit Kumar, and Vinod Singh Rajput

4.1 INTRODUCTION

Additive manufacturing (AM) is a process of creating a product from three-dimensional (3D) model data in a layer fashion. It was first recognized in the 1980s to create models and prototypes. This technology was established for the assistance of engineers to develop 3D parts. AM was known as rapid prototyping earlier. This technique has brought new advancements, including time and price savings, increased labor involvement, thereby resulting in a shorter production cycle, and complex shape production [1, 2]. AM allows scientists and students to rapidly develop and analyze models for comprehensive research [3]. Medical practitioners may develop a model of damaged anatomy for better analysis and planning for surgery using AM techniques [4]. AM is a generative production technique that generates 2D layers and then stacks those layers to create a 3D item. Figure 4.1 represents the fundamental process steps involved in product development using AM. The basic steps of AM processes are explained in detail.

The initial step is to model the part with computer-aided design (CAD) software. The tessellated representation of the part is contained in the Standard Tessellation Language (STL) file that is exported from the CAD model. Tessellation creates the data of triangle vertices and surface normal, approximating the part's surface using a variety of triangles. These STL files are examined for faults, and defect-free STL files are then utilized with specific software for slicing and parameter adjustments. The next step involves slicing the model, orienting it, creating support structures as

DOI: 10.1201/9781003686941-4

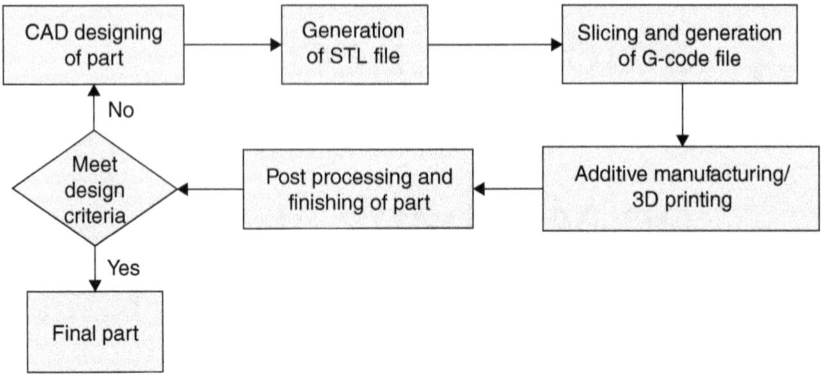

FIGURE 4.1 Fundamental process steps involved in AM.

needed, and other tasks using the slicing program. For high-quality printing, the process parameters must be established. Also, product's size and shape, kind of material, etc. all affect the parameter settings. The support structure is essential when printing hollow constructions or overhanging sections. Once the model's data has been recorded, the program creates the toolpath in the form of G-codes. To print the model, this G-code file is loaded into the AM machine. In a further step, the printed pieces are cleaned to make them functional. The supporting structures and excess material are removed. However, surface smoothing is also carried out for a superb appearance. Typically, larger objects are printed in separate sections and then joined; therefore, this process also includes assembly. Examining the product and recommending any necessary adjustments is the final phase.

AM process may be categorized into many groups. The AM techniques are categorized based on the level of material aggregation, such as liquid, solid, and powder-based. A few other procedures are categorized based on the energy employed, such as thermal energy from laser or electron beams that are used to melt, sinter, or bond metals and polymers. However, AM may be classified into seven categories, according to ISO/ASTM 52900, published by the International Standardization Organization (ISO) and the American Society for Testing and Materials (ASTM) [5]. Among the seven techniques, fused deposition modeling (FDM) is the most uncomplicated and broadly used process. It extrudes the material through the nozzle and deposits it over the bed. Choosing the right material and setting the right parameters are crucial to achieving the qualities needed for the specific application through FDM. It takes a variety of combinations and multiple testing runs for operators to understand how process parameters affect the developed product. This chapter describes the basic principle of FDM process, its process parameters, and materials that are generally used for biomedical applications. Additionally, it mostly discusses how process parameters affect materials' mechanical strengths and also recognizes the optimization techniques for process parameters. Finally, it provides a framework for more research and discussion.

4.2 FUSED DEPOSITION MODELING

FDM is an extrusion-based technique. Also known as fused layer modeling (FLM), it was used by Scott Crump in 1988 and patented in 1992. He founded Stratasys Inc. and introduced its first machine, named 3D Modeler, in the year 1992. An FDM machine generally consists of two controlled extrusion nozzles and a build platform. Extrusion nozzles are the key elements of the FDM machine. The thermoplastic polymer filaments that are wounded on the spool are used as input material. The roller arrangements are provided to feed the filaments toward nozzles (Figure 4.2). These rollers are spaced according to the filament diameter and moved in opposite directions. When the filament is loaded through this drive mechanism, it passes through the heating chamber or reservoir, where it gets liquefied. Heating coils are wrapped around the chamber, used to ideally heat and melt the polymer in the semi-liquid state. Then, the nozzle deposits the semi-molten form of material to fabricate the first layer according to the geometry of the model. Depending on the slice thickness, the build plate descends in the Z-direction following the deposition of the first layer. The subsequent layer cold-welds and deposits onto the preceding layer. Once the material extrudes through the nozzle, it must solidify and bond together. The bonding between the layers and near regions occurs due to sufficient residual heat. This process goes on until the model is completed. After the completion of the model, finishing can be done if required.

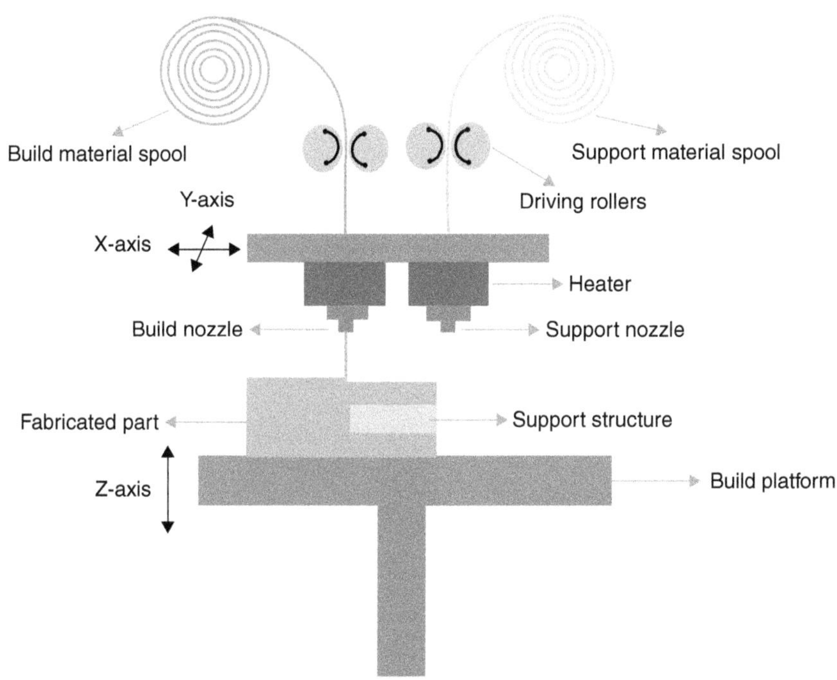

FIGURE 4.2 Schematic diagram for fused deposition modeling.

4.3 MATERIALS

There are various thermoplastic polymer materials for the extrusion-based machine. These materials are pre-formed in filaments and wounded on a spool to be used on FDM machine. Amorphous thermoplastic polymers are an ideal material for the extrusion process due to their low processing temperatures and coefficient of thermal expansion. Amorphous thermoplastic polymers have a viscosity that varies gradually with temperature. The extrusion of material can be precisely controlled by adjusting the nozzle's temperature and pressure.

On the other hand, when crystalline polymers fuse, a clear phase transition occurs. In this case, it is hard to control a homogeneous extrusion process. FDM has been used with a broad variety of polymers for various applications, including biopolymers, acrylonitrile butadiene styrene (ABS), polylactic acid (PLA), nylon, polycarbonate (PC), and polyether ether ketone (PEEK). There is a continuous improvement in the properties of materials. Some important properties to be known are mechanical strengths, thermal properties, viscosity, and density. The list of some polymers with advantages, drawbacks, and applications is presented in Table 4.1.

Other than pure thermoplastic polymers, composite materials, and metals are the present trends for extrusion materials. Different manufacturers are producing filaments with a combination of more than one polymer. The blends of binary and tertiary polymers are also being used for extrusion. Other than this, thermoplastic polymers are being reinforced with fibers and ceramic powders. The fibers and powders are fine enough to extrude out from the diameter of the nozzles.

TABLE 4.1
List of Polymers

Polymers	Advantage	Disadvantage	Application
PLA	Biodegradable, biocompatible, good mechanical and thermal properties	Low heat resistance, low melting point, slow degradation, low toughness.	Tissue engineering, drug delivery
PCL	Highly processable	Low degradation, low melting point, hydrophobic,	Tissue engineering, medical devices
PVA	Biodegradable, water soluble	not easily processable, high cost	Tissue engineering, drug delivery
PCs	Chemistry-dependent mechanical properties	Slow degradation	Tissue engineering, drug delivery, fixation devices
Polyurethanes	Mechanically strong	Do not degrade often	Prostheses and tissue engineering

4.4 PROCESS PARAMETER

Process parameters are those that pertain to the FDM process and have an effect on both the production process and the parts that are produced. FDM technique has various variables that influence the properties of the build and production time which include machine parameters, printing parameters, build orientation, and other affecting parameters, as presented in Figure 4.3a. Optimizing part properties like surface property, accuracy of dimensions, mechanical characteristics, and thermal characteristics require choosing the correct combination of process parameters. Some of the standard parameters are listed in Table 4.2 with a description.

4.5 PROCESS PARAMETER OPTIMIZATION

The numerous process variables included in the fabrication of 3D products using FDM, the mechanical properties, and other characteristics of the part may differ on the change of the values of process parameters. Extensive research has been conducted to determine the effect of process variables on mechanical behavior such as tensile, impact, compressive, and flexural strength of the part.

4.5.1 Layer Thickness

The layer thickness influences the heat dissipation and impacts the component's strength. The count of layers is lowered, the distortion is lessened, and the strength is increased if the layer height increases. However, adding more layer thickness results in a staircase effect that lowers accuracy and produces a poorer texture. Rather than this, a reduction in layer thickness will increase the count of layers, which will improve heat transfer from the top to bottom layers and raise the temperature at the contact of bonding. Consequently, strength is increased through proper diffusion between neighboring rasters. More layers, however, result in a high number of cycles of heating and cooling overall, which raises residual stress and increases the risk of part debonding, alteration, and cracking in interlayers.

The thickness of the layer has been optimized for mechanical strengths by several researchers. The highest strength was observed at a layer height of 0.16 mm in a work on the compression strength of lattice-based cellular PLA structures with biomedical applications [6]. Furthermore, tensile strengths at layer heights of 0.05, 0.10, 0.15, and 0.20 mm were assessed for PLA specimens. It was discovered that as layer height rises, tensile strength falls [7].

Similarly, the tensile, compression, and flexural strength of PEEK was investigated at three layer heights: 200, 300, and 400 µm. The layer thickness of 300 µm showed significant mechanical strengths with a raster angle of 0/90°. It was concluded that tensile strength was influenced more than compression and flexural strength by layer thickness [6]. It was also observed that due to reduction in layer height, the strength increases and elongation decreases due to an increase in cooling rate for nylon material. Further, when layer height was raised from 0.15 to 0.55 mm, elongation at break was enhanced by 125–251% [7].

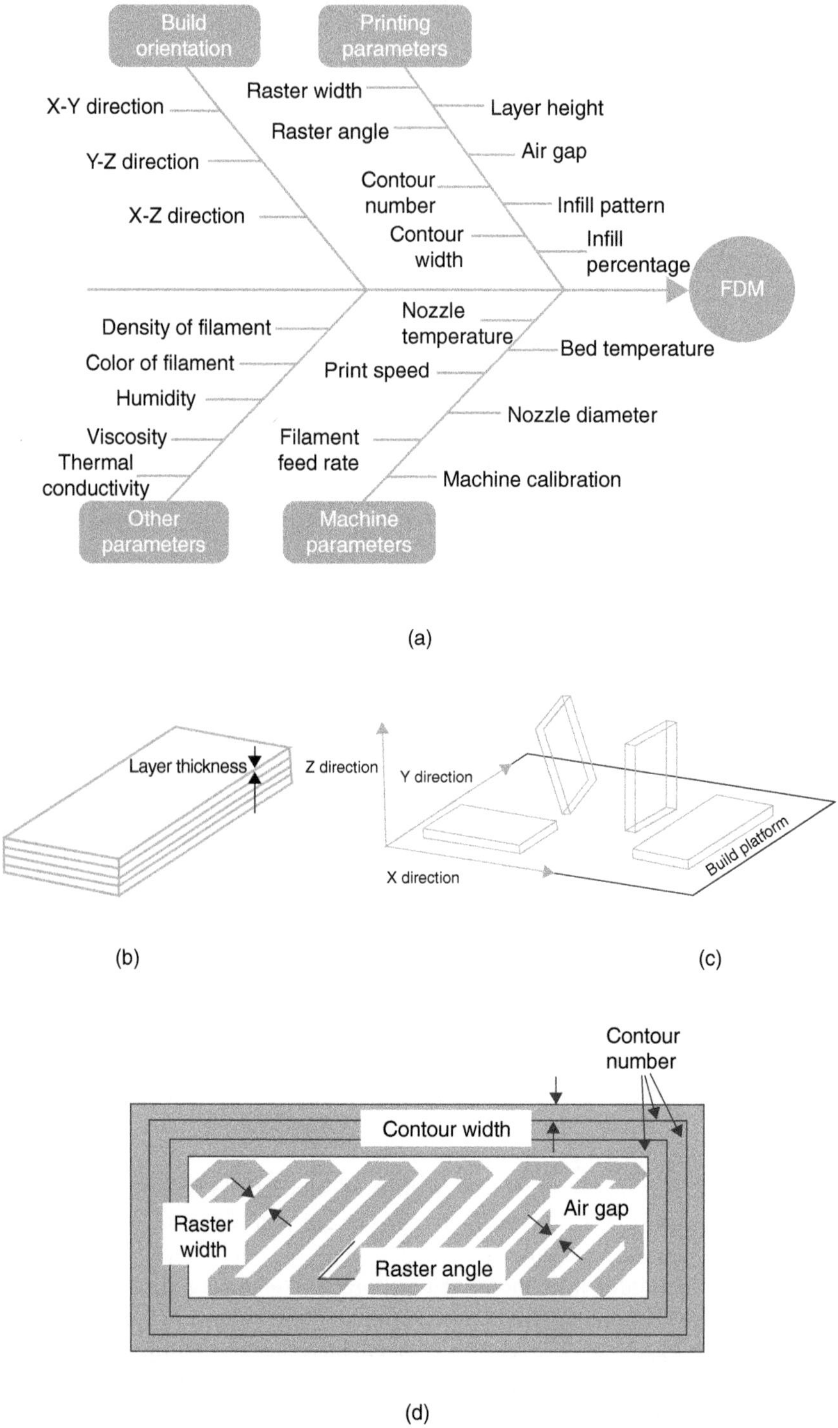

FIGURE 4.3 (a) Cause and effect diagram of process parameters of FDM, (b) layer thickness, (c) build orientations, (d) other process parameters.

TABLE 4.2
FDM Process Parameters with Description

S. No.	Parameter	Description
1.	Layer thickness	Layer thickness is the height between the build platform and nozzle tip during extrusion in the Z direction (refer Figure 4.3b).
2.	Build orientation	It is the orientation (angle) of part with build platform concerning the X, Y, and Z direction of machine (refer Figure 4.3c).
3.	Raster angle	It is the bead's angle concerning the x-axis on the build platform (refer Figure 4.3d). It varies between 0°and 90°.
4.	Raster width	It is the thickness of bead along the nozzle path. It depends on nozzle diameter.
5.	Raster gap	It is the space between the two adjacent deposited beads (or raster) on a single layer. It may be positive, zero, or negative (if the raster overlaps).
6.	Infill percentage	The printed 3D parts are solid from the outside, but the internal structure is not necessarily solid. This internal structure is known as infill, and it varies in terms of shape, size, and volume. Infill percentage is the volume of the material deposited inside the outer solid layer of the building part. It is also known as infill density.
7.	Infill pattern	Infill pattern is defined as shape of infill of the building part. Diamond, hexagonal, linear, and circular are some common infill patterns.
8.	Contour width	It is the thickness of the boundary which surrounds the raster.
9.	Contour number	It is the count of boundary layers to surround the infill raster of part.
10.	Print speed	It is the printing speed of the extrusion nozzles in the X–Y plane. It directly affects the printing time of the part.
11.	Extrusion temperature	It is the temperature of the material which extrudes out of the heated nozzles. It is different for different materials.
12.	Bed temperature	It is the temperature of the build platform during extrusion. It also depends upon the material used. It is necessary for sticking of initial layers and to avoid the temperature gradient between layers.

Layer height depends directly on nozzle tip diameter. Theoretically, the layer height can be 0.01 mm of thickness even if the nozzle diameter is around 0.5 mm, but this will increase the build time which is not desirable. It has been observed that approximately 85% of the FDM part accuracy depends upon layer thickness. The majority of work has concluded that 0.1 mm thickness is considerable over 0.2 mm thickness for obtaining desirable tensile strength [8]. A poly(*p*-dioxanone) biodegradable stent has also been fabricated at an optimum layer height of 0.1 mm for maximum modulus and yield strength. The layer height of 0.1 mm has lesser and smaller voids between the extruded raster, thus providing strong interfacial bonding as shown in Figure 4.4 [9].

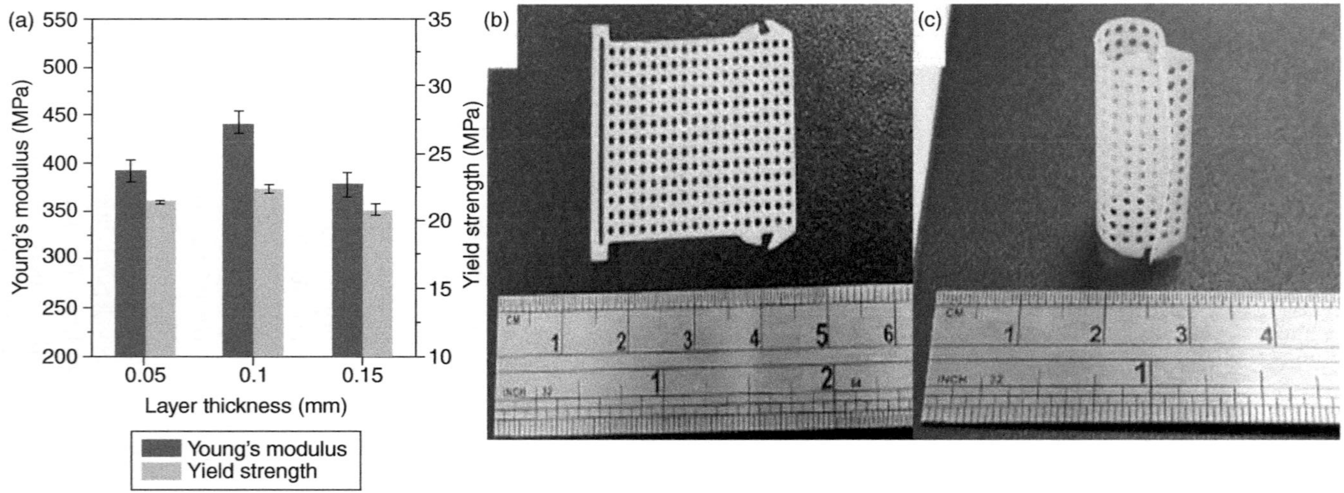

FIGURE 4.4 (a) Tensile strength of poly(*p*-dioxanone) at different layer thicknesses, (b) stent at unfolded state, and (c) stent at folded state [9].

4.5.2 Build Orientation

The build platform informs the zone of printing of part, that is, X–Y plane and Z direction indicates the layer height. In the context of the tensile strength of FDM printed part having biomedical application is mostly influenced by build orientation. The lower angles, that is, 0°–15° angle, of build orientation are best suited for tensile strength [10]. The optimal orientation for an FDM-fabricated PLA specimen to maximize tensile strength is with a 0° angle, either lying flat or on the edge of the build platform when compared to 45° or 90° build orientation. This orientation aligns the deposited raster with the applied load direction. The vertical direction of printing reduced the tensile strength by approximately 36% as compared to flat orientation [11]. Instead of this, when the part is exposed to cyclic tensile load fabricated at 45° build orientation, it has higher fatigue strength compared to 0° angle orientation [12]. A study investigated that the ultimate tensile strength of PLA was reduced with the decrease in build orientation angle (90°–0°). The largest gap between angles 90° and 0° for tensile strength is 52.29% at 0.1 mm layer height [13]. The flexural strength of PLA-printed sample showed an enhanced value at 30° build orientation when compared to 0° and 60° angles. The raster length at 0° is smaller; therefore, it provides less resistance to bending. On the other hand, 60° orientation has too long a raster length that slightly decreases the flexural strength [14]. The Design of Experiments (DOE) technique has shown that on the increase of build orientation angle, the compression strength increased and it is higher for 90° orientation angle for poly methyl methacrylate porous structure (refer Figure 4.5) [15].

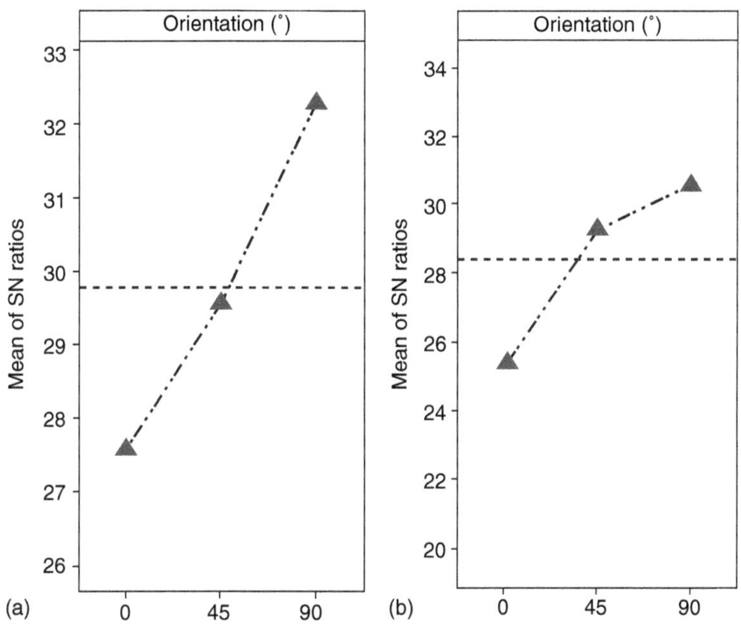

FIGURE 4.5 Main effect (S/N) plot for PMMA lattice structure. (a) Compressive strength, (b) compressive strength/density [15].

4.5.3 Extrusion Temperature

The temperature is a critical parameter for achieving desirable mechanical strengths. The optimal extrusion temperature maintains the flowability of material, which ultimately affects part characteristics. The cooling of extruded filaments below the glass transition temperature results in a fusion between the new and previous layers for improved mechanical properties. The longer the cooling period, the stronger the bonding effect between layers. The desirable temperature gradient can be maintained by adjusting the build platform temperature and layer thickness. The tensile, flexural, and impact strengths were studied at various extrusion temperatures (180°C, 200°C, 220°C, 240°C) for PLA material. It was detected that tensile and flexural strengths were enhanced with a rise in temperature, but impact strength was decreased. This may be because less viscosity at higher temperatures reduces the thickness of the part and hence decreases the impact strength [16]. In addition to this, PLA/carbon fiber composite also had a higher tensile strength of 23.56 MPa at 225°C and the lowest at 205°C of value 21.47 MPa [17]. The diffusion duration is prolonged by the greater nozzle temperature, which improves bonding and overall mechanical strengths [18]. PLA/alumina composite for biomedical applications showed that flexural strength improved with an upsurge in nozzle temperature, but tensile strength was first increased and then decreased just opposite to the compression strength trend [19]. The triply periodic minimal surface (TPMS) structured PLA/graphene oxide (GO) scaffolds have been printed based on the optimized printing temperature of base PLA polymer (refer Figure 4.6a and 4.6b). The tensile strength of the sample is first raised and then it falls opposite to the compression strength trend. The rise in temperature ensures proper bonding among layers to obtain high tensile strength, but the over-melting of the polymer decreases the tensile strength due to the rise in fluidity of material. Instead, compression strength first decreased and then increased, as shown in Figure 4.6c [20].

4.5.4 Infill Pattern and Infill Percentage

The mechanical performance of FDM parts is affected by the infill pattern and ratio. The infill pattern affects the interaction between the filaments when they are loaded. When the infill pattern density inside the part increases, the load-holding capacity of the part will also rise. Research showed that the compression strength of the samples at 80–100% infill density for various infill patterns showed higher strength-to-weight ratios. Also, the 2D infill patterns (grid, cross, and triangular) performed better than the 3D patterns based on strength-to-weight ratio and at the topmost load for the compression strength test. This is due to 2D structures being aligned to the cylinder's axes and the principal stress field [21]. Similarly, the infill pattern (triangle, grid, quarter cubic, and tri-hexagon) affected the compressive strength of PLA component. The triangular infill structure showed higher absorbed energy and stiffness at 40% porosity. This recommends its application in implant and tissue engineering. Also, the triangular pattern has fewer voids and ratchet lines, which leads to improved strength. The grid pattern depicted the maximum compressive strength of 72 MPa in the load direction [22]. Similarly, the effect of grid, triangle, and honeycomb infill patterns

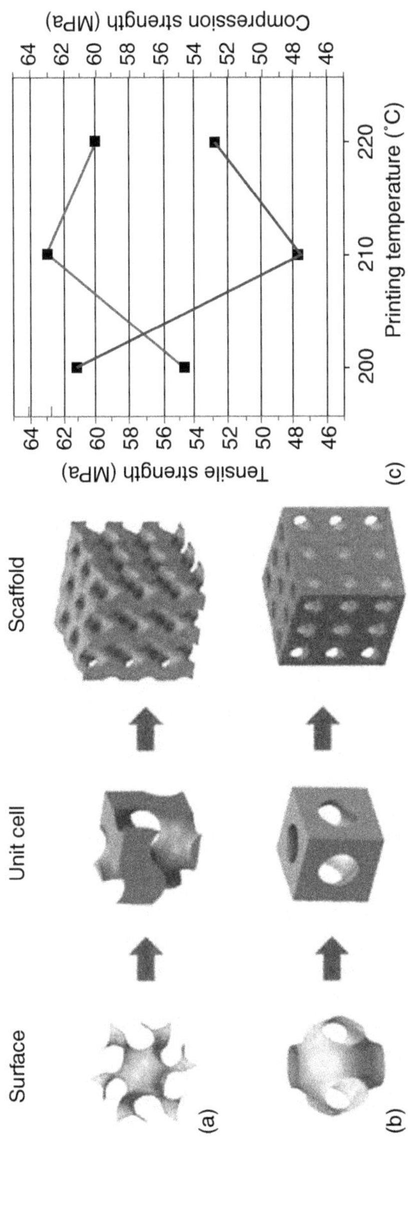

FIGURE 4.6 TPMS scaffold. (a) Gyroid, (b) Schwarz-P, (c) printing temperature effect on PLA composite [20].

was observed in flexural and tensile loads. The triangular pattern was found to have the maximum strength under both bending and tensile stress, followed by grid and honeycomb patterns. This could be because necking happened in a honeycomb and triangular pattern that failed to have a brittle fracture [23]. In addition to this, concentric patterns in a study showed the highest tensile strength and modulus followed by Hilbert, rectilinear, and honeycomb patterns at all filling percentages (20%, 50%, and 100%). The flexural strength is also higher for concentric infill patterns of about 102 MPa at 100% filling percentage. The drop in infill percentage reduced the load-bearing capacity [24]. Similarly, the tensile strength of PLA was investigated at different infill patterns using Taguchi DOE. It showed that the square infill pattern verified the highest tensile strength and energy absorption, followed by the triangular pattern, as shown in Figure 4.7a and 4.7b [25]. Additionally, the tensile modulus, strength, and yield of PLA were also increased by a rise in the filling density from 20% to 100%. The ultimate tensile strength was calculated based on cross-section area previously, but when the total area is not filled, only the infilled area carries the load [26]. In another work, tensile, impact, and flexural strength specimens showed higher values at higher infill percentages as shown in Figure 4.7c [27].

4.5.5 Raster Angle

The material deposition angle, that is, raster angle, is generally calculated with respect to the X-axis and varies from 0º to 90º. The compressive, flexural, and tensile strengths for PEEK samples were tested at different raster angles, that is, 0°, 30°, and 45°. The optimized mechanical strengths were found at 0° and 300 μm layer height. At 0°/90° angle, filaments are oriented in the load direction and produce strongly bonded samples when subjected to tensile loads. While in other orientations, the samples were exposed to tensile and shear loads, which failed to lower tensile strength. In compression and bending loads, interlayer bonding plays a major role in predicting the strengths [6]. The carbon fiber-reinforced polyamide was investigated for its tensile strength at various layer heights and raster angles. It was found that 0° and 90° raster angles have tensile strengths of 58.9 and 91.53 MPa, respectively. This depicted that the tensile strength increases with an increase in raster angle (refer Figure 4.8) [28]. Similarly, the flexural strengths of the PLA sample showed a higher average value at a low level of deposition angle, that is, 0° [29]. Similarly, ABS filament demonstrated that the yield, ultimate, and tensile strength are maximum at 0° raster and has lower compression strength at 45° raster angle [30].

4.5.6 Print Speed

The mechanical behavior of parts has been studied for print head speed ranging mostly from 30 to 70 mm/s. The print head velocity affects the building time of the part. A faster build rate deteriorates the surface finish and dimensional accuracy of the part. The compression strength of the PLA composite experienced a decrease from 34.3% to 14.6 % on the rise of printing speed from 30 to 70 mm/s [31]. In addition to this, the PLA/hydroxyapatite (HA) composite tube also showed better compression

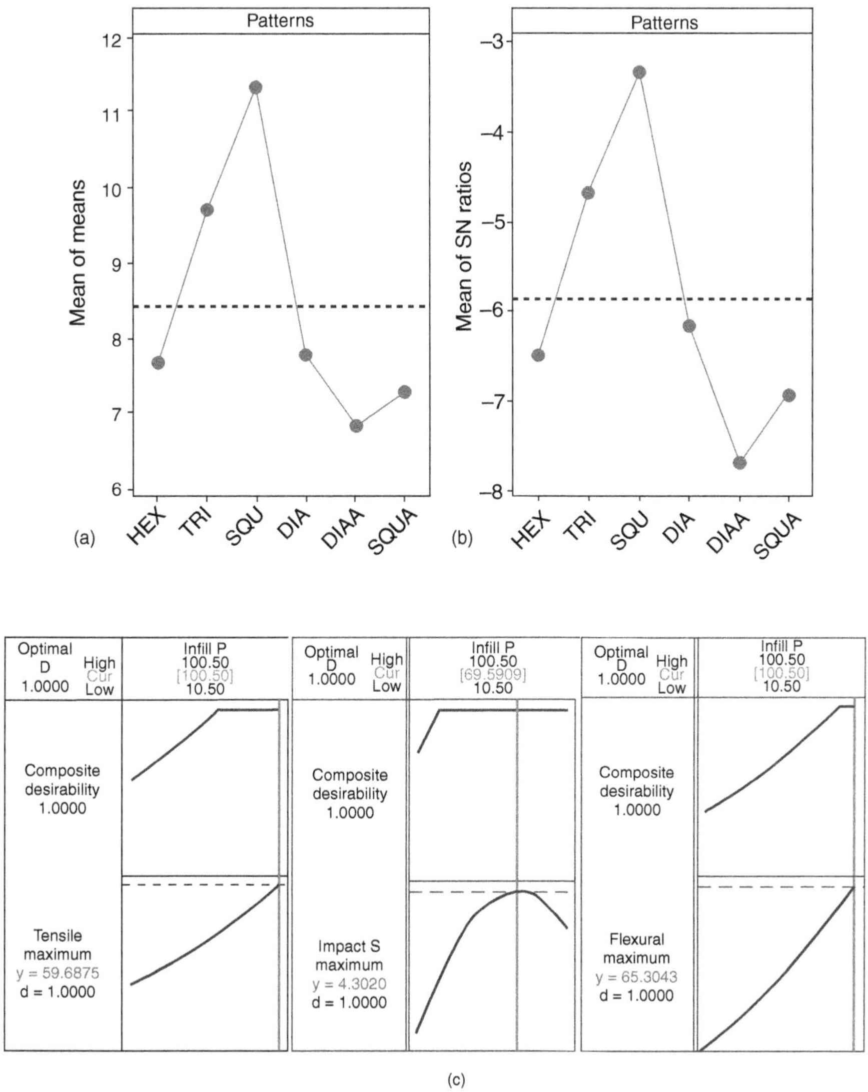

FIGURE 4.7 Mean effect plot for infill pattern effect on tensile strength for (a) means, (b) S/N ratio [25], and (c) infill percentage effect on tensile, impact, and flexural strength of PLA polymer [27].

strength at a lower speed of printing (10 mm/s) [32] (refer Figure 4.9). Similarly, the tensile strength of PLA specimens decreases with an upsurge in printing speed from 20 to 100 mm/s [33]. Contrary to this, the average flexural force for PLA material increased with a rise in extrusion speed from 40 to 60 mm/s [29]. Similarly, the flexural strength of PLA was analyzed by response surface methodology (RSM) and particle swarm optimization (PSO) methods, and it was observed that with the rise

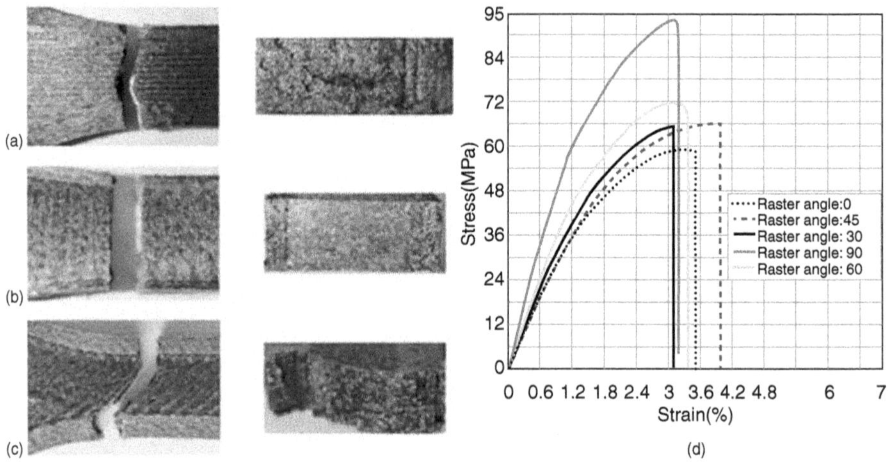

FIGURE 4.8 Fracture tensile specimen. (a) 90° raster angle, (b) 0° raster angle, (c) 45° raster angle, and (d) stress–strain curve at different raster angles [28].

in print speed, flexural strength increases. The print speed of 60 mm/sec achieved 88.045 and 86.54 MPa flexural strength for the experimental and regression models, respectively [34]. The printing speed of more than 70 mm/s for different materials such as PLA, PEEK, ABS, and polyethylene terephthalate glycol (PETG) has been analyzed using the DOE technique. It has been observed that Youngs' modulus of PLA changes approximately by 20% at a speed between 70 and 110 mm/s. PEEK-printed parts get weakened by an increase in speed as the thickness of the raster width gets reduced at higher speeds. Similarly, the flexural and tensile strengths decreased with a rise in speeds for ABS and PETG [35].

4.5.7 Raster Gap and Raster Width

The gap between the raster may be zero, positive, or negative. The deposited materials come in contact when there is no gap. A structure with a negative air gap is denser than one with a positive air gap, which is more loosely packed. Raster width acts opposite to the raster gap, that is, larger raster gap, will give lower raster width. DOE technique has been adopted to examine the impact of raster gap and width along with other process parameters on PLA part for tensile, flexural, and impact strengths. It has been observed that zero air gap provided higher impact strength for PLA, although the change is not significant. The flexural strength at zero gap was the lowest and tensile strength was not affected. The larger raster width (0.5 mm) provided higher tensile, flexural, and impact strengths [36]. Similarly, raster width and raster angle have been evaluated to investigate the tensile strength of PLA. It has been depicted that an upsurge in raster width of up to 600 μm provides the optimum tensile strength, because at a large width, the raster will have higher thermal energy, which ultimately provides proper bonding and strength [37]. Instead, for ABS polymer, positive air gap increased the tensile and flexural strength at lower layer height showing an increase

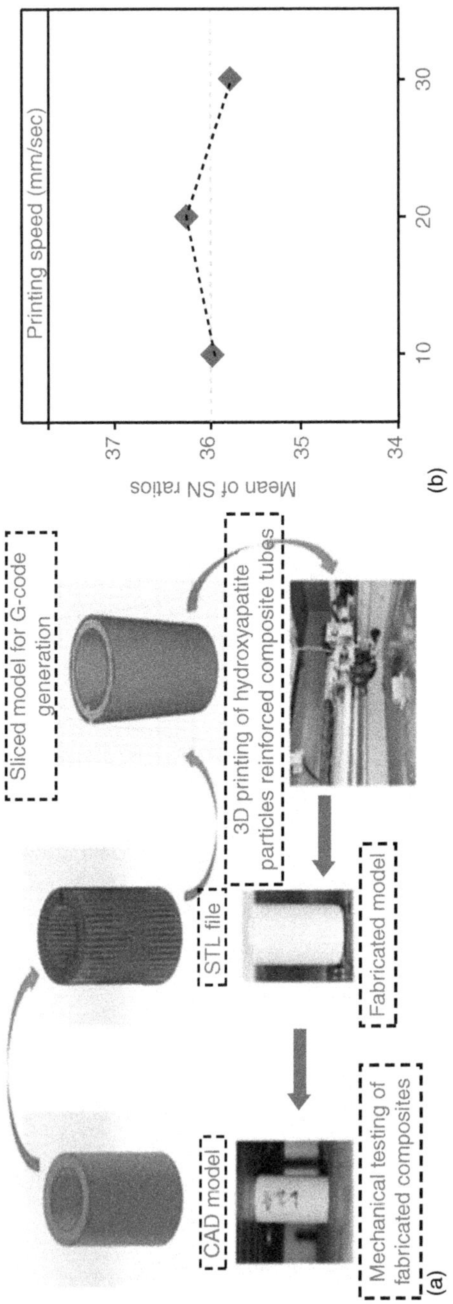

FIGURE 4.9 (a) PLA/hydroxyapatite tube, and mean S/N ratios plot for tube compression strength [32].

in strength with air gaps [38]. Additionally, the compressive strength of printed part has been influenced by the combined effect of the air gap, layer height, contour count, raster angle, raster width, and part orientation [39].

Other process variables such as contour count, contour width, and bed temperature have been studied less, and also, they have less effect on mechanical performances of different polymers or their composites having biomedical applications.

4.6 OPTIMIZATION TECHNIQUES

Numerous strategies from the DOE field and optimization methods have been combined to obtain the best configuration of the FDM process parameters. Various statistical methods, such as factorial design, and face-centered central composite design, are popular DOE techniques used for designing small-number experiments. In case several objectives are involved, the multi-objective optimization (MOO) approach is adopted instead of a single objective. The optimization through advanced techniques, such as machine learning, merges the FDM through artificial intelligence (AI). This highlights the importance of AI in achieving precision and innovation through the optimization of parameters in advanced manufacturing techniques. These methods are employed to discover the impactful process variables, the relationship between variables, and the individual influence of each FDM parameter on the output variable. In FDM, to attain high mechanical properties, various optimization techniques such as high dimensional accuracy, low surface roughness, lower defect, low build-in time and cost, etc. have been used in literature and some of them are listed in Table 4.3.

TABLE 4.3
Popular Optimization Techniques

Techniques	Linear/ Non-Linear Model	Accuracy	Interaction between Factors	References
One-factor-at-a-time (OFAT) method	Linear	Low	No	[40]
Taguchi design method	Linear	Low	Yes	[14, 32, 36, 41]
Factorial design method	Linear	Moderate	Yes	[17, 42]
Response surface method	Both	Very High	Yes	[27, 34, 38]
Fuzzy logic	Both	High	No	[43]
Gray relational analysis	Linear	Moderate	Yes	[44, 45]
Bacterial foraging optimization (BFO)	Both	High	No	[46]
Particle swarm optimization (PSO)	Both	Very High	No	[34, 46, 47]
Genetic algorithm	Both	High	No	[48, 49]
ANN	Both	Highest	No	[28, 47, 48, 50, 51]

4.7 CONCLUSION AND FUTURE SCOPE

FDM is an extrusion-based 3D printing technique, which is the most cost-effective among the other 3D printing techniques. It uses thermoplastic polymers or their composites for the development of products. The products' strength and accuracy depend upon the setting of process variables. Numerous industries, including automotive, fashion, aerospace, sports, and medicine, have extensive uses for FDM. This chapter focuses on the study of the impact of FDM process parameters on the mechanical behavior of materials that have medical applications. The effect of different variables has been studied mostly on tensile, flexural, compression, and impact strengths. It is evident that the layer height, infill pattern, build orientation, infill percentage, nozzle temperature, raster angle, and raster width are the most influential variables affecting mechanical strength. Researchers mostly optimized the process parameters for PLA, ABS, PEEK, and their composites keeping focused on their biomedical applications. The chapter also listed various popular optimization techniques, such as Taguchi, RSM, factorial designs, and new AI-based optimizations like artificial neural networks (ANNs). A balanced approach had to be developed by the researchers to determine the best settings for the various operating conditions.

Researchers are more engrossed in the optimization of key parameters, but there are still several parameters like contour number, contour width, and feed rate that need to be explored as they may affect the mechanical strengths of the products. It has been noted that there is more room for research on biomedical materials because the effects of other elements, such as moisture, temperature, and noise levels, are also not well supported by the literature. In addition to variables, there is room to expand the range of materials with potential uses in biomedicine. There will be a future need to use AI and ML-based optimization techniques as they correlate to the best outcome and provide multi-parameter optimization.

REFERENCES

1. Kumar, M., Sharma, V. (2021) Additive manufacturing techniques for the fabrication of tissue engineering scaffolds: a review. *Rapid Prototyping Journal*, 27: 1230–1272.
2. Durga Prasad Reddy, R., Sharma, V. (2020) Additive manufacturing in drug delivery applications: a review. *International Journal of Pharmaceutics*, 589: 119820.
3. Choudhary, N., Sharma, V., Kumar, P. (2021) Reinforcement of polylactic acid with bioceramics (alumina and YSZ composites) and their thermomechanical and physical properties for biomedical application. *Journal of Vinyl and Additive Technology*, 27: 612–625.
4. Reddy, R.D.P., Elgazzar, H., Sharma, V. (2022) Investigations of personalized and sustainable approach of oral drug delivery systems through additive manufacturing. *Rapid Prototyping Journal*, 28: 855–872.
5. Durga Prasad Reddy, R., Sharma, V. (2023) Investigations of hybrid infill pattern in additive manufactured tablets: a novel approach towards tunable drug release. *Journal of Biomedical Materials Research Part B: Applied Biomaterials*, 111: 1869–1882.
6. Wu, W., Geng, P., Li, G., Zhao, D., Zhang, H., Zhao, J. (2015) Influence of layer thickness and raster angle on the mechanical properties of 3D-printed PEEK and a comparative mechanical study between PEEK and ABS. *Materials*, 8: 5834–5846.

7. Moradi, M., Aminzadeh, A., Rahmatabadi, D., Rasouli, S.A. (2021) Statistical and experimental analysis of process parameters of 3D nylon printed parts by fused deposition modeling: response surface modeling and optimization. *Journal of Materials Engineering and Performance*, 30: 5441–5454.
8. Mengesha Medibew, T. (2022) A comprehensive review on the optimization of the fused deposition modeling process parameter for better tensile strength of PLA-printed parts. *Advances in Materials Science and Engineering*, 2022: 1–11.
9. Chen, E., Xiong, Z., Cai, X., Liu, S., Qin, X., Sun, J., Jin, X., Sun, K., (2023) Bioresorbable PPDO sliding-lock stents with optimized FDM parameters for congenital heart disease treatment. *Journal of the Mechanical Behavior of Biomedical Materials*, 138: 105609.
10. Abdelrhman, A.M., Gan, W., Kurniawan, D. (2019) Effect of part orientation on dimensional accuracy, part strength, and surface quality of three dimensional printed part. *IOP Conference Series: Materials Science and Engineering*, 694: 012048.
11. Eryildiz, M. (2021) Effect of build orientation on mechanical behaviour and build time of FDM 3D-printed PLA parts: an experimental investigation. *European Mechanical Science*, 5: 116–120.
12. Afrose, M.F., Masood, S.H., Iovenitti, P., Nikzad, M., Sbarski, I. (2016) Effects of part build orientations on fatigue behaviour of FDM-processed PLA material. *Progress in Additive Manufacturing*, 1: 21–28.
13. Yao, T., Deng, Z., Zhang, K., Li, S. (2019) A method to predict the ultimate tensile strength of 3D printing polylactic acid (PLA) materials with different printing orientations. *Composites Part B: Engineering*, 163: 393–402.
14. Bardiya, S., Jerald, J., Satheeshkumar, V. (2021) The impact of process parameters on the tensile strength, flexural strength and the manufacturing time of fused filament fabricated (FFF) parts. *Materials Today: Proceedings*, 39: 1362–1366.
15. Sivakumar, N.K., Palaniyappan, S., Sekar, V., Alodhayb, A., Braim, M. (2023) An optimization approach for studying the effect of lattice unit cell's design-based factors on additively manufactured poly methyl methacrylate cranio-implant. *Journal of the Mechanical Behavior of Biomedical Materials*, 141: 105791.
16. Jatti, V.S., Jatti, S.V., Patel, A.P., Jatti, V.S. (2019) A study on effect of fused deposition modeling process parameters on mechanical properties. *International Journal of Scientific and Technology Research*, 8: 689–693.
17. Durga Prasada Rao, V., Rajiv, P., Navya Geethika, V. (2019) Effect of fused deposition modelling (FDM) process parameters on tensile strength of carbon fibre PLA. *Materials Today: Proceedings,* 18: 2012–2018.
18. Zhou, X., Hsieh, S.J., Sun, Y. (2017) Experimental and numerical investigation of the thermal behaviour of polylactic acid during the fused deposition process. *Virtual and Physical Prototyping*, 12: 221–233.
19. Choudhary, N., Sharma, V., Kumar, P. (2023) Polylactic acid-based composite using fused filament fabrication: process optimization and biomedical application. *Polymer Composites*, 44: 69–88.
20. Guo, W., Yang, Y., Liu, C., Bu, W., Guo, F., Li, J., Wang, E., Peng Z., Mai, H., You, H., Long, Y., (2023) 3D printed TPMS structural PLA/GO scaffold: process parameter optimization, porous structure, mechanical and biological properties. *Journal of the Mechanical Behavior of Biomedical Materials*, 142: 105848.
21. Pernet, B., Nagel, J.K., Zhang, H. (2022) Compressive strength assessment of 3D printing infill patterns. *Procedia CIRP*, 105: 682–687.
22. Aloyaydi, B., Sivasankaran, S., Mustafa, A. (2020) Investigation of infill-patterns on mechanical response of 3D printed poly-lactic-acid. *Polymer Testing*, 87: 106557.

23. Chadha, A., Ul Haq, M.I., Raina, A., Singh, R.R., Penumarti, N.B., Bishnoi, M.S. (2019) Effect of fused deposition modelling process parameters on mechanical properties of 3D printed parts. *World Journal of Engineering*, 16: 550–559.
24. Akhoundi, B., Behravesh, A.H. (2019) Effect of filling pattern on the tensile and flexural mechanical properties of FDM 3D printed products. *Experimental Mechanics*, 59: 883–897.
25. John, J., Devjani, D., Ali, S., Abdallah, S., Pervaiz, S. (2023) Optimization of 3D printed polylactic acid structures with different infill patterns using Taguchi-grey relational analysis. *Advanced Industrial and Engineering Polymer Research*, 6: 62–78.
26. Alafaghani, A., Qattawi, A., Alrawi, B., Guzman, A. (2017) Experimental optimization of fused deposition modelling processing parameters: a design-for-manufacturing approach. *Procedia Manufacturing*, 10: 791–803.
27. Jatti, V.S., Sapre, M.S., Jatti, A.V., Khedkar, N.K., Jatti, V.S. (2022) Mechanical properties of 3D-printed components using fused deposition modeling: optimization using the desirability approach and machine learning regressor. *Applied System Innovation*, 5: 112.
28. Muhamedagic, K., Berus, L., Potočnik, D., Cekic, A., Begic-Hajdarevic, D., Cohodar Husic, M., Ficko, M., (2022) Effect of process parameters on tensile strength of FDM printed carbon fiber reinforced polyamide parts. *Applied Sciences*, 12: 6028.
29. Luzanin, O., Guduric, V., Ristic, I., Muhic, S. (2017) Investigating impact of five build parameters on the maximum flexural force in FDM specimens – a definitive screening design approach. *Rapid Prototyping Journal*, 23: 1088–1098.
30. Popescu, D., Zapciu, A., Amza, C., Baciu, F., Marinescu, R. (2018) FDM process parameters influence over the mechanical properties of polymer specimens: a review. *Polymer Testing*, 69: 157–166.
31. Yang, T.-C., Yeh, C.-H. (2020) Morphology and mechanical properties of 3D printed wood fiber/polylactic acid composite parts using fused deposition modeling (FDM): the effects of printing speed. *Polymers*, 12: 1334.
32. Mathiazhagan, N., Palaniyappan, S., Sivakumar, N.K. (2023) Effect of fused filament fabrication parameters on crashworthiness studies of hydroxyapatite particle reinforced PLA composite thin-walled tubes. *Journal of the Mechanical Behavior of Biomedical Materials*, 138: 105611.
33. Miazio, Ł. (2019) Impact of print speed on strength of samples printed in FDM technology. *Agricultural Engineering*, 23: 33–38.
34. Saad, M.S., Mohd Nor, A., Zakaria, M.Z., Baharudin, M.E., Yusoff, W.S. (2021) Modelling and evolutionary computation optimization on FDM process for flexural strength using integrated approach RSM and PSO. *Progress in Additive Manufacturing*, 6: 143–154.
35. Algarni, M., Ghazali, S. Comparative study of the sensitivity of PLA, ABS, PEEK, and PETG's mechanical properties to FDM printing process parameters. *Crystals*, 2021, 11: 995.
36. Liu, X., Zhang, M., Li, S., Si, L., Peng, J., Hu, Y. (2017) Mechanical property parametric appraisal of fused deposition modeling parts based on the gray Taguchi method. *The International Journal of Advanced Manufacturing Technology*, 89: 2387–2397.
37. Rajpurohit, S.R., Dave, H.K. (2018) Flexural strength of fused filament fabricated (FFF) PLA parts on an open-source 3D printer. *Advances in Manufacturing*, 6: 430–441.
38. Sood, A.K., Ohdar, R.K., Mahapatra, S.S. (2010) Parametric appraisal of mechanical property of fused deposition modelling processed parts. *Materials and Design*, 31: 287–295.

39. Mishra, S.B., Abhishek, K., Satapathy, M.P., Mahapatra, S.S. (2017) Parametric appraisal of compressive strength of FDM build parts. *Materials Today: Proceedings*, 4: 9456–9460.
40. Sandanamsamy, L., Mogan, J., Rajan, K., Harun, W.S.W., Ishak, I., Romlay, F.R.M., Samykano, M., Kadirgama, K., (2023) Effect of process parameter on tensile properties of FDM printed PLA. *Materials Today: Proceedings*, 109: 1–6.
41. Maldonado-García, B., Pal, A.K., Misra, M., Gregori, S., Mohanty, A.K. (2021) Sustainable 3D printed composites from recycled ocean plastics and pyrolyzed soy-hulls: optimization of printing parameters, performance studies and prototypes development. *Composites Part C: Open Access*, 6: 100197.
42. Myers, D., Abdel-Wahab, A., Hafeez, F., Kovacev, N., Essa, K. (2022) Optimisation of the additive manufacturing parameters of polylactic acid (PLA) cellular structures for biomedical applications. *Journal of the Mechanical Behavior of Biomedical Materials*, 136: 105447.
43. Sai, T., Pathak, V.K., Srivastava, A.K. (2020) Modeling and optimization of fused deposition modeling (FDM) process through printing PLA implants using adaptive neuro-fuzzy inference system (ANFIS) model and whale optimization algorithm. *Journal of the Brazilian Society of Mechanical Sciences and Engineering*, 42: 617.
44. Dusanapudi, S., Krupakaran, R.L., Srinivas, A., Nikhil, K.S., Vamshi, T. (2023) Optimization and experimental analysis of mechanical properties and porosity on FDM based 3D printed ABS sample. *Materials Today: Proceedings.*
45. Almuflih, A.S., Abas, M., Khan, I., Noor, S. (2024) Parametric optimization of FDM process for PA12-CF parts using integrated response surface methodology, grey relational analysis, and grey wolf optimization. *Polymers*, 16: 1508.
46. Raju, M., Gupta, M.K., Bhanot, N., Sharma, V.S. (2019) A hybrid PSO–BFO evolutionary algorithm for optimization of fused deposition modelling process parameters. *Journal of Intelligent Manufacturing*, 30: 2743–2758.
47. Seyedzavvar, M. (2023) A hybrid ANN/PSO optimization of material composition and process parameters for enhancement of mechanical characteristics of 3D-printed sample. *Rapid Prototyping Journal*, 29: 1270–1288.
48. Deshwal, S., Kumar, A., Chhabra, D. (2020) Exercising hybrid statistical tools GA-RSM, GA-ANN and GA-ANFIS to optimize FDM process parameters for tensile strength improvement. *CIRP Journal of Manufacturing Science and Technology*, 31: 189–199.
49. Singh, B., Kumar, R., Chohan, J.S.. (2021) Multi-objective optimization of 3D printing process using genetic algorithm for fabrication of copper reinforced ABS parts. *Materials Today: Proceedings*, 48: 981–988.
50. Khusheef, A.S., Hashemi, R., Shahbazi, M. (2024) Optimizing FDM process parameters: predictive insights through Taguchi, regression, and neural networks. *Physica Scripta*, 99: 066005.
51. Giri, J., Shahane, P., Jachak, S., Chadge, R., Giri, P. (2021) Optimization of FDM process parameters for dual extruder 3D printer using artificial neural network. *Materials Today: Proceedings*, 43: 3242–3249.

5 Parametric Optimization of FDM Process Using Heat Transfer Search Algorithm

Nikhil Aditya and Siba Sankar Mahapatra

5.1 INTRODUCTION

Nowadays, industries depend on advanced technologies to meet more exacting processes with better quality to address the recent competitive market. Competition in all industrial sectors, like automobiles, nuclear, electronics, aeronautics, aerospace, chemical, oil, and gas, pressures the manufacturers to improve the performance of their machining processes. Many researchers are trying to improve the machining processes using artificial intelligence and machine learning (AI/ML). In the last few years, additive manufacturing (AM) techniques have come into existence owing to the limitations of traditional methods, such as material wastage in the form of chips and the inability to produce complex and intricate shapes with good dimensional accuracy [1]. Contradictory to traditional subtractive manufacturing techniques, AM is the process that creates products via layer-by-layer deposition using a three-dimensional (3D) model as described by the American Society for Testing and Materials Standard (ASTM) [2]. The seven types of AM, as defined by ASTM, are powder bed fusion, VAT photopolymerization, material extrusion, direct energy deposition, sheet lamination, and binder jetting [3]. Fused deposition modeling (FDM) comes under the category of material extrusion. In the material extrusion method, a composite or thermoplastic material filament is used to make a 3D product. Acrylonitrile butadiene styrene (ABS), polylactic acid (PLA), and polycarbonate (PC) are the most frequently used thermoplastic base materials in the FDM process, which are made to reach a semi-liquid stage by heating before deposition on the build platform. After heating, the material's deposition takes place layer-wise to obtain the desired 3D part [4]. Compared to alternative 3D printing techniques, the FDM process is more affordable. However, poor surface finish and the need for a support structure is a major disadvantage [5]. Furthermore, several input variables are associated with the FDM process that are responsible for the quality of the final produced part. These parameters include print speed, infill density, nozzle temperature, part orientation, layer thickness, raster width, air gap, and raster angle [6]. A variety of Design of Experiments (DOE) methods have been used to increase the FDM process's efficiency [7]. DOE methods like Taguchi's experimental design and response surface

DOI: 10.1201/9781003686941-5

methodology were widely used in the FDM process to design the experiments and for statistical analysis. Ahmad et al. [8] carried out parametric optimization for oil palm fiber-reinforced composites made by FDM. The impact of variables like printing speed, infill density, orientation, and layer thickness on responses like tensile strength and flexural strength was ascertained by the study using the Taguchi method. Moradi et al. [9] looked into the impact of the number of contours, infill percentage, and layer thickness on failure loads, build time, and part weight of 3D-printed nylon through FDM using response surface method (RSM). The effects of bed temperature, infill percentage, and layer height on the density, surface roughness, and micro-hardness of 3D-printed ABS by FDM have been studied by Venkatraman and Raghuraman [10]. A total of 15 experiments have been conducted based on Box–Behenken Design (BBD). Chohan et al. [11] studied the impact of a chemical-based finishing process on the FDM-produced part surface finish using Taguchi and the technique for order of preference by similarity to the ideal solution (TOPSIS) approach. Heidari-Rarani et al. [12] used the Taguchi method to optimize the FDM's input parameters for PLA. Khalid and Peng [13] analyzed the impact of five input parameters, such as percentage infill, infill type, build orientation, height of the layer, and shell number, on the sustainability of the FDM process. The Taguchi method has been used to optimize the input process variables. To enhance the part quality, sustainability, and productivity, Camposeco-Negrete [14] optimized the FDM process using the desirability approach and Taguchi's method. Zaman et al. [15] performed the optimization of the FDM's process parameter using the Taguchi method to maximize the compressive strength of the built part. Lestari et al. [16] optimized the FDM process for prosthetics manufacturing. The main objective is to minimize the printing time and socket weight using DOE techniques such as the Taguchi method and RSM. Ambade et al. [17] used the BBD method to analyze the impact of different input variables on the tensile strength, percentage elongation, and maximum force of the part made by the FDM process using PLA. Sahoo et al. [18] used Taguchi's L9 orthogonal array to analyze the effect of print speed, single layer thickness, and infill percentage on the strength and hardness of the FDM-printed part. The results of the study have concluded that the infill percentage and the thickness of the layer are the major contributing factors affecting the strength and hardness of the part. To enhance the mechanical characteristics and dimensional accuracy of the FDM-printed part using PLA, Popović et al. [19] analyzed the impact of the speed of the printing and the nozzle temperature on tensile strength and surface roughness of the build part.

Apart from various DOE techniques, metaheuristics have been widely used to obtain quality solutions for different parameters of the FDM process [20]. Most of the FDM optimization methods are nature-inspired algorithms belonging to AI. Chohan et al. [21] used the Taguchi technique and genetic algorithm (GA) to optimize the FDM process's process variables. Saad et al. [22] used a hybrid approach consisting of particle swarm optimization (PSO) and RSM to maximize the flexural strength of the FDM-produced part. Deshwal et al. [23] used a combination of DOE and metaheuristic algorithms to determine the optimum parameters of the FDM process for achieving maximum tensile strength. They hybridized the optimization techniques such as adaptive neuro-fuzzy inference system (ANFIS), RSM, GA, and artificial neural network (ANN). The mechanical characteristics of the FDM-produced specimens, including flexural, impact, tensile strength, and surface roughness, have been optimized by Chinchanikar et al. [24]. They have used optimization techniques such as non-dominated sorting genetic algorithm (NSGA), TOPSIS, RSM, and gray relational analysis (GRA). The hybrid approach using GA with RSM has proven to be better than other optimization techniques. Yadav et al. [25] studied the impact of different input variables, such as jerk, printing speed, and acceleration, on the surface roughness of the final part produced by FDM. The three input parameters are monitored to minimize the surface roughness. Sai et al. [26] used ANFIS and whale optimization algorithms (WOAs) to obtain the best configuration of infill density, raster angle, layer height, and internal structure for an FDM-produced implant for biomedical applications. The thorough literature analysis leads to the conclusion that high dimensional accuracy, improved mechanical qualities, surface smoothness, and build time can only be achieved by optimizing the FDM process [27]. Apart from the input variables related to the FDM process, Chinchanikar et al. [28] analyzed the impact of number of neurons and hidden layers of ANN on the output response of the FDM process. Tura et al. [29] analyzed the influence of different input variables on the tensile strength of the part made by the FDM process. ANN and fuzzy logic have been deployed to get a prediction model. The results concluded that the infill density is a more significant input variable compared to extruder temperature and print speed. Ulkir and Akgun [30] aimed to minimize the surface roughness of the part produced by the FDM process. Apart from the Box–Bhenken design approach, ANN is used to determine the relationship between input and output variables, which is further optimized using GA. Several DOE techniques have been used to achieve the optimal value for different process parameters, such as RSM, BBD, and GRA. GA and PSO have been the first choice in the metaheuristic algorithms category for the optimization of the input variables of the FDM process. Some hybrid techniques, such as PSO-RSM, GA-RSM, and GA-ANN, have been used recently. The industry welcomes any improvement in the manufacturing process as the advancement in manufacturing helps in the growth of the industry. However, despite having a minimal number of user-defined parameters and simple architecture [31], there is a lack of literature evaluating the performance of the heat transfer search (HTS) algorithm on FDM process optimization. Furthermore, the HTS algorithm has outperformed some well-performing algorithms, such as teaching–learning-based optimization (TLBO) and differential evolution (DE) on constrained optimization [32]. Therefore, the present study uses the HTS algorithm to evaluate its performance by optimizing five objective functions related to the FDM process and comparing them fairly against the state-of-the-art algorithms.

5.2 HTS ALGORITHM

The laws of heat transfer and thermodynamics serve as the motivation for the HTS algorithm, which is a population-based metaheuristic optimization algorithm [32]. The HTS algorithm imitates the thermal interaction between the surroundings and the system, where thermal equilibrium can be established via three heat transfer modes: radiation, convection, and conduction. The HTS method compares the population of candidate solutions to the molecules in a system, while the fitness function and decision variables stand in for the system's energy and temperature. The search for the optimal solution in HTS occurs by giving equal probability to each mode of heat transfer. The range of probabilities (R) for the three phases of heat transfer is 0–0.3333, 0.3333–0.6666, and 0.6666–1. The values for the probability are the random numbers generated from the uniform distribution. After executing the three modes of heat transfer, the greedy selection is performed to keep the improved solutions, and elite solutions replace the two worst solutions. The detailed search mechanism of the HTS algorithm is given below:

Let the population vector be given as:

$$X = \left\{x_1^1, x_2^1, \ldots, x_d^1, \ldots\ldots x_n^m\right\} \tag{5.1}$$

In Equation (5.1), X denotes the population of size m and dimension n. The conduction phase is executed when the R value lies between 0 and 0.3333, and only a single dimension of each candidate solution is updated. The conduction phase occurs in two phases. The first phase of conduction is executed when $iter \leq \frac{T_{max}}{CDF}$, where $iter$ represents the number of the current iteration, T_{max} denotes the limiting number of iterations, and CDF is the conduction factor whose value is kept at 2 to maintain the intensification and diversification behavior of the algorithm [32]. The solutions in phase one of the conduction are updated according to Equation (5.2):

$$\left.\begin{array}{ll} x_{i,j}^{new} = x_{k,j}^{old} + CDS_1, & f(x_i) > f(x_k) \\ x_{k,j}^{new} = x_{i,j}^{old} + CDS_2, & f(x_k) > f(x_i) \end{array}\right\}, \; iter \leq \frac{T_{max}}{CDF} \tag{5.2}$$

In Equation (5.2), i denotes the candidate solutions ($i \in 1, 2, \ldots, n$), whereas k represents the randomly selected solution such that $i \neq k$. However, for each candidate solution, only the jth dimension is updated. CDS_1 and CDS_2 regulate the step length for randomly selected variable j. The mathematical equations for CDS_1 and CDS_2 are given below:

$$CDS_1 = -R^2 x_{k,j}^{old} \tag{5.3}$$

$$CDS_2 = -R^2 x_{i,j}^{old} \tag{5.4}$$

In Equations (5.3) and (5.4), R varies between 0 and 0.3333. Therefore, R^2 gives a smaller quantity than R, which leads to a refined search or intensification. The second phase of the conduction phase is executed when $iter \geq \frac{T_{max}}{CDF}$. In the second phase of conduction, the updation equations for the particular dimension are the same as Equation (5.2). However, step length (CDS_3 and CDS_4) is defined to promote exploration. Mathematically, CDS_3 and CDS_4 can be defined as follows:

$$CDS_3 = -rx_{k,j}^{old} \tag{5.5}$$

$$CDS_4 = -rx_{i,j}^{old} \tag{5.6}$$

In Equations (5.5) and (5.6), the value of r varies between 0 and 1, providing more step length than phase one of the conduction. Therefore, phase two of the conduction process assists exploration.

The convection phase in the heat transfer search algorithm is executed when the R-value lies between 0.6666 and 1. In the convection phase, each dimension of every candidate solution is updated by moving it toward the solution with the best fitness value in the population. The procedure for the same is given below:

During stage one, when $iter \le \frac{T_{max}}{COF}$, where COF (convection factor) is set to 10 [32], the equations for updating the solutions are as follows:

$$\left.\begin{aligned} x_{i,j}^{new} &= x_{i,j}^{old} + COS \\ COS &= R\left(x_s - x_{ms} \times TCF\right) \\ TCF &= \left|R - r\right| \end{aligned}\right\}, \; iter \le \frac{T_{max}}{COF} \tag{5.7}$$

In Equation (5.7), i varies from 1 to m, and j ranges from 1 to n to update all dimensions of the population. x_s is the dimension corresponding to the best solution, whereas x_{ms} is the mean value of the particular dimension of the population. During the second stage of convection, TCF is updated as given below in Equation (5.8):

$$TCF = round\left(1 + r\right)\}, \; iter \ge \frac{T_{max}}{COF} \tag{5.8}$$

TCF controls the diversification and intensification in the convection phase as it maintains a value between 0 and 1 in the first stage and is allotted a value of either 1 or 2 in the second stage.

The radiation phase in the HTS algorithm is executed once R attains a value between 0.3333 and 0.6666. In the radiation heat transfer phase, a particular solution interacts with the best solution and the other candidate solutions in the population. In the first stage of radiation, when $iter \le \frac{T_{max}}{RDF}$, where RDF (radiation factor) is set to 2 [32], the equations for updating the solutions are given below:

$$\left.\begin{array}{r} x_{i,j}^{new} = x_{i,j}^{old} + RDS_1,\ f(x_k) < f(x_i) \\ x_{i,j}^{new} = x_{i,j}^{old} + RDS_2,\ f(x_i) < f(x_k) \\ RDS_1 = R\left(x_{k,j}^{old} - x_{i,j}^{old}\right) \\ RDS_2 = R\left(x_{i,j}^{old} - x_{k,j}^{old}\right) \end{array}\right\},\ iter \le \frac{T_{max}}{RDF} \tag{5.9}$$

In Equation (5.9), k is the randomly chosen solution from the population ($k \in 1, 2, \ldots, m$), which interacts with the rest of the solutions ($i = 1, 2, \ldots, m$) in each dimension ($j = 1, 2, \ldots, n$), unlike the conduction phase. Similarly, in the second phase of radiation, the solutions are updated according to Equation (5.10):

$$\left.\begin{array}{r} x_{i,j}^{new} = x_{i,j}^{old} + RDS_3,\ f(x_k) < f(x_i) \\ x_{i,j}^{new} = x_{i,j}^{old} + RDS_4,\ f(x_i) < f(x_k) \\ RDS_3 = r\left(x_{k,j}^{old} - x_{i,j}^{old}\right) \\ RDS_4 = r\left(x_{i,j}^{old} - x_{k,j}^{old}\right) \end{array}\right\},\ iter \ge \frac{T_{max}}{RDF} \tag{5.10}$$

Three phases are necessary for the HTS algorithm's search process to function: radiation, conduction, and convection. Each phase of the heat transfer is further subdivided into two stages based on iteration number, and the search process in each stage depends upon *CDF*, *COF*, *RDF*, CDS_1, CDS_2, $CDS_{3,}$ $CDS_{4,}$ *TCF*, $RDS_{1,}$ RDS_2, RDS_3, and RDS_4. These factors have proven to provide the algorithm with balanced intensification and diversification characteristics [33]. After executing any one of the heat transfer phases, greedy selection is performed to keep quality solutions, and the two best solutions replace the two worst solutions in the population. Therefore, the present study uses the HTS algorithms' search mechanism to obtain the optimal values of the input variables of the FDM process.

5.3 RESULTS AND DISCUSSION

The performance of the HTS algorithm is assessed in this study by solving the five problems related to FDM process optimization. There are no user-defined parameters in the HTS algorithm except the population size (m) and maximum number of iterations ($iter_{max}$). However, an unbiased comparison is made against the state-of-the-art methods by maintaining the m and $iter_{max}$ so that the maximum value of the function evaluations is the same as that of competing methods for a particular problem. The five problems with their description and results of the HTS algorithm are given belowwhere the performance of the best algorithm compared to the state of the art algorithms is highlighted in bold.

5.3.1 Problem 1

Peng et al. [34] performed the parametric optimization of the FDM process to obtain optimal values of width of the filament 'X_1' (mm), velocity of extrusion 'X_2' (mm/s), and filling 'X_3' (mm/s), and thickness of layer 'X_4' (mm). The output responses include wrap deformation, build time, and dimensional error. A fuzzy inference system is used to produce a combined response (y), and second-order RSM is used to ascertain the link between the single response and the input variables. The problem statement is given in Equation (5.11):

$$\begin{aligned} Max\, y = {} & 806 - 33763.1 \times X1 + 225.30 \times X_2 + 14.81 \times X_3 \\ & - 2759.88 \times X_4 + 67560.93 \times X_1^2 - 3.57 \times X_2^2 + 0.17 \times X_3^2 \\ & + 10607.87 \times X_4^2 + 17.16 \times X_1 \times X_2 + 172.22 \times X_1 \times X_3 \\ & - 1.86 \times X_2 \times X_3 + 5.57 \times X_2 \times X_4 - 61.53 \times X_3 \times X_4 \end{aligned} \quad (5.11)$$

In Equation (5.11), X_1 lies beween 0.1700 and 0.2500, X_2 and X_3 are bounded between 20 and 30, whereas the search bound for X_4 lies in the range of 0.1500 to 0.3000. Rao and Rai [35] found the solution to this problem using TLBO. They maintained an upper limit on the function evaluations, equal to 2000, equal to the function evaluations maintained by Peng et al. [34] while solving the problem with the GA toolbox. Therefore, the present work initializes the HTS algorithm with the population size (m) of 50 while running it for the maximum iterations ($iter_{max}$) of 20. The algorithm is run 25 times independently to obtain the average performance. The best results obtained by GA, TLBO, and HTS algorithms are given in Table 5.1.

From Table 5.1, it can be noted that the HTS algorithm produces better results than GA and matches the performance of TLBO in a limited number of iterations. The mean values obtained by the TLBO and HTS algorithms with standard deviation (Std) are given in Table 5.2.

Table 5.2 demonstrates that TLBO maintains better stability than the HTS algorithm by achieving a mean value of 334.65 and Std of 0. However, the HTS reaches the

TABLE 5.1
Best Results Obtained by GA, TLBO, and HTS Algorithm

Parameters	GA	TLBO	HTS
x_1	0.1702 mm	0.25 mm	0.25 mm
x_2	22.4908 mm/s	21.8523 mm/s	21.85286 mm/s
x_3	23.896 mm/s	40 mm/s	40 mm/s
x_4	0.2875 mm	0.15 mm	0.15 mm
y	192.0682	**334.65**	**334.657**

Note: The bold entries indicate the comparable or better performance of the algorithms in comparison with the state of the art algorithms.

TABLE 5.2
Mean Value Comparison of TLBO and HTS Algorithms

Algorithm	Best	Worst	Mean	Std
TLBO	**334.65**	**334.65**	**334.65**	**0**
HTS	**334.657**	294.334	329.023	9.324

Note: The bold entries indicate the comparable or better performance of the algorithms in comparison with the state of the art algorithms.

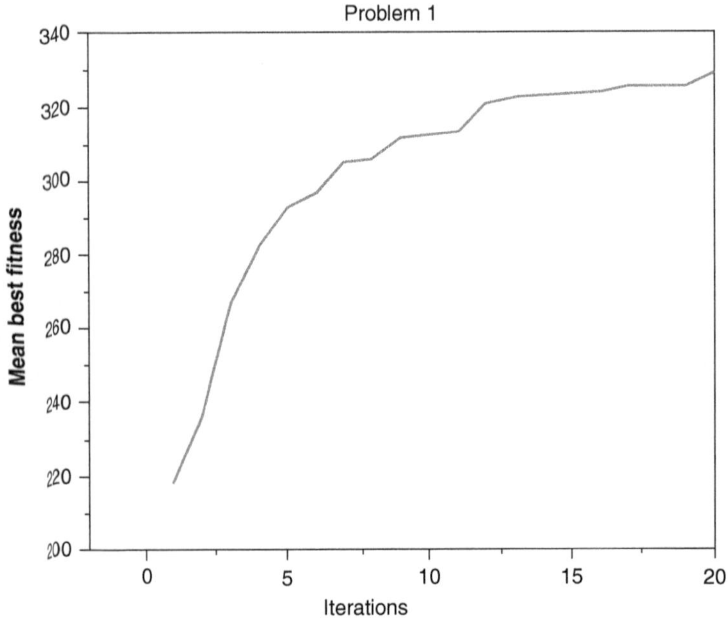

FIGURE 5.1 Search history of the HTS algorithm for problem 1.

same best value as TLBO but lacks the precision to maintain the best value during all 25 runs. The convergence history of the HTS for the problem is shown in Figure 5.1.

From Figure 5.1, it can be seen that the HTS provides competitive results within 20 iterations; however, further increments in the number of iterations may improve the mean fitness value obtained by the algorithm.

5.3.2 Problem 2

Mohanty et al. [36] optimized the FDM's process variables, such as the thickness of the single layer 'X_1' (mm), part orientation 'X_2' (°), raster angle 'X_3' (°) and width 'X_4' (mm) and air gap 'X_5' (mm) to maintain the build part's dimensional accuracy (y).

The dimensional accuracy of the build part is assessed by horizontal length, vertical length, angle of inclination, and hole diameter. A regression equation is established between output responses and input parameters, which is given below:

$$\begin{aligned} Min\, y = & -5.99 + 0.26 \times X_1 - 0.01418 \times X_2 - 0.00573 \times X_3 \\ & + 28.65 \times X_4 - 46.2 \times X_5 + 2.52 \times X_1^2 + 0.000195 \times X_2^2 \\ & + 0.000104 \times X_3^2 - 31.56 \times X_4^2 + 5408 \times X_5^2 - 60.3 \times X_1 \\ & \times X_5 + 0.413 \times X_2 \times X_5 - 0.061 \times X_3 \times X_5 - 6.2 \times X_4 \times X_5 \end{aligned} \tag{5.12}$$

The variable X_1 is restricted between 0.127 and 0.254, X_2 and X_3 share a common lower bound value of 0. However, the upper bound for X_2 and X_3 is 30 and 60, respectively. The variable X_4 is confined between 0.4064 and 0.5064, whereas X_5 is searched in the range of 0 to 0.008. Apart from the experimental investigation, Mohanty et al. [36] used 10 metaheuristic approaches, namely GA, forensic-based investigation (FBI) optimization, PSO, simulated annealing (SA), moth-flame optimization (MFO), Lichtenberg optimization algorithm (LOA), grey wolf optimization (GWO), sunflower optimization algorithm (SOA), Jaya algorithm, and WOA. Different algorithms have converged at different number of iterations. However, the current study maintains a population size (m) of 30 with maximum iterations ($iter_{max}$) of 200. The comparison study of best values obtained by the HTS algorithm is given in Table 5.3. The results of the state-of-the-art algorithms are taken from Mohanty et al. [36].

Regarding the minimization, the best values are obtained by MFO, WOA, Jaya algorithm, SOA, FBI, and LOA. The HTS algorithm competes strongly with the state-of-the-art algorithms and matches their performance with limited function evaluations. The mean value of the output response obtained by the HTS algorithm is evaluated by running it independently 25 times. The preliminary statistical analysis of the HTS algorithm by reporting best, worst, mean, and Std is mentioned in Table 5.4.

From Table 5.4, it follows that the HTS algorithm cannot maintain the best value of 0.070711 for all 25 runs, therefore attaining a mean value of 0.079908 and a Std of 0.012544. The search history for the problem is given in Figure 5.2.

The search history depicted in Figure 5.2 shows the progressive improvement in the minimization of the objective function, and convergence is obtained within 200 iterations.

5.3.3 Problem 3

Dev and Srivastava [37] performed an experimental investigation using ABS material to maximize the flexural strength (y) of the FDM build part. The study identified the significant parameters affecting the flexural strength using an analysis of variance (ANOVA). These significant parameters are the layer thickness 'X_1' (mm), nozzle temperature 'X_2' (°), and print head speed 'X_3' (mm/s). The relationship between the input parameters and flexural is established using RSM, which is used as a fitness

TABLE 5.3
Comparison of the HTS Algorithm with the State-of-the-Art Algorithms

Parameters	GA	PSO	SA	MFO	GWO	WOA	Jaya	SOA	FBI	LOA	HTS
x_1	0.127	0.127	0.127	**0.127**	0.127	**0.127**	**0.127**	**0.127**	**0.127**	**0.127**	**0.127**
x_2	30	29.99	28.48	**30**	30	**30**	**30**	**30**	**29.99**	**30**	**30**
x_3	28.79	28.81	27.91	**28.8**	29.33	**28.8**	**28.8**	**28.8**	**28.84**	**28.83**	**28.803**
x_4	0.5064	0.5064	0.5064	**0.5064**	0.5064	**0.5064**	**0.5064**	**0.5064**	**0.5064**	**0.5064**	**0.5064**
x_5	0.004	0.004	0.004	**0.004**	0.004	**0.004**	**0.004**	**0.004**	**0.004**	**0.004**	**0.004**
y	0.070715	0.070712	0.07762	**0.070711**	0.070743	**0.070711**	**0.070711**	**0.070711**	**0.707**	**0.070711**	**0.070711**

Note: The bold entries indicate the comparable or better performance of the algorithms in comparison with the state of the art algorithms.

TABLE 5.4
Average Performance of the HTS Algorithm

Algorithm	Best	Worst	Mean	Std
HTS	0.070711	0.10337	0.079908	0.012544

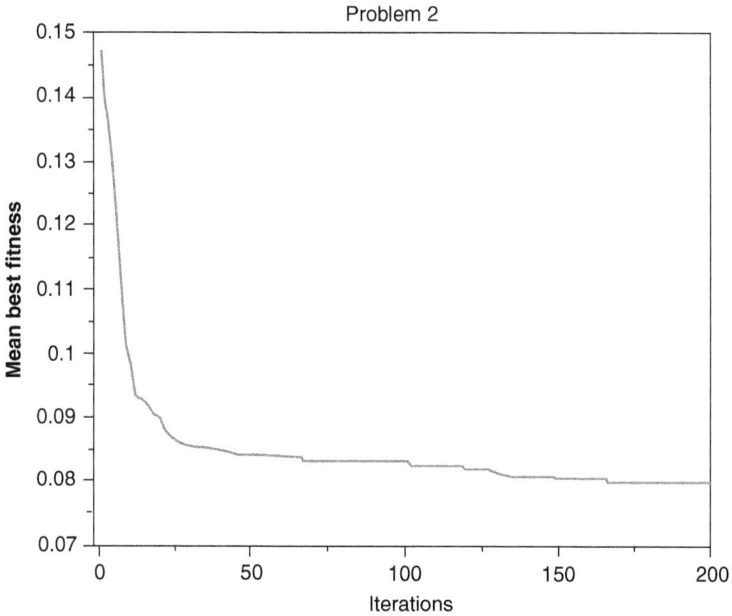

FIGURE 5.2 Search history of the HTS algorithm for problem 2.

function in GA. The resulting methodology is named as RSM-GA. Equation (5.13) ascertains the link between the output response and the input variables:

$$\begin{aligned} Max\,y = & -1545 - 671 \times X_1 + 14.66 \times X_2 - 0.2013 \times X_3 \\ & + 349.4 \times X_1^2 - 0.03306 \times X_2^2 + 2.142 \times X_1 \times X_2 \end{aligned} \tag{5.13}$$

The variable X_1 lies between 0.1 and 0.3, whereas the search bound for the X_2 variable is 220 and 240. Furthermore, the X_3 variable is confined between 30 and 70. The present work uses m equal to 30 and $iter_{max}$ equal to 500. The best results furnished by the HTS algorithm and RSM-GA are given in Table 5.5.

From Table 5.5, it can be noticed that the HTS algorithm produces better results than RSM-GA during maximization of flexural strength. However, the average value obtained by RSM-GA is superior to the HTS algorithm, as RSM-GA produces a

TABLE 5.5
Best Results Obtained by RSM-GA and HTS Algorithm

Parameters	RSM-GA	HTS
x_1	0.120 mm	**0.1 mm**
x_2	224.958°	**224.9577°**
x_3	30.356 mm/s	**30 mm/s**
Y	58.3862 MPa	**58.38756 MPa**

Note: The bold entries indicate the comparable or better performance of the algorithms in comparison with the state of the art algorithms.

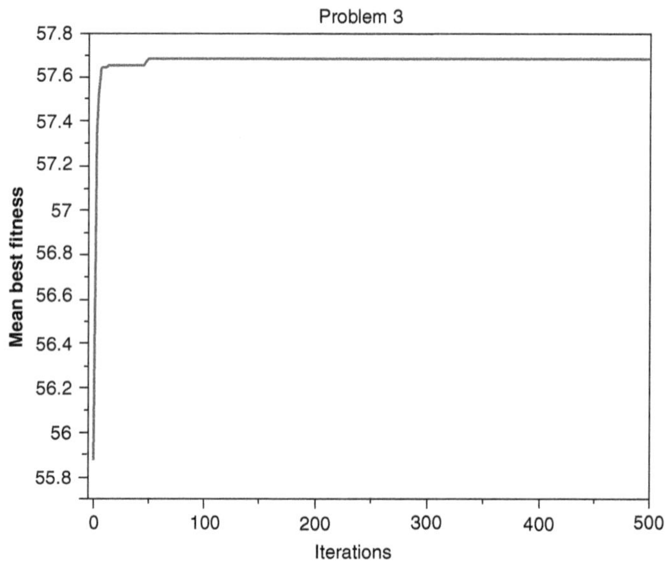

FIGURE 5.3 Search history of the HTS algorithm for problem 3.

mean value of 58.3791, whereas the HTS algorithm reports a mean value of 57.6852. The convergence curve for the problem is given in Figure 5.3. From Figure 5.3, it can noticed that the HTS algorithm converges to the optimal value of 57.6852 within 50 iterations.

5.3.4 Problem 4

Rayegani and Onwubolu [38] analyzed the impact of input variables, such as the orientation of the part 'X_1' (°), raster angle 'X_2' (°), raster width 'X_3' (mm), and air gap 'X_4' (mm) on tensile strength (y) of the part made by the FDM process. A non-linear relationship is established between output and input variables based on the group method for data handling (GMDH), given in Equation (5.14):

TABLE 5.6
Best Results Obtained by DE and HTS Algorithm

Parameters	DE	HTS
x_1	**0°**	**0°**
x_2	**50°**	**45°**
x_3	**0.2034 mm**	**0.2034 mm**
x_4	**–0.0025 mm**	**–0.0025 mm**
Y	**36.8604 MPa**	**36.8604 MPa**

Note: The bold entries indicate the comparable or better performance of the algorithms in comparison with the state of the art algorithms.

$$\begin{aligned} Max\, y = {} & 33.8477 + 0\times X_1 + 0\times X_2 - 1.6447\times X_3 - 45.5050 \\ & \times X_4 - 0.0011\times X_1\times X_2 + 0.0206\times X_1\times X_3 + 0.2293\times X_1 \\ & \times X_4 - 0.0671\times X_2\times X_3 - 0.0732\times X_2\times X_4 + 33.1809\times X_3 \\ & \times X_4 - 0.0013\times X_1^2 + 0.0020\times X_2^2 - 4.6791\times X_3^2 - 23.8441\times X_4^2 \end{aligned} \quad (5.14)$$

The lower bound for the X_1 and X_2 is 0, whereas the upper bound is 90 and 45, respectively. The search range for X_3 lies between 0.2034 and 0.5588, whereas the variable X_4 is searched between –0.0025 and 0.5588. Rayegani and Onwubolu [38] obtained the best input variable values to get the tensile strength's maximal value using DE. The population size (m) and maximum iterations for the DE are kept at 50 and 1000 [38]. Therefore, the HTS algorithm is initialized with a population size (m) of 50 and run for the maximum iterations of 1000. The best results obtained by DE and HTS algorithms are given in Table 5.6.

The HTS algorithm achieves the maximum tensile strength value of 36.8604 MPa, matching the DE algorithm's performance. However, the mean value obtained by the HTS algorithm for the 25 runs is 36.85851, which is a bit inferior to the mean value reported by the DE algorithm (36.8604). The convergence curve in Figure 5.4 shows the convergence efficiency of the HTS algorithm to reach the optimal solution in the limited number of iterations.

5.3.5 Problem 5

The base material's quality significantly impacts the characteristics of parts created by FDM. To boost the mechanical characteristics of the ABS–polymer blend filament (ABS/PC), Kumar et al. [39] investigated the impact of the speed of screw 'X_1' (wt.%), the composition of ABS/PC 'X_2' (RPM), and temperature of barrel 'X_3' (°C) on the filaments's tensile property (y). Apart from DOE, GA and ANN are combined with RSM (GA-RSM, GA-ANN). The expression relating to the input and output parameters is given in Equation (5.15):

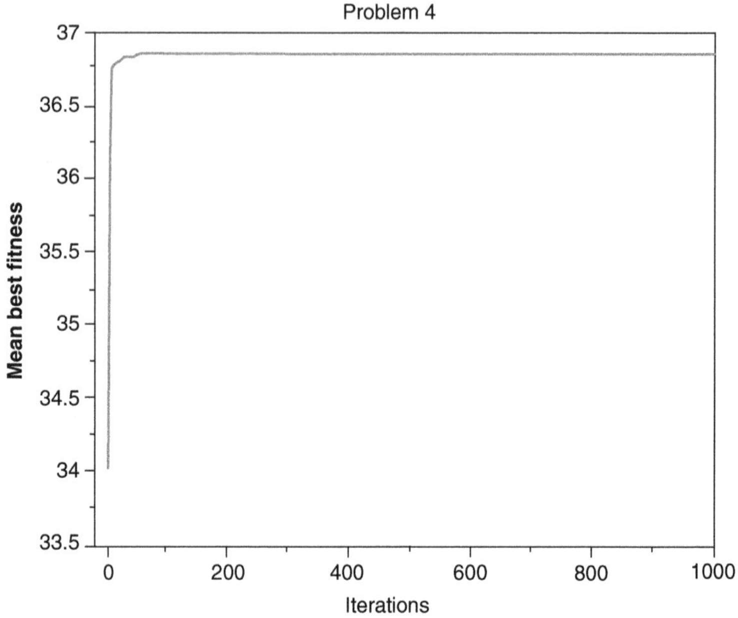

FIGURE 5.4 Search history of the HTS algorithm for problem 4.

$$Max\,y = -855.37075 + 10.31479 \times X_1 - 1.31512 \times X_2 + 8.50281 \times X_3 - 0.009393 \times X_1 \times X_2 - 0.027292 \times X_1 \times X_3 + 0.007236 \times X_2 \times X_3 - 0.386537 \times X_1^2 - 0.020469 \times X_3^2 \quad (5.15)$$

The search range for X_1 lies between 1 and 9, whereas for X_2 it is 10 and 50. The lower and upper bounds for X_3 are 210 and 226, respectively. Kumar et al. [39] ran their hybrid models for a maximum iteration of 20 with a population size of 100. Therefore, the HTS algorithm is run for the maximum iterations ($iter_{max}$) of 20 while maintaining a population size (m) of 100. The best values for different parameters obtained by HTS, GA-RSM, and GA-ANN are presented in Table 5.7.

The HTS algorithm obtains better results than GA-RSM, as observed in Table 5.7. However, GA-ANN (50.112 MPa) provides better results than the HTS algorithm and GA-RSM. The average performance of the HTS algorithm (mean value equals 49.40021 MPa) is better than GA-RSM (mean value equals 49.3906 MPa). However, GA-ANN produces a better mean value (50.6739 MPa) than the HTS algorithm. The convergence curve in Figure 5.5 shows that there is still a scope for improvement in the quality of the solution. Therefore, it can be concluded that the competitive results are obtained at lower iterations. However, the quality of the solutions can be enhanced by increasing the number of iterations in the HTS algorithm.

TABLE 5.7
Best Values Obtained by GA-RSM, GA-ANN, and HTS Algorithm

Parameters	GA-RSM	GA-ANN	HTS
x_1	5 RPM	**4 RPM**	5 RPM
x_2	50.054/49.946%	**50.223/49.777%**	50/50%
x_3	213°C	**210°C**	212.9435°C
Y	49.404 MPa	**50.112 MPa**	49.41369 MPa

Note: The bold entries indicate the comparable or better performance of the algorithms in comparison with the state of the art algorithms.

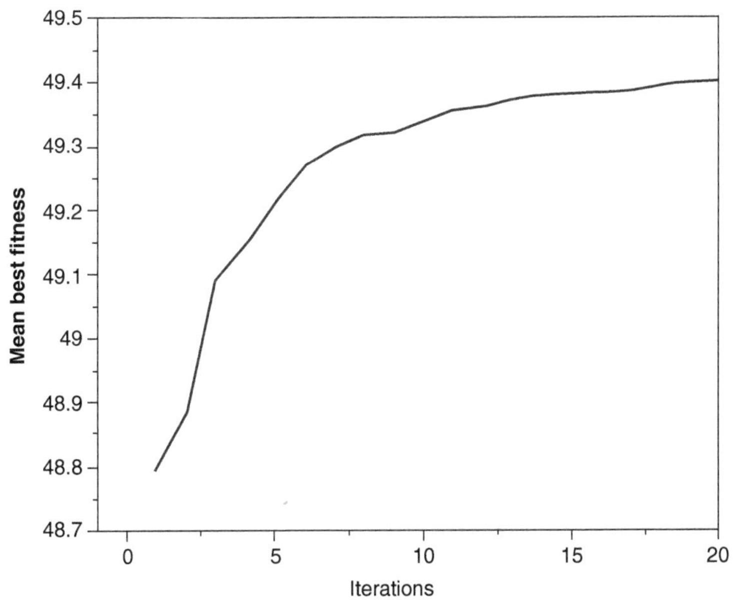

FIGURE 5.5 Search history of the HTS algorithm for problem 5.

5.4 CONCLUSION

AM processes like FDM are widely used in different manufacturing industries and are gaining popularity due to reduced wastage of materials and the ability to produce complex-shaped products. The present study determines the potential of the HTS algorithm during the optimization of different objectives related to the FDM process. The comparison is performed with the state-of-the-art algorithms on five different problems. The HTS algorithm matches the best-found solution by TLBO, MFO, WOA, Jaya, SOA, FBI, LOA, and DE. Furthermore, the results obtained by the HTS algorithm are superior to GA, PSO, SA, GWO, and GA-RSM. However, the algorithm's stability can be improved to obtain better mean values than TLBO, DE, GA-RSM, and GA-ANN. The convergence analysis proves that the HTS algorithm can produce competitive results at lower iterations, but for some problems, it leads to a premature solution due to lower exploration. Further clarification on the comparison study can be introduced by comparing different statistics of the results, such as best, median, worst, and average values of the solutions. The future scope of the present work can be extended by proposing a multi-objective HTS algorithm for optimizing two or more objectives simultaneously related to the FDM process. Furthermore, some strategies related to constrained optimization can be developed to handle the objectives with equality and inequality constraints.

REFERENCES

1. Jiménez M, Romero L, Domínguez IA, Espinosa MD, Domínguez M (2019) Additive manufacturing technologies: an overview about 3D printing methods and future prospects. *Complexity*, 2019(1):9656938. https://doi.org/10.1155/2019/9656938
2. Jadhav A, Jadhav VS (2022) A review on 3D printing: an additive manufacturing technology. *Materials Today: Proceedings*, 62:2094–2099. https://doi.org/10.1016/j.matpr.2022.02.558
3. Pragana JPM, Sampaio RFV, Bragança IMF, Silva C.M.A., Martins P.A.F. (2021) Hybrid metal additive manufacturing: a state-of-the-art review. *Advances in Industrial and Manufacturing Engineering*, 2:100032. https://doi.org/10.1016/j.aime.2021.100032
4. Wang C., Tan X.P., Tor S.B., Lim C.S. (2020) Machine learning in additive manufacturing: state-of-the-art and perspectives. *Additive Manufacturing*, 36:101538. https://doi.org/10.1016/j.addma.2020.101538
5. Rajan K, Samykano M, Kadirgama K, Harun WS, Rahman MM (2022) Fused deposition modeling: process, materials, parameters, properties, and applications. *International Journal of Advanced Manufacturing Technology*, 120(3):1531–1570. https://doi.org/10.1007/s00170-022-08860-7.
6. Galetto M, Verna E, Genta G (2021) Effect of process parameters on parts quality and process efficiency of fused deposition modeling. *Computers & Industrial Engineering*, 156:107238. https://doi.org/10.1016/j.cie.2021.107238.
7. Mengesha Medibew T (2022) A comprehensive review on the optimization of the fused deposition modeling process parameter for better tensile strength of PLA-printed parts. *Advances in Material Sciences and Engineering*, 2022(1):5490831. https://doi.org/10.1155/2022/5490831
8. Ahmad MN, Ishak MR, Taha M, Mustapha F, Leman Z, Anak Lukista DD, Ghazali I (2022) Application of Taguchi method to optimize the parameter of fused deposition modeling (FDM) using oil palm fiber reinforced thermoplastic composites. *Polymers*, 14 (11):2140.
9. Moradi M, Aminzadeh A, Rahmatabadi D, Rasouli SA (2021) Statistical and experimental analysis of process parameters of 3D nylon printed parts by fused deposition modeling: response surface modeling and optimization. *Journal of Materials Engineering Performance*, 30(7):5441–5454. https://doi.org/10.1007/s11665-021-05848-4

10. Venkatraman R, Raghuraman S (2021) Experimental analysis on density, micro-hardness, surface roughness and processing time of acrylonitrile butadiene styrene (ABS) through fused deposition modeling (FDM) using Box Behnken design (BBD). *Materials Today Communication*, 27:102353. https://doi.org/10.1016/j.mtcomm.2021.102353
11. Chohan JS, Kumar R, Singh TB, Singh S, Sharma S, Singh J, Mia M, Pimenov DY, Chattopadhyaya S, Dwivedi SP, Kapłonek W (2020) Taguchi S/N and TOPSIS based optimization of fused deposition modelling and vapor finishing process for manufacturing of ABS plastic parts. *Materials*, 13(22):1–15. https://doi.org/10.3390/ma13225176
12. Heidari-Rarani M, Ezati N, Sadeghi P, Badrossamay MR (2022) Optimization of FDM process parameters for tensile properties of polylactic acid specimens using Taguchi design of experiment method. *Journal of Thermoplastic Composite Materials*, 35(12): 2435–2452. https://doi.org/10.1177/0892705720964560
13. Khalid M, Peng Q (2021) Investigation of printing parameters of additive manufacturing process for sustainability using design of experiments. *Journal of Mechanical Design*, 143(3):032001. https://doi.org/10.1115/1.4049521
14. Camposeco-Negrete C (2020) Optimization of FDM parameters for improving part quality, productivity and sustainability of the process using Taguchi methodology and desirability approach. *Progress in Additive Manufacturing*, 5(1):59–65. https://doi.org/10.1007/s40964-020-00115-9
15. Zaman UK, Boesch E, Siadat A, Rivette M, Baqai AA (2019) Impact of fused deposition modeling (FDM) process parameters on strength of built parts using Taguchi's design of experiments. *International Journal of Advanced Manufacturing Technology*, 101:1215–1226. https://doi.org/10.1007/s00170-018-3014-6
16. Lestari WD, Adyono N, Faizin AK, Haqiyah A, Sanjaya KH, Nugroho A, Kusmasari W, Ammarullah MI (2024) Optimization of 3D printed parameters for socket prosthetic manufacturing using the Taguchi method and response surface methodology. *Results in Engineering*, 21:101847. https://doi.org/10.1016/j.rineng.2024.101847
17. Ambade V, Rajurkar S, Awari G, Yelamasetti B, Shelare S (2023) Influence of FDM process parameters on tensile strength of parts printed by PLA material. *International Journal on Interactive Design and Manufacturing*, 2023:1–12. https://doi.org/10.1007/s12008-023-01490-7
18. Sahoo S, Sutar H, Senapati P, Mohanto BS, Dhal PR, Baral SK (2023) Experimental investigation and optimization of the FDM process using PLA. *Materials Today: Proceedings*, 74:843–847. https://doi.org/10.1016/j.matpr.2022.11.208
19. Popović M, Pjević M, Milovanović A, Mladenović G, Milošević M (2023) Printing parameter optimization of PLA material concerning geometrical accuracy and tensile properties relative to FDM process productivity. *Journal of Mechanical Science and Technology*, 37 (2):697–706. https://doi.org/10.1007/s12206-023-0113-6
20. Datta S, Raza MS, Das AK, Saha P, Pratihar DK (2020) Experimental investigations and parametric optimization of laser beam welding of NiTinol sheets by metaheuristic techniques and desirability function analysis. *Optics & Laser Technology*, 124:105982. https://doi.org/10.1016/j.optlastec.2019.105982
21. Chohan JS, Kumar R, Yadav A, Chauhan P, Singh S, Sharma S, Li C, Dwivedi SP, Rajkumar S (2022) Optimization of FDM printing process parameters on surface finish, thickness, and outer dimension with ABS polymer specimens using Taguchi orthogonal array and genetic algorithms. *Mathematical Problems in Engineering*, 2022:2698845. https://doi.org/10.1155/2022/2698845

22. Saad MS, Mohd Nor A, Zakaria MZ, Baharudin ME, Yusoff WS (2021) Modelling and evolutionary computation optimization on FDM process for flexural strength using integrated approach RSM and PSO. *Progress in Additive Manufacturing*, 6:143–154. https://doi.org/10.1007/s40964-020-00157-z
23. Deshwal S, Kumar A, Chhabra D (2020) Exercising hybrid statistical tools GA-RSM, GA-ANN and GA-ANFIS to optimize FDM process parameters for tensile strength improvement. *CIRP Journal of Manufacturing Science and Technology*, 31:189–199. https://doi.org/10.1016/j.cirpj.2020.05.009
24. Chinchanikar S, Shinde S, Shaikh S, Gaikwad V, Ambhore NH (2024) Multi-objective optimization of FDM using hybrid genetic algorithm-based multi-criteria decision-making (MCDM) techniques. *Journal of the Institution of Engineers (India): Series D*, 105(1):49–63. https://doi.org/10.1007/s40033-023-00459-w
25. Yadav K, Rohilla S, Ali A, Yadav M, Chhabra D (2023) Effect of speed, acceleration, and jerk on surface roughness of FDM-fabricated parts. *Journal of Materials Engineering and Performance*, 33(14):6261–6305. https://doi.org/10.1007/s11665-023-08476-2
26. Sai T, Pathak VK, Srivastava AK (2020) Modeling and optimization of fused deposition modeling (FDM) process through printing PLA implants using adaptive neuro-fuzzy inference system (ANFIS) model and whale optimization algorithm. *Journal of the Brazilian Society of Mechanical Sciences and Engineering*, 42:1–19. https://doi.org/10.1007/s40430-020-02699-3
27. Mohamed OA, Masood SH, Bhowmik JL (2015) Optimization of fused deposition modeling process parameters: a review of current research and future prospects. *Advances in Manufacturing*, 3:42–53. https://doi.org/10.1007/s40436-014-0097-7
28. Chinchanikar S, Shinde S, Gaikwad V, Shaikh A, Rondhe M, Naik M (2024) ANN modelling of surface roughness of FDM parts considering the effect of hidden layers, neurons, and process parameters. *Advances in Materials and Processing Technology*, 10(1):22–32. https://doi.org/10.1080/2374068X.2022.2091085
29. Tura AD, Lemu HG, Mamo HB, Santhosh AJ (2023) Prediction of tensile strength in fused deposition modeling process using artificial neural network and fuzzy logic. *Progress in Additive Manufacturing*, 8(3):529–539. https://doi.org/10.1007/s40964-022-00346-y
30. Ulkir O, Akgun G (2023) Predicting and optimising the surface roughness of additive manufactured parts using an artificial neural network model and genetic algorithm. *Science and Technology of Welding and Joining*, 28(7):548–557. https://doi.org/10.1080/13621718.2023.2200572
31. Maharana D, Kotecha P (2016) Simultaneous heat transfer search for computationally expensive numerical optimization. *IEEE Congress on Evolutionary Computation CEC*, 2016:2982–2988. https://doi.org/10.1109/CEC.2016.7744166
32. Patel VK, Savsani VJ (2015) Heat transfer search (HTS): a novel optimization algorithm. *Information Sciences*, 324:217–246. https://doi.org/10.1016/j.ins.2015.06.044
33. Raja BD, Patel V, Jhala RL (2017) Thermal design and optimization of fin-and-tube heat exchanger using heat transfer search algorithm. *Thermal Science and Engineering Progress*, 4:45–57. https://doi.org/10.1016/j.tsep.2017.08.004
34. Peng A, Xiao X, Yue R (2014) Process parameter optimization for fused deposition modeling using response surface methodology combined with fuzzy inference system. *International Journal of Advanced Manufacturing Technology*, 73:87–100. https://doi.org/10.1007/s00170-014-5796-5
35. Rao RV, Rai DP (2016) Optimization of fused deposition modeling process using teaching-learning-based optimization algorithm. *Engineering Science and*

Technology, an International Journal, 19(1):587–603. https://doi.org/10.1016/j.jestch.2015.09.008

36. Mohanty A, Nag KS, Bagal DK, Barua A, Jeet S, Mahapatra SS, Cherkia H (2022) Parametric optimization of parameters affecting dimension precision of FDM printed part using hybrid Taguchi-MARCOS-nature inspired heuristic optimization technique. *Materials Today: Proceedings*, 50:893–903. https://doi.org/10.1016/j.matpr.2021.06.216
37. Dev S, Srivastava R (2021) Optimization of fused deposition modeling (FDM) process parameters for flexural strength. *Materials Today: Proceedings*, 44:3012–3016. https://doi.org/10.1016/j.matpr.2021.02.436
38. Rayegani F, Onwubolu GC (2014) Fused deposition modelling (FDM) process parameter prediction and optimization using group method for data handling (GMDH) and differential evolution (DE). *International Journal of Advanced Manufacturing Technology*, 73:509–519. https://doi.org/10.1007/s00170-014-5835-2
39. Kumar P, Gupta P, Singh I (2023) Performance analysis of acrylonitrile–butadiene–styrene–polycarbonate polymer blend filament for fused deposition modeling printing using hybrid artificial intelligence algorithms. *Journal of Materials Engineering and Performance*, 32(4):1924–1937. https://doi.org/10.1007/s11665-022-07243-z

6 Parametric Optimization of the FDM Process Parameters

A MOORA-Driven Multi-Objective Decision-Support Approach

Arpan Paul, Manik Chandra Das, and Siba Sankar Mahapatra

6.1 INTRODUCTION

Additive manufacturing (AM) is a highly efficient production process, well-suited for creating low-volume and customized products with complex geometries [1]. Its advantages, such as high part accuracy, cost-effectiveness, minimal material waste, and strong mechanical properties, have made AM technology an increasingly flexible solution for producing a diverse array of components [2]. This technology has found applications across various sectors, including prototyping, tooling, aerospace, automotive, construction, and biomedical industries.

The concept of AM technology dates back to the 19th century, initially referred to as generative manufacturing or rapid prototyping [3]. During its early stages, it was primarily used to create prototypes of original designs. However, advancements in the field have enabled fused deposition modeling (FDM)-processed parts to serve as both prototypes and end-use components.

In the FDM process, raw materials are supplied in the form of filaments, which are heated and extruded through a nozzle into a semi-molten state. This material is then deposited layer by layer onto a build platform, creating the desired three-dimensional (3D) object. The design process begins with a model created in 3D computer-aided design (CAD) software, which is subsequently converted into a stereolithography (STL) file format for compatibility with the printing machine. The nozzle operates along the X and Y axes, while the build platform moves along the Z axis. Initially, the platform starts at its highest position for the first layer. Afterward, it lowers incrementally by a distance equal to the layer thickness for each subsequent layer, continuing this cycle until the complete part is formed.

 DOI: 10.1201/9781003686941-6

FIGURE 6.1 An actual FDM process and its machining components.

Figure 6.1 illustrates an actual FDM process along with its key components. For this study, experiments were conducted using IdeaMaker simulation software and followed a full factorial design methodology. The response variables, production time and cost, were collected and subsequently analyzed using the multi-objective optimization on the basis of ratio analysis (MOORA) method, a powerful multi-criteria decision-making (MCDM) tool. This approach facilitated the comprehensive evaluation and optimization of the production parameters.

6.2 LITERATURE REVIEW

Researchers have performed time and cost analyses of AM processes using various process parameters. Hopkinson and Dickens [4] conducted a cost analysis of the AM process by comparing it with the conventional injection molding process, showing a better performance of the AM process. Different types of AM technologies, including the FDM process, have become increasingly popular because of their ability to enable mass customization.

Several studies have been carried out by researchers to determine the key optimization parameters and their levels that influence the production efficiency of FDM processes. Nancharaiah [5] utilized three parameters in their study, which was structured using the Design of Experiments (DoE) approach. Their findings revealed that layer thickness and air gap play a significant role in affecting processing time. Similarly, Zhanh and Peng [6] used the Taguchi method integrated with fuzzy comprehensive evaluation to optimize four critical FDM process parameters: wire-width compensation, extrusion velocity, filling velocity, and layer height. Their study revealed that wire-width compensation had the most significant influence on the process. Ali et al. [7] selected Taguchi DoE algorithms. Seven parameters (slice height, air gap, road width, number of contours, raster angle, STL deviation, and angle) were utilized to evaluate the signal-to-noise ratio. Their results revealed that minimizing build time and material consumption is desirable. Baich et al. [8] performed a cost sensitivity analysis on infill patterns and identified the double-dense infill design as the most expensive. They also found that low-density infill offers the greatest cost savings, while sparse-built infill requires the shortest amount of time to complete.

Liu et al. [9] identified raster width, raster gap, layer thickness, deposition direction, and deposition style as crucial variables for the Taguchi technique and analysis of variance (ANOVA). In their research, the optimal setting parameters were chosen to optimize the mechanical qualities of FDM products. Rathee et al. [10] employed response surface methodology for their analysis, working on raster angle, slice height, orientation, contour width, raster width, and air gap. They discovered that the construction time was significantly impacted by spatial orientation. Padhi et al. [11] employed a combined fuzzy inference system and Taguchi approach to optimize the operational parameters of the FDM process. They focused on parameters such as layer height, orientation, raster width, raster angle, and air gap. Wu [12] considered four factors for his research: layer height, printing supplies, print time, and print size accuracy. Their investigation met the print quality with the shortest print time. A creative approach was employed by Jaiswal et al. [13] to determine the ideal build orientation that reduces errors related to both geometry and material composition. In their study, Nandernezhad et al. [14] used differential scanning calorimetry to analyze how changes in the mesostructure of 3D-printed components influence their mechanical and thermal properties. They focused on three parameters: layer height, infill pattern, and infill percentage. Papazetis and Vosniakos [15] applied the Taguchi method, ANOVA, and artificial neural networks for their analysis, taking into account variables like printing speed, orientation, layer thickness, and flow adjustments. They found factor windows in their research where parts free of defects were created while maintaining material deposition stability. Luca et al. [16] identified the operating factors for calculating build time and overall build cost using a single-parameter build orientation. Surface area and support volume were the two parameters that were used by Fritz and Kim [17], who found that structural compliance decreased when AM cost and duration increased. Nayak et al. [18] explored the effect of operational parameters like layer thickness, build orientation, laser power, and scan speed on the mechanical properties of the FDM process and the direct metal laser sintering (DMLS) process. Dev and Srivastava [19] employed a hybrid approach of response surface methodology and genetic algorithm considering three process parameters. They studied flexural mechanical properties by varying layer thickness, nozzle temperature, and print head speed and found some parameters with optimal values. A comparative inspection of five processing constraints, that is, air gap, raster width, raster angle, part orientation, and layer thickness, was performed by Mohanty et al. [20]. They applied 10 different nature-inspired algorithms and reported build orientation as the most substantial parameter. Using the Taguchi method, Kam et al. [21] investigated the effect of layer thickness, extruder temperature, filling structure, and occupancy rate on tensile and impact strength of 3D-printed Polyamide12 samples. They found the layer thickness to be the most influential factor for enhancing the mechanical properties. Paul and Das [22] used printing time and printing cost as response parameters to study the effect of layer height, print speed, and part orientation using Taguchi's DoE. They showed that all these three parameters significantly impact the printing time and cost.

The reviewed studies highlight the growing importance of AM and its optimization through various methodologies. Researchers have extensively analyzed time, cost, and quality aspects of AM processes, particularly focusing on the FDM technique. Key insights include:

1. *Process Parameters and Optimization Techniques:*
    - Commonly optimized parameters include layer thickness, print speed, build orientation, air gap, raster width, and infill patterns.
    - Optimization methodologies such as Taguchi's DoE, response surface methodology, fuzzy inference systems, and nature-inspired algorithms were widely employed.
2. *Key Findings:*
    - Parameters like layer thickness and build orientation significantly affect processing time, cost, and mechanical properties.
    - Studies reveal that minimizing infill density or adopting sparse infill patterns can reduce costs and production time without compromising quality.
    - Techniques such as hybrid genetic algorithms and artificial neural networks have been successfully utilized to identify optimal settings for improving efficiency and part quality.
3. *Applications and Case Studies:*
    - Multiple studies explored the impact of AM parameters on mechanical properties, such as tensile strength and thermal stability, using innovative approaches like differential scanning calorimetry.
    - Comparative studies of process constraints indicate that build orientation and layer thickness remain among the most critical factors for both cost and quality optimization.

Overall, the literature emphasizes the importance of MCDM tools and innovative optimization techniques to enhance the efficiency and performance of AM processes. The findings provide valuable guidance for tailoring AM parameters to specific production requirements, enabling cost-effective and high-quality manufacturing.

This chapter aims to determine the operational parameter settings for the optimization of the responses. For this purpose, an integrated MCDM tool of fuzzy analytic hierarchy process (F-AHP) and MOORA has been used. The F-AHP method was employed to assign subjective weights to the criteria, while the MOORA technique was used to compute the performance scores for each experimental dealing combination.

The remainder of the chapter is structured as follows: Section 6.3 details the process parameter settings and experimental methodologies. Section 6.4 describes the optimization approach utilizing the MOORA method. Section 6.5 presents the results and provides a discussion. Finally, Section 6.6 concludes the study.

6.3 PROCESS PARAMETERS SETTINGS AND EXPERIMENTAL STRATEGIES

From the literature study, it has been found that the layer thickness and print speed affect the printing time as well as the quality of the part being produced [7, 14]. Also, these two factors affect the mechanical and thermal properties of the printed part [10, 15]. Build orientation is another factor that significantly impacts the build time [12, 13]. So, for the experiment of the current study, we have chosen three parameters: (1) layer thickness, (2) print speed, and (3) build orientation. The term "layer thickness" refers to the distance between the nozzle tip and the build platform, where the material is extruded. "Print speed" refers to the nozzle's velocity in the X–Y plane during the part's printing across the build platform. "Build orientation" is the part's rotation with respect to the Z-axis before the commencement of printing. These three factors and their three levels of values have been finalized from the literature, and the printing machine specifications are shown in Table 6.1. Table 6.2 provides the main technical specifications of the reference FDM machine.

TABLE 6.1
Selected Parameters and Their Levels for the Experiment

	Levels		
Selected Parameters	**–1**	**0**	**1**
Layer thickness (mm)	0.1ss5	0.25	0.35
Print speed (mm/s)	30	65	100
Build orientation (degree)	0	45	90

TABLE 6.2
Key Technical Specifications of the FDM Machine

Items	Technical Specifications
Machine ID	Raise3D Pro2
Maximum capacity	$305 \times 305 \times 300$ mm^3
Positioning resolution	0.78125 μm on the *X/Y* axis
Supported nozzle diameters	0.2, 0.4, 0.6, 0.8, and 1.0 mm
Filament diameter	1.75 mm
Supported materials	ABS, TPU, PLA, NYLON, HIPS, PC, PETG, etc.
Nozzle speed range	30–150 mm/s

6.4 RESEARCH METHODOLOGY

6.4.1 Design of Experiment

The experimental design has been performed using full factorial design following the DoE. This method can be used to adjust multiple input factors or parameters in studies to find out how they affect the desired response. DoE includes full factorial design, in which two or more inputs are examined for optimization. Each input has unique levels, and every possible combination is taken into account. For a full factorial design, the number of experiments can be calculated as m^n, where n is the number of factors considered, and m is the number of levels of value. In this case, three factors are being used, each with three levels. Thus, a total of 27 (3^3) tests are carried out.

6.4.2 Material Selection

The selection of appropriate printing material is a further important aspect. The most common and extensively used material for FDM components' manufacturing is polymer. Polylactic acid (PLA) material was utilized in the experiment simulation because of its superior mechanical properties and ease of availability. PLA is often referred to as biodegradable plastic due to its biodegradable nature. Specifications of this material are described below:

- Printing temperature (°C): 180–230
- Build platform temperature (°C): 20–60
- Heat resistance: Low
- Flexibility: Brittle
- Strength: High
- Moisture absorption: Yes

6.4.3 Optimization Using MOORA and F-AHP

Over the past 50 years, numerous scholars have contributed to the field of MCDM by inventing novel techniques, including AHP, ANP, DEMATEL, ELECTRE, TOPSIS, PROMETHEE, VIKOR, etc., to aid in decision-making [23–28]. The subjective weights for total production time and printing cost were assigned using the Fuzzy AHP method for optimizing the response variables [29]. The MOORA method has been employed to calculate the performance scores of experiments with various parameter configurations [30, 31]. This technique was selected because of its reliability and ease of use. The steps involved in this process are described below.

Step 1: Create the decision matrix $D = \left[X_{ij}\right]_{m \times n}$, where X_{ij} represents the experiment's result for the ith alternative in comparison to the jth criterion. In this case, n= number of criteria and m= number of alternatives.

$$D = \left[X_{ij}\right]_{m\times n} = \begin{bmatrix} x_{11} & x_{12} & \cdots & \cdots & x_{1n} \\ x_{21} & x_{22} & \cdots & \cdots & x_{2n} \\ \cdots & \cdots & \cdots & \cdots & \cdots \\ \cdots & \cdots & \cdots & \cdots & \cdots \\ x_{m1} & x_{m2} & \cdots & \cdots & x_{mn} \end{bmatrix}$$

Step 2: Apply Equation (6.1) to normalize the choice matrix.

$$x_{ij}^{*} = \frac{x_{ij}}{\sqrt{\sum_{i=1}^{m} x_{ij}^{2}}} \quad \text{(For } j = 1, 2, 3, \ldots, n\text{)} \tag{6.1}$$

Step 3: Compute the algebraic sum of the weighted normalized elements of the decision matrix using Equation (6.2) to determine the performance score (y_i).

$$y_i = \sum_{j=1}^{g} w_j x_{ij}^{*} - \sum_{j=g+1}^{n} w_j x_{ij}^{*} \tag{6.2}$$

where $j = 1, 2, 3, \ldots, g$ corresponds to the outputs that need to be maximized (higher is the better), and $j = g+1, g+2, \ldots, n$ corresponds to the outputs that need to be minimized (lower is the better). The alternatives are subsequently ranked in descending order according to their performance score (y_i).

Chang [29] developed the fuzzy AHP extent analysis approach, which was used to compute the subjective weights (w_j) for total production time and printing cost [29]. Opinions from three academic experts have been considered, upon which the fuzzy AHP method was applied. The criteria weights for the production time and printing cost have been obtained as 0.5 and 0.5. The best nonfuzzy performances (BNPs) of the weights are shown in Table 6.3.

TABLE 6.3
BNP of the Weights for the Response Variables

Response Variable	Geometric Mean	Fuzzy Weights	*BNP*	Normalized *BNP*
Production time	(1.817, 2.076, 2.305)	(0.422, 0.391, 0.365)	0.393	0.5
Printing cost	(0.434, 0.482, 0.550)	(0.365, 0.391, 0.422)	0.393	0.5

6.4.4 Preparation of CAD Model

A CAD model of the single slider crank mechanism was created in the SolidWorks platform to experiment with FDM technology, as shown in Figure 6.2. Next, the model is converted into a .stl file format, as 3D printing software is compatible with this format. After that, the transformed model is loaded into the IdeaMaker program to be sliced, after which additional printing-related operations are carried out. Table 6.4 displays the machining parameters that remained constant throughout the experiment.

Based on the experimental results, a decision matrix of 27 experiments and two response variables has been established, as shown in Table 6.5. The performance of each experiment has been decided using the integrated fuzzy AHP–MOORA method.

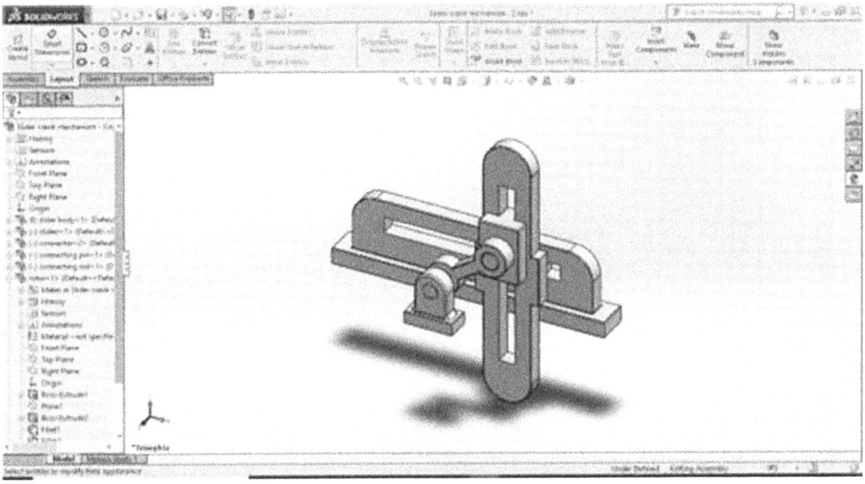

FIGURE 6.2 Single slider crank mechanism prepared in SolidWorks.

TABLE 6.4
Additional Parameters Settings for IdeaMaker Software

Fixed Parameters	Values
Infill density	30%
Extrusion width	0.40 mm
Number of extruders	2
Infill pattern type	Grid
Support infill type	Honeycomb
Support material printing speed	100 mm/s
Retraction speed	40 mm/s
Platform addition	Raft only
Maximum shell overlap	50%
Heated bed temperature	60•C
Extruder temperature	205•C

TABLE 6.5
Decision Matrix for Experimental Runs in Terms of Performance Evaluation Criteria

	Coded Parameter Values			Performance Assessment Criteria	
Experimental Run Order	Layer Thickness	Print Speed	Build Orientation	Printing Time (Hours)	Printing Cost (₹)
1	−1	−1	−1	11.76	11842.5
2	−1	−1	0	9.61	9655.1
3	−1	−1	1	13.66	13733.1
4	−1	0	−1	8.12	8234.1
5	−1	0	0	6.63	6704.3
6	−1	0	1	10.03	10133.6
7	−1	1	−1	7.22	7346.9
8	−1	1	0	6.13	6204.2
9	−1	1	1	10.49	10593.1
10	0	−1	−1	7.43	7551.6
11	0	−1	0	5.90	5980.7
12	0	−1	1	8.41	8534.7
13	0	0	−1	5.25	5398.9
14	0	0	0	4.12	4214.1
15	0	0	1	6.51	6653.2
16	0	1	−1	4.71	4866.1
17	0	1	0	3.82	3915.0
18	0	1	1	6.23	6377.9
19	1	−1	−1	6.39	6529.2
20	1	−1	0	4.99	5076.3
21	1	−1	1	7.12	7252.1
22	1	0	−1	4.55	4707.2
23	1	0	0	3.66	3759.3
24	1	0	1	5.53	5678.6
25	1	1	−1	4.10	4255.6
26	1	1	0	3.24	3342.4
27	1	1	1	5.29	5447.9

6.5 CALCULATION OF PRODUCTION TIME AND PRINTING COST

The two variables, printing time and cost, have been selected as response variables for each experiment in the current work as described below:

1. The total time spent on each experiment comprises the time spent on actual constructing or printing, as well as pre- and post-processing. The pre-processing time, which is approximately 10 minutes, is almost the same for all of the studies. The size and complexity of the part are the primary determinants of the post-processing time. The single slider crank mechanism (dimension: $125 \times 117.65 \times 49.63$ mm^3) requires an approximate post-processing time of 20 minutes. The experiment has yielded the actual printing time, which varies depending on the parameter values.
2. The cost of printing is calculated as the product of the actual printing time (hours) and the machine hour rate (₹/hour). The total of the hourly costs for labor, power, tools, and finance has been used to calculate the machine hour rate. As calculated below, the computed machine hour charge for the FDM machine made by Raise3D comes to ₹ 990.23/hour.

The total cost of the product is calculated as the summation of material cost and printing cost. Printing cost is the multiplication of printing time and machine hour rate. The formula for calculating the machine hour rate is

Machine hour rate = Financial cost + Power cost + Tooling cost + Labor cost

1. Financial cost is the purchase cost of the machine used in the printing process.

   Purchase cost = Present worth + Depreciation (18%) + Interest (10%) + Insurance

   = ₹ (4,30,462 + 77,483 + 43,046 + 60,564)

   = ₹ 6,11,555/year = ₹ 254.81/hour.

2. Power cost is calculated as the multiplication of the motor power of the 3D printing machine and the current unit price, which is ₹ 4.89, according to CESC Limited. As the motor is 600 W and if it runs for 1 hour, it will consume 0.6 kW • 1 hour = 0.6 kW-hour of power.

   Power cost = (0.6 × ₹ 4.89) = ₹ 2.93/hour.

3. Tooling cost can be calculated from the nozzle used as an extruder in the printing from its life. Assuming a brass nozzle of price ₹ 1660 and having a lifespan of 1000 hours,

   Tooling cost = ₹ 3.32/hour.

4. Assuming the working time of the supervisor is 6 days a week and 8 hours a day,

   Labor cost is assumed as ₹ 35000/month = ₹ 729.17/hour.

   So, machine hour rate = ₹ (254.81 + 2.93 + 3.32 + 729.17)
   = ₹ 990.23

A decision matrix with the experimental results having 27 experiments, three input parameters, and two performance evaluation criteria is shown in Table 6.5.

Subsequently, using Equation (6.1), the normalized values for each alternative are determined, and the associated performance score is calculated as a weighted algebraic sum of the normalized elements, applying Equation (6.2). The same data is shown in Table 6.6.

6.6 RESULTS AND DISCUSSION

The performance scores of the parametric combinations, as outlined in Table 6.6, have been ranked using the MOORA method, with the results displayed in Table 6.7. These rankings provide a clear indication of the efficacy of each parameter configuration. A higher performance score denotes superior outcomes across the chosen parameters, irrespective of individual responses. Notably, the optimal performance score is achieved in experiment 26. This experiment utilized parameters set to a layer thickness of 0.35 mm, a print speed of 100 mm/s, and a build orientation of 45° (level 0).

These settings yielded the best results for the FDM process using PLA material to manufacture the single slider crank mechanism arrangement. The findings underscore the importance of selecting the appropriate combination of input parameters to achieve optimal results in terms of performance metrics. Such optimization can serve as a valuable reference for professionals and researchers striving to improve the efficiency, quality, and cost-effectiveness of FDM-based AM processes.

Table 6.8 displays the average performance score (y_i) for each process parameter level. The performance score was determined by computing the average of all scores corresponding to that specific level. Based on the calculation, the average y_i range for the FDM parameters at various levels is 0.109 for layer thickness, 0.067 for print speed, and 0.072 for build orientation. As layer thickness has the highest score, or 0.109, it can be stated that layer thickness is the most important controllable parameter in this FDM method, followed by print speed and build orientation. Thus, it is evident that the parameter combination with the highest performance score aligns with the optimal operational parameter configurations, resulting in the best responses. Table 6.8 displays the average performance score (y_i) for each process parameter level. The performance score was determined by computing the average of all scores corresponding to that specific level. Based on the calculation, the average y_i range for the FDM parameters at various levels is 0.109 for layer thickness, 0.067 for print speed, and 0.072 for build orientation. As layer thickness has the highest score of 0.109, it can be stated that layer thickness is the most important controllable

TABLE 6.6
Normalization Matrix for Printing Time and Cost

Experimental Run Order	Coded Parameter Values			Performance Assessment Criteria		
	Layer Thickness	Print Speed	Build Orientation	Normalized Time	Normalized Cost	Performance Scores (y_i)
1	–1	–1	–1	0.308	0.311	–0.310
2	–1	–1	0	0.254	0.253	–0.254
3	–1	–1	1	0.355	0.360	–0.358
4	–1	0	–1	0.217	0.218	–0.218
5	–1	0	0	0.180	0.177	–0.178
6	–1	0	1	0.265	0.267	–0.266
7	–1	1	–1	0.195	0.195	–0.195
8	–1	1	0	0.167	0.164	–0.165
9	–1	1	1	0.276	0.279	–0.278
10	0	–1	–1	0.200	0.200	–0.200
11	0	–1	0	0.162	0.158	–0.160
12	0	–1	1	0.224	0.226	–0.225
13	0	0	–1	0.146	0.145	–0.145
14	0	0	0	0.117	0.112	–0.114
15	0	0	1	0.177	0.177	–0.177
16	0	1	–1	0.132	0.131	–0.131
17	0	1	0	0.110	0.105	–0.107
18	0	1	1	0.170	0.170	–0.170
19	1	–1	–1	0.174	0.174	–0.174
20	1	–1	0	0.139	0.135	–0.137
21	1	–1	1	0.192	0.193	–0.192
22	1	0	–1	0.128	0.127	–0.127
23	1	0	0	0.106	0.101	–0.103
24	1	0	1	0.152	0.152	–0.152
25	1	1	–1	0.117	0.115	–0.116
26	1	1	0	0.095	0.090	–0.092
27	1	1	1	0.147	0.146	–0.146

TABLE 6.7
Ranks and Performance Scores for Every Combination of Parameters

Experimental Run Order	Coded Parameter Values			Performance Scores (y_i)	Ranks
	Layer Thickness	Print Speed	Build Orientation		
1	–1	–1	–1	–0.310	26
2	–1	–1	0	–0.254	23
3	–1	–1	1	–0.358	27
4	–1	0	–1	–0.218	21
5	–1	0	0	–0.178	17
6	–1	0	1	–0.266	24
7	–1	1	–1	–0.195	19
8	–1	1	0	–0.165	13
9	–1	1	1	–0.278	25
10	0	–1	–1	–0.200	20
11	0	–1	0	–0.160	12
12	0	–1	1	–0.225	22
13	0	0	–1	–0.145	9
14	0	0	0	–0.114	4
15	0	0	1	–0.177	16
16	0	1	–1	–0.131	7
17	0	1	0	–0.107	3
18	0	1	1	–0.170	14
19	1	–1	–1	–0.174	15
20	1	–1	0	–0.137	8
21	1	–1	1	–0.192	18
22	1	0	–1	–0.127	6
23	1	0	0	–0.103	2
24	1	0	1	–0.152	11
25	1	1	–1	–0.116	5
26	1	1	0	–0.092	1
27	1	1	1	–0.146	10

TABLE 6.8
Performance Score Response Table for Each Level

Process Parameters	Average Performance Score (y_i)			Range (Max–Min)
	Level −1	Level 0	Level 1	
Layer thickness	−0.247	−0.159	−0.138	0.109
Print speed	−0.223	−0.165	−0.156	0.067
Orientation	−0.180	−0.146	−0.218	0.072

TABLE 6.9
Estimated Model Coefficients for Means for Printing Time

Term	Coefficients	SE Coefficients	*T*	*P*
Constant	6.70000	0.1209	55.420	0
Layer thickness – 0.15	2.59444	0.1710	15.175	0
Layer thickness – 0.25	−0.88000	0.1710	−5.147	0
Print speed – 30	1.66333	0.1710	9.729	0
Print speed – 65	−0.65556	0.1710	−3.834	0.001
Build orientation – 0	−0.08556	0.1710	−0.500	0.622
Build orientation – 45	−1.35556	0.1710	−7.928	0

SE, standard error.

Also, $S = 0.6282$, $R^2 = 95.5\%$, $R^2_{adj} = 94.1\%$ are obtained.

parameter in this FDM method, followed by print speed and build orientation. Thus, it is evident that the parameter combination with the highest performance score aligns with the optimal operational parameter configurations, resulting in the best responses. This finding emphasizes the critical role of layer thickness in determining the overall performance of the FDM process. It also highlights the need for carefully selecting parameter levels to maximize efficiency and achieve the desired quality in AM.

6.7 TAGUCHI ANALYSIS

The impact of different process parameters on a multi-objective response known as MPI has been investigated through Taguchi's parameter design methodology utilizing Minitab® 21.2 (64-bit) software. Taguchi analysis for both the output responses, printing time (hour) and printing cost (INR) versus layer thickness, print speed, and build orientation is performed, and valuable insights are provided.

6.7.1 Taguchi Analysis for Printing Time

The estimated model coefficients for means for printing time are tabulated in Table 6.9. Process parameters with p-values below 0.05 are considered significant

TABLE 6.10
ANOVA for Means for Printing Time

Source	DF	Seq SS	Adj SS	Adj MS	*F*	*P*
Layer thickness	2	94.004	94.004	47.0019	119.10	0
Print speed	2	37.908	37.908	18.9542	48.03	0
Build orientation	2	35.295	35.295	17.6474	44.72	0
Residual error	20	7.893	7.893	0.3946		
Total	26	175.100				

Adj MS, adjusted mean square; Adj SS, adjusted sum of squares; DF, degrees of freedom; Seq SS, sequential sum of squares.

[32]. So, the regression model appears to be highly significant, as the indicated *p*-values are very low for the overall model (except for "Build orientation – 0"). The *p*-value of 0.622 associated with "Build orientation – 0" suggests that the coefficient may not be statistically significant. Also, the model accounts for a significant portion of the variance in the dependent variable, with R^2 at 95.5%. So, the model appears to be well-fitted and statistically significant, but further diagnostics on residuals and potential model refinement may be warranted. These diagnostics could include examining the normality of residuals and checking for multicollinearity among independent variables to ensure model accuracy. Additionally, validating the model using an independent dataset would strengthen confidence in its predictive capabilities.

The ANOVA for means for print speed is performed, and the results are shown in Table 6.10. Referring to this table, it seems that all three factors (layer thickness, print speed, and build orientation) have a substantial effect on the measured outcome, as indicated by the low *p*-values. The *F*-values obtained are quite large, indicating that the variance between the groups (due to these factors) is much larger than the variance within the groups (due to random error). The probability associated with these *F*-values (the *p*-values, which are all close to zero) suggests that these results are statistically significant.

In practical terms, this means that layer thickness, print speed, and build orientation all significantly impact the measured outcome. Additionally, one observation (observation 3), as shown in Table 6.11, appears to be an outlier based on its standardized residual.

The response table for means for printing time, as shown in Table 6.12, likely represents the mean values of the response variable (printing time) at various levels of the factors under consideration. This table is significant as it offers a detailed understanding of how the response variable is influenced by changes in the parameter levels. By presenting mean response values for each level of the factors, it facilitates a direct comparison and highlights trends or patterns in the data.

This analysis provides critical insights into the relationships between each factor and the printing time, aiding in the identification of optimal settings. The results

TABLE 6.11
Unusual Observations for Means

Observation	Means	Fit	SE Fit	Residual	St Residual
3	13.660	12.399	0.320	1.261	2.33 R

SE, standard error; St, standardized.

TABLE 6.12
Response Table for Means for Printing Time

Factor	Level 1	Level 2	Level 3	Delta	Rank
Layer thickness	9.294	5.820	4.986	4.309	1
Print speed	8.363	6.044	5.692	2.671	3
Build orientation	6.614	5.344	8.141	2.797	2

indicate that among the factors studied, layer thickness has the most substantial impact on printing time, followed by print speed and build orientation. Such insights are invaluable for process optimization, as they help prioritize parameters that significantly influence the response. Moreover, this approach enhances decision-making in parameter selection to achieve reduced printing times while maintaining the desired quality and cost efficiency.

The main effects plots for means for printing time are depicted in Figure 6.3. From the main effects plots, we can gain insights into how each factor individually affects the mean of the response variable. The following insights can be gained from the main effects plot.

- Increasing layer thickness and print speed decreases the mean of the response variable, suggesting an inverse relationship between them. Adjusting layer thickness and print speed to lower levels may result in higher mean response values, potentially indicating improved product quality.
- The effect of build orientation on the mean response variable is less straightforward compared to layer thickness and print speed. There seems to be a nonlinear relationship between the build orientation and the mean response value. It provides valuable information for optimizing the manufacturing process.
- Build orientation may have a more complex relationship with the mean response, possibly requiring further investigation to determine the optimal orientation for maximizing the response variable.

Figure 6.4 displays the residual plots for the mean printing time. The normal probability plot indicates that most points lie along the straight line, implying that the data is roughly normally distributed. The residuals versus fitted values plot indicates

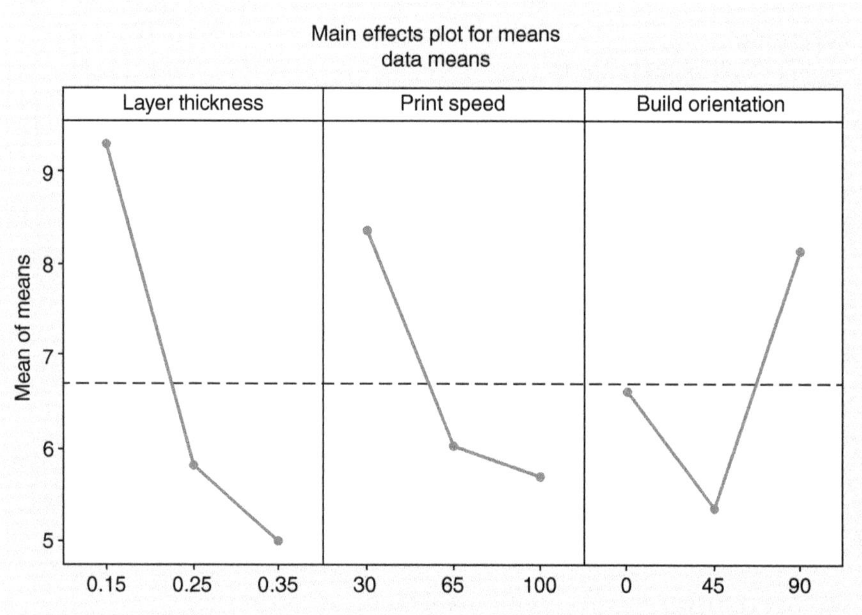

FIGURE 6.3 Main effect plots for means for printing time.

that the regression model's assumptions are met and the model is appropriate for the data. In other words, there is no systematic pattern in the residuals, indicating that the model effectively captures the relationship between the variables. The histogram of residuals suggests an approximate normal distribution as the residuals are symmetrically distributed around zero. The residuals versus observation order with a zigzag pattern indicates that there is a presence of autocorrelation (residuals from adjacent observations are correlated) or heteroscedasticity (variability of the residuals changes as a function of the predictor variable or observation order) in the data.

6.7.2 Taguchi Analysis for Printing Cost

Table 6.13 presents the estimated coefficients for the mean printing cost. The regression model demonstrates high significance, supported by very low *p*-values for most predictors (excluding "Build orientation – 0"). However, the *p*-value of 0.699 for "Build orientation – 0" suggests a potential lack of statistical significance for its coefficient. The model accounts for a considerable portion of the variance in the dependent variable, with an R^2 of 95.5%, and it appears well-fitted and statistically significant; additional diagnostics on residuals and potential model refinement may be necessary.

A large absolute value of the *t*-statistic indicates a substantial effect of the predictor variable on the response variable, regardless of whether the coefficient is positive or negative. In the case of "Build orientation 0," the *t*-statistic is –0.392, which suggests that the coefficient estimate is not statistically significant.

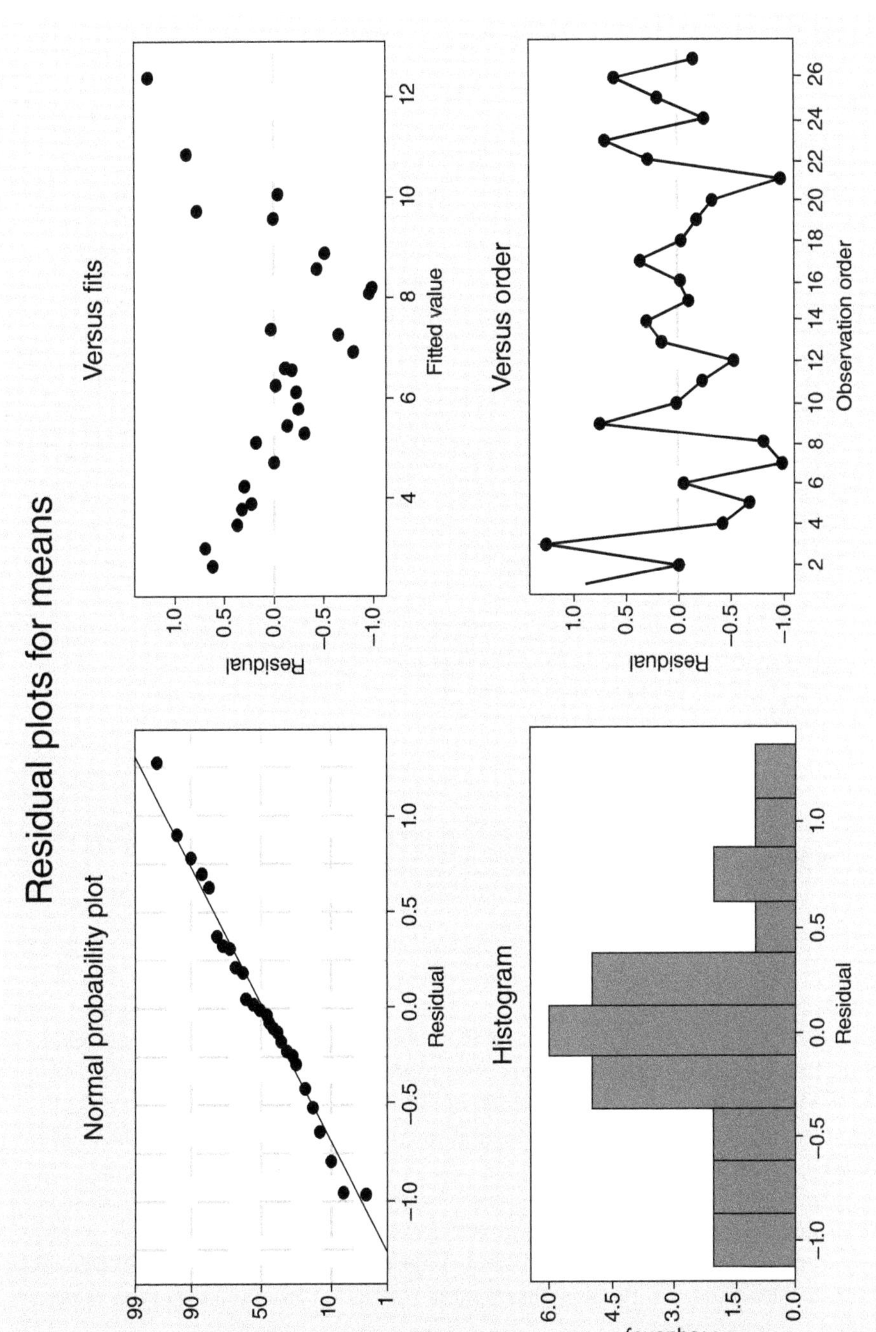

FIGURE 6.4 Residual plots for mean for print speed.

TABLE 6.13
Estimated Model Coefficients for Means for Printing Cost

Term	Coefficients	SE Coefficients	*T*	*P*
Constant	6814.36	119.8	56.904	0
Layer thickness – 0.15	2568.63	169.4	15.167	0
Layer thickness – 0.25	−870.78	169.4	−5.142	0
Print speed – 30	1647.34	169.4	9.727	0
Print speed – 65	−649.55	169.4	−3.835	0.001
Build orientation – 0	−66.35	169.4	−0.392	0.699
Build orientation – 45	−1386.43	169.4	−8.186	0

SE, standard error.

Also, $S = 622.3$, $R^2 = 95.5\%$, $R^2_{adj} = 94.2\%$ are obtained.

TABLE 6.14
ANOVA for Means for Printing Cost

Source	DF	Seq SS	Adj SS	Adj MS	*F*	*P*
Layer thickness	2	92149262	92149262	46074631	118.99	0
Print speed	2	37181109	37181109	18590555	48.01	0
Build orientation	2	36334183	36334183	18167091	46.92	0
Residual error	20	7743962	7743962	387198		
Total	26	173408516				

Adj MS, adjusted mean square; Adj SS, adjusted sum of squares; DF, degrees of freedom; Seq SS, sequential sum of squares.

The ANOVA analysis for printing cost reveals significant effects for all three factors: layer thickness, print speed, and build orientation, as indicated by their low p-values in Table 6.14. The obtained F-values are notably large, signifying that the variance between groups (attributed to these factors) exceeds the variance within groups (resulting from random error). The close-to-zero p-values associated with these F-values underscore the statistical significance of these findings. Practically, this implies that layer thickness, print speed, and build orientation significantly influence the measured outcome.

Furthermore, Table 6.15 highlights one observation identified as an outlier based on its standardized residual. Observation 3 has a mean value that deviates significantly from the fitted mean, with a relatively large residual. The standardized residual, 2.33 *R*, further indicates that this observation's deviation from the expected value is notable.

The response table for means for printing cost is shown in Table 6.16. The table provides valuable information about how variations in layer thickness, print speed,

TABLE 6.15
Unusual Observations for Means

Observation	Means	Fit	SE Fit	Residual	St Residual
3	13733.100	12483.104	316.835	1249.996	2.33 *R*

SE, standard error; St, standardized.

TABLE 6.16
Response Table for Means for Printing Cost

Factor	Level 1	Level 2	Level 3	Delta	Rank
Layer thickness	9383	5944	5117	4266	1
Print speed	8462	6165	5817	2645	3
Build orientation	6748	5428	8267	2839	2

and build orientation affect the mean response variable. This information can be used to optimize these factors to achieve desired outcomes in the process being studied. Insights from this table are given below.

- The mean response variable decreases as layer thickness increases, indicating an inverse relationship, which suggests that thinner layers may lead to higher mean response values.
- The mean response variable also decreases as print speed increases, indicating an inverse relationship, which implies that slower print speeds may lead to higher mean response values.
- The mean response variable increases as build orientation changes from level 2 to level 3, suggesting that certain orientations may lead to higher mean response values than others.

The main effects plot for means for printing cost, shown in Figure 6.5, illustrates the variations of the factors – layer thickness, print speed, and build orientation – on the mean of the response variable (printing cost), similar to the trends observed for printing time. This plot reveals an inverse relationship between layer thickness and print speed, indicating that as layer thickness increases, printing cost tends to decrease with higher print speeds. Additionally, the plot suggests a more complex and non-linear relationship for build orientation, where the effect of this parameter on the printing cost appears to be influenced by other factors and does not follow a simple linear pattern.

These insights highlight the intricate interplay between the parameters and their effects on the cost, providing useful guidance for process optimization. Understanding

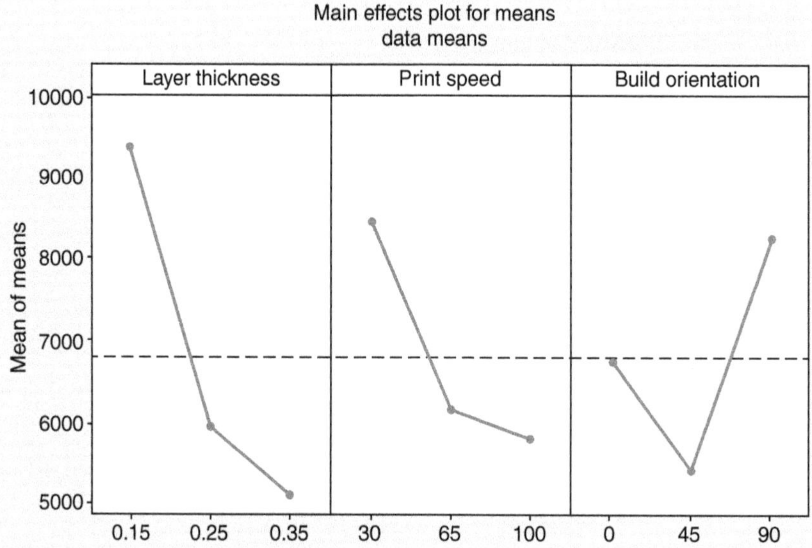

FIGURE 6.5 Main effects plot for means for printing cost.

these relationships allows for better control over both printing time and cost, ensuring efficient use of resources while maintaining high-quality outcomes. Further investigation into the combined effects of these parameters could help refine the process even further and lead to more precise and cost-effective AM strategies.

The residual plots for the means of printing cost, as shown in Figure 6.6, exhibit patterns identical to those observed for printing time, offering a consistent view of the model's performance across both response variables. The insights gathered from the various diagnostic plots indicate the appropriateness of the regression model for the data. The normal probability plot shows that most of the points closely align with the straight line, suggesting that the residuals are approximately normally distributed, which is a crucial assumption for the validity of the model. Additionally, the residuals versus fitted values plot demonstrates that the residuals are evenly distributed around zero, indicating that the assumption of homoscedasticity is satisfied and that the model fits the data well.

The histogram of residuals further confirms this by displaying a symmetric distribution centered around zero, supporting the hypothesis of approximate normality. However, the residuals versus observation order plot reveals a noticeable zigzag pattern, which may indicate potential issues such as autocorrelation or heteroscedasticity in the data. This suggests that while the model performs well overall, further diagnostics and possibly adjustments may be necessary to address these issues and improve the model's robustness.

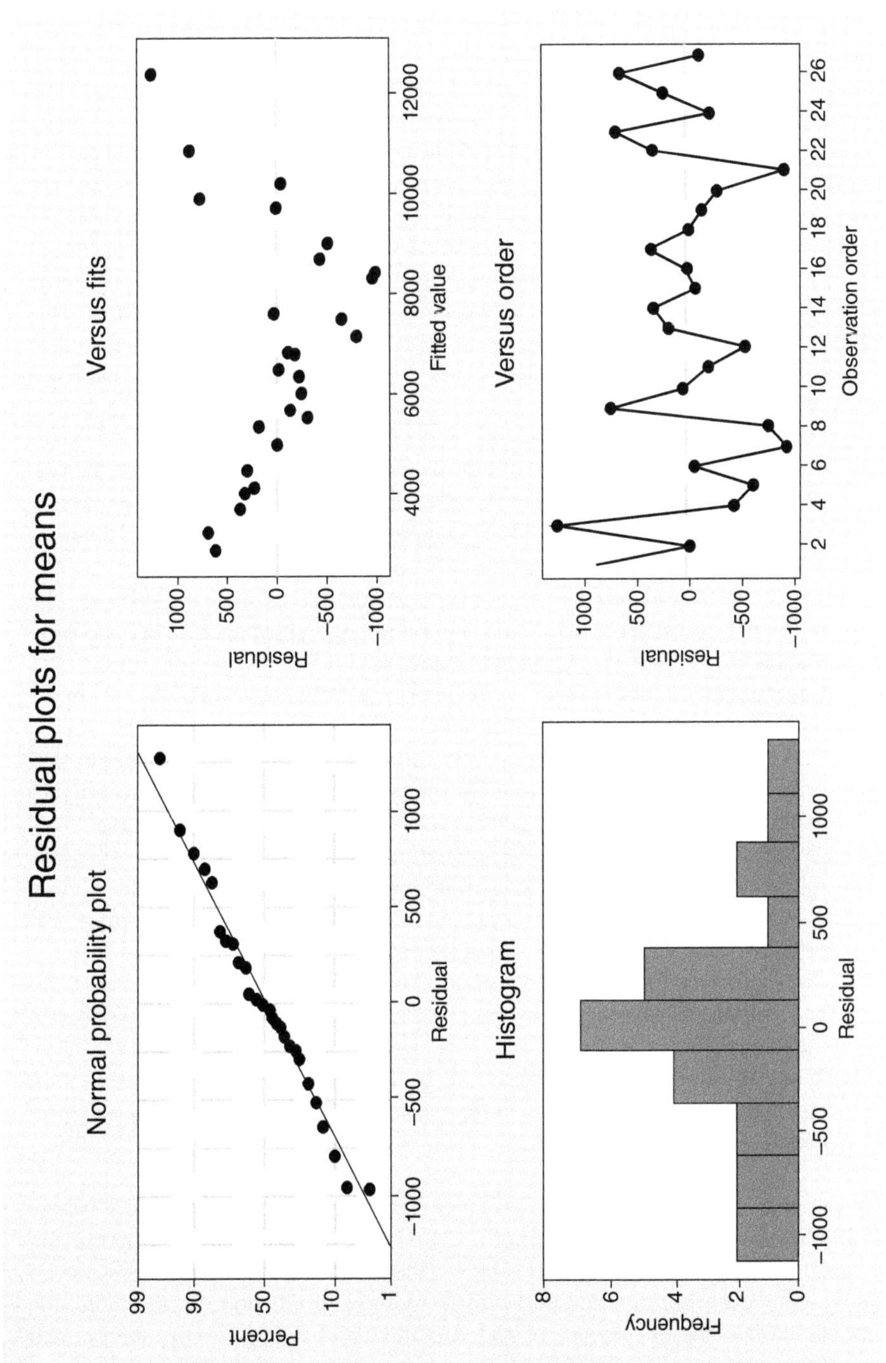

FIGURE 6.6 Residual plots for means for printing cost.

6.8 CONCLUSION

In the present work, the F-AHP and MOORA methods have been employed to address the multi-response parametric optimization problem using Taguchi's DoE framework. This approach ensures a systematic evaluation of input parameters and their impact on the desired outcomes. The input parameters considered for this study include layer thickness, print speed, and build orientation, which are critical factors influencing the AM process. The response variables selected for evaluation are printing time and printing cost, reflecting the practical significance of these factors in optimizing production efficiency and cost-effectiveness. The test specimen chosen for this research is a single slider crank mechanism arrangement, providing a realistic scenario for assessing the influence of the input parameters. The findings from this study not only highlight the effectiveness of the adopted methodologies but also offer valuable insights for improving AM processes. A detailed discussion of the outcomes is presented below:

1. From the performance score, it is determined that layer thickness of 0.35 mm, print speed of 100 mm/s, and orientation of 45° are the optimal parameter settings. Additionally, the analysis demonstrates that layer thickness emerges as the most crucial attribute of the FDM process.
2. The study also highlights that the most impactful parameters, ranked by the difference between the maximum and minimum performance scores, are layer thickness, print speed, and build orientation.
3. Layer thickness, print speed, and build orientation all have a significant impact on both printing time and printing cost.
4. There is an inverse relationship between printing time, layer thickness, and print speed. Thinner layers and slower print speeds tend to result in higher mean response values, potentially indicating improved product quality.
5. The relationship between build orientation and the mean response variable is more complex. While specific orientations may lead to higher mean response values compared to others, further investigation is needed to determine the optimal orientation for maximizing the response variable.
6. The regression models for both printing time and printing cost are well-fitted and statistically significant, explaining a substantial portion of the variance in the dependent variables.
7. While the current models appear to be appropriate, further diagnostics on residuals and potential model refinement may be warranted to improve predictive accuracy.

In summary, the Taguchi analysis provides significant insights into the key factors that influence printing time and printing cost, which are crucial for optimizing the manufacturing process. By focusing on these factors, it becomes possible to improve both the efficiency and cost-effectiveness of production. Furthermore, the study emphasizes the importance of applying this approach to identify the optimal parameters for various aspects of AM. To enhance the overall quality and performance of the manufacturing process, professionals, engineers, and researchers in the field of AM can broaden the use of this parametric optimization model. This could lead to more refined processes, reduced costs, and improved final product quality. Additionally, the insights gained from this study can serve as a valuable resource, guiding future research and practical applications in the field.

REFERENCES

1. Thakur, A., Vates, U.K., and Mishra, S., 2023, "Parametric optimization of 3D-printed PLA part using response surface methodology for mechanical properties." *Smart Science,* **12**(1), pp. 91–104.
2. Gupta, N.K., Rakesh, P.K., Rastogi, V., and Singh, I., 2024, "Process parametric optimization of fused deposition modeling for manufacturing of acrylonitrile butadiene styrene parts." *Journal of Micromanufacturing*, **7**(1), pp. 90–96.
3. Gebhardt, A., 2012, "*Understanding Additive Manufacturing: Rapid Prototyping – Rapid Tooling – Rapid Manufacturing*." Munich, Germany: Carl Hanser Verlag.
4. Hopkinson, N., and Dickens, P., 2000, "A comparison between stereolithography and aluminium injection moulding tooling." *Rapid Prototyping Journal*, **6**(4), pp. 253–258.
5. Nancharaiah, T., 2011, "Optimization of process parameters in FDM process using design of experiments." *International Journal on Emerging Technologies*, **2**(1), pp. 100–102.
6. Zhang, J.W., and Peng, A.H., 2012, "Process-parameter optimization for fused deposition modeling based on Taguchi method." *Advanced Materials Research*, **538**, pp. 444–447.
7. Ali, F., Chowdary, B.V., and Maharaj, J., 2014, "Influence of some process parameters on build time, material consumption, and surface roughness of FDM processed parts: Inferences based on the Taguchi design of experiments." In *Proceedings of The 2014 IACJ/ISAM Joint International Conference*, Orlando, FL, USA, September 2014.
8. Baich, L., Manogharan, G., and Marie, H., 2015, "Study of infill print design on production cost-time of 3D printed ABS parts." *International Journal of Rapid Manufacturing*, **5**(3–4), pp. 308–319.
9. Liu, X., Zhang, M., Li, S., Si, L., Peng, J., and Hu, Y., 2017, "Mechanical property parametric appraisal of fused deposition modeling parts based on the gray Taguchi method." *The International Journal of Advanced Manufacturing Technology*, **89**(5), pp. 2387–2397.
10. Rathee, S., Sandeep, M., Maheshwari, S., and Siddiquee, A.N., 2017, "Effect of varying spatial orientations on build time requirements for FDM process: a case study." *Defence Technology*, **13**(2), pp. 92–100.
11. Padhi, S.K., Sahu, R.K., Mahapatra, S.S., Das, H.C., Sood, A.K., Patro, B., and Mondal, A.K., 2017, "Optimization of fused deposition modeling process parameters using a fuzzy inference system coupled with Taguchi philosophy." *Advances in Manufacturing*, **5**, pp. 231–242.
12. Wu, J., 2018, "Study on optimization of 3D printing parameters." *IOP Conference Series: Materials Science and Engineering*, **392**(6), pp. 062050.
13. Jaiswal, P., Patel, J., and Rai, R., 2018, "Build orientation optimization for additive manufacturing of functionally graded material objects." *The International Journal of Advanced Manufacturing Technology*, **96**(1), pp. 223–235.

14. Nandernezhad, N., Unal, S., Khani, N., and Koc, B., 2019, "Material extrusion-based additive manufacturing of structurally controlled poly (lactic acid)/carbon nanotube nanocomposites." *The International Journal of Advanced Manufacturing Technology*, **102**(5), pp. 2119–2132.
15. Papazetis, G., and Vosniakos, G.C., 2019, "Mapping of deposition-stable and defect-free additive manufacturing via material extrusion from minimal experiments." *The International Journal of Advanced Manufacturing Technology*, **100**(9), pp. 2207–2219.
16. Luca, D.A., Stefano, P.D., Dolatnezhadsomarin, A., Guardiani, E., and Khorram, E., 2020, "A reliable build orientation optimization method in additive manufacturing: the application to FDM technology." *The International Journal of Advanced Manufacturing Technology*, **108**(1), pp. 263–276.
17. Fritz, K., and Kim, Y., 2020, "Simultaneous topology and build orientation optimization for minimization of additive manufacturing cost and time." *International Journal for Numerical Methods in Engineering*, **121**(5), pp. 3442–3481.
18. Nayak, P.K., Sahu, A.K., and Mahapatra, S.S., 2020, "Effect of process parameters on the mechanical behavior of FDM and DMLS build parts." *Materials Today: Proceedings*, **22**, pp. 1443–1451.
19. Dev, S., and Srivastava, R., 2021, "Optimization of fused deposition modeling (FDM) process parameters for flexural strength." *Materials Today: Proceedings*, **44**, pp. 3012–3016.
20. Mohanty, A., Nag, K.S., Bagal, D.K., Barua, A., Jeet, S., Mahapatra, S.S., and Cherkia, H., 2022, "Parametric optimization of parameters affecting dimension precision of FDM printed part using hybrid Taguchi-MARCOS-nature inspired heuristic optimization technique." *Materials Today: Proceedings*, **50**, pp. 893–903.
21. Kam, M., Ipekci, A. and Şengül, Ö., 2023, "Investigation of the effect of FDM process parameters on mechanical properties of 3D printed PA12 samples using Taguchi method." *Journal of Thermoplastic Composite Materials*, **36**(1), pp. 307–325.
22. Paul, A., and Das, M., 2024, "A decision support system for the selection of FDM process parameters using MOORA." *Management Science Letters*, **14**(3), pp. 181–188.
23. Saaty, T.L., 1990, "How to make a decision: the analytic hierarchy process." *European Journal of Operational Research*, **48**(1), pp. 9–26.
24. Saaty, T.L., 1996, "*Decision Making with Dependence and Feedback: The Analytic Network Process*." Pittsburgh: RWS Publications, **4992**(2).
25. Hwang, C.L., Lai, Y.J., and Liu, T.Y., 1993, "A new approach for multiple objective decision making." *Computers & Operations Research*, **20**(8), pp. 889–899.
26. Opricovic, S., and Tzeng, G.H., 2007, "Extended VIKOR method in comparison with outranking methods." *European Journal of Operational Research*, **178**(2), pp. 514–529.
27. Roy, B., 1968, "Classement et choix en présence de points de vue multiples." *Revue française d'informatique et de recherche opérationnelle*, **2**(8), pp. 57–75.
28. Brans, J.P., 1982, "L'ingenierie de la decision, l'laboration d'instruments d'aidea la decision." In *Colloque sur l'Aidea la Decision*. Faculte des Sciences de l'Administration, Universite Laval.
29. Chang, D.Y., 1996, "Applications of the extent analysis method on fuzzy AHP." *European Journal of Operational Research*, **95**(3), pp. 649–655.
30. Brauers, W.K., 2004, "Multi-objective optimization (MOO) in privatization." *Journal of Business Economics and Management*, **5**(2), pp. 59–65.

31. Brauers, W.K., and Zavadskas, E.K., 2006, "The MOORA method and its application to privatization in a transition economy." *Control and Cybernetics*, **35**(2), pp. 445–469.
32. Sood, A.K., Chaturvedi, V., Datta, S., and Mahapatra, S.S., 2011, "Optimization of process parameters in fused deposition modeling using weighted principal component analysis." *Journal of Advanced Manufacturing Systems*, **10**(2), pp. 241–259.

7 Revolutionizing Application Sectors of Fused Deposition Modeling

From Rapid Prototyping to Customization

Pankaj Kumar, Sanhita Mandal, Rohit Anand, and Ganesh Sahadeo Meshram

7.1 INTRODUCTION

Fused deposition modeling (FDM) is the leading three-dimensional (3D) printing technology that has made it revolutionize manufacturing and designing in almost all industries. Developed by S. Scott Crump in the late 1980s, FDM is one of the newest ideas that have grown from an innovation to a basic requirement in making complex, strong, and accurate parts (Zindani and Kumar 2019; Prasad *et al.* 2024). This chapter comprises the fundamental concepts of FDM, its advantages, and its novel applications in the medical, aerospace, automotive, and scientific fields. FDM is a method in which thermoplastic materials are extruded via a heated nozzle, stacking the material to fabricate a 3D object. The process initiates with a digital model of a 3D object that is sliced into thin cross-sectional layers. The schematic of FDM process is shown in Figure 7.1. FDM machine follows the pathway of these layers, where it deposits the thermoplastic material, which solidifies on cooling to give the final structure. The commonly used materials in FDM are acrylonitrile butadiene styrene (ABS), polylactic acid (PLA), and high-performance thermoplastics like polyether ether ketone (PEEK). There has been a significant increase in the number of published articles in recent years, which indicates vigorous research efforts in the field of FDM. Figure 7.2 shows the statistics of publications pertaining to the "application of FDM" for the period from 2000 to 2023.

There are several advantages driving the adoption of FDM. First, FDM is very flexible and capable of generating parts with complex geometries that cannot be made in traditional manufacturing methods. Its capability for making intricate custom components has been most helpful when producing highly complicated parts (Belter and Dollar 2015; Mandal, Basak, *et al.* 2024). Second, it is relatively inexpensive to apply, especially for low-run production and prototyping. The relatively low cost of

DOI: 10.1201/9781003686941-7

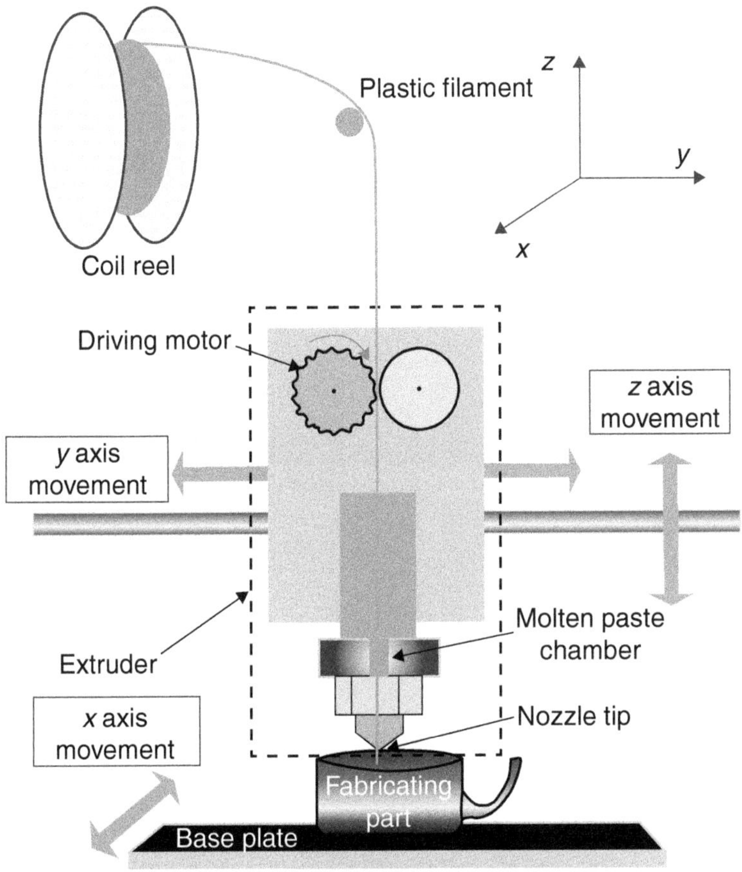

FIGURE 7.1 Schematic of fused deposition modelling process. (From Jin *et al.* (2015).)

the material and the avoidance of the expensive molds and tooling reduce the total cost of manufacturing (Zhao *et al.* 2018; Mandal *et al.* 2023). Third, FDM facilitates fast prototyping, where designers and engineers can create and test various iterations of a design within a very short period, speeding up the development cycle (Liao *et al.* 2018, Mandal, Kumar, *et al.* 2024). Furthermore, the technology provides a wide variety of available materials, which include high-strength thermoplastics and biocompatible materials that make it versatile for most applications as illustrated in Figure 7.3, and its detailed application is shown in Table 7.1 (Singh *et al.* 2018).

In the health sector, FDM has greatly improved patient care and medical science. A custom prosthetic or an orthotic product is a product that is fashioned to fit specific patients better, thus bringing comfort and functional ability. Using FDM, patient-specific surgical guides and anatomical models facilitate surgeons in doing complex surgical procedures with precision. This technology allows for the manufacturing of personal medical devices and implants, including cranial plates and dental prosthetics that fit properly with the body of a patient (Talinungsang *et al.* 2019). Moreover,

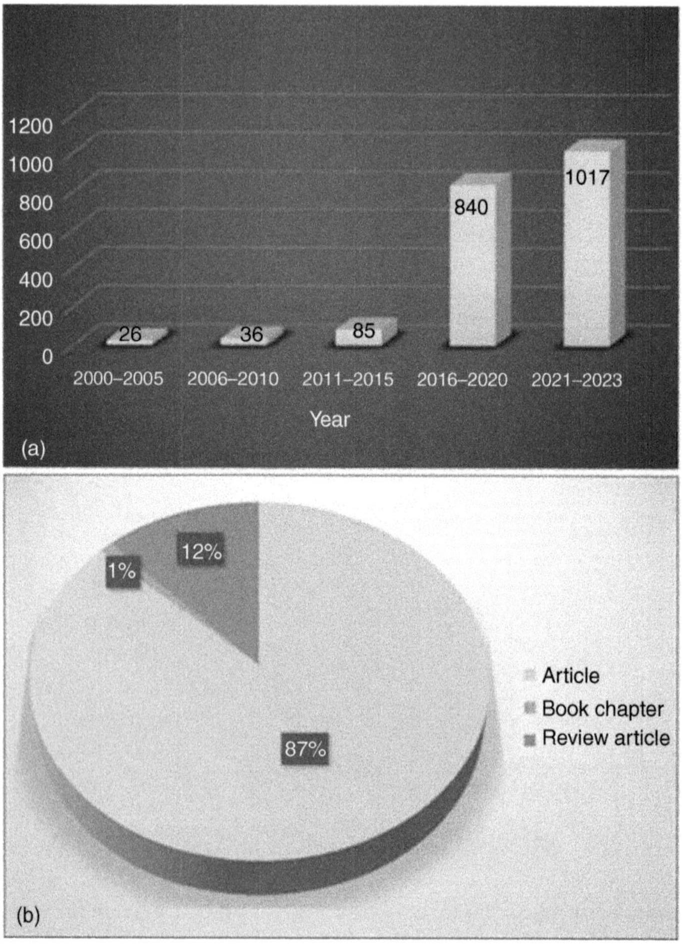

FIGURE 7.2 (a) The number of publications in the span of 2000–2023 related to the application of fused deposition modeling. (b) Different types of publications.(From Web of Science, Keyword: "Application of Fused Deposition Modeling," Time: 2000–2023.)

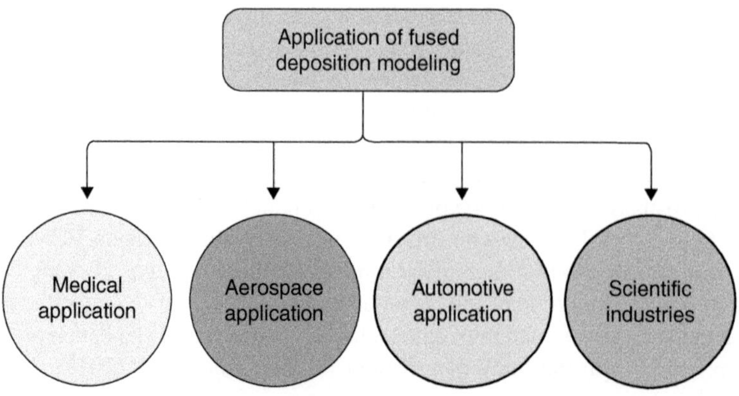

FIGURE 7.3 Application of fused deposition modeling.

TABLE 7.1
Application Areas of FDM and Its Specific Uses

Application Area	Specific Use	Reference
Medical	Drug delivery system	Goyanes, Buanz, *et al.* (2015), Goyanes, Robles Martinez, *et al.* (2015), Khaled *et al.* (2015)
	3D bioprinting	Daminabo *et al.* (2020), Hazrat Ali and Abilgaziyev (2021)
	Tissue engineering	Lee *et al.* (2010), Kovalcik *et al.* (2020)
	Medical devices and implants	Meaney and Goyen (2007), Murphy and Atala (2014)
	Prosthetics and orthotics	Teixeira *et al.* (2019), Xu *et al.* (2023)
Aerospace	Lightweight component manufacturing	Najmon *et al.* (2019), Alami *et al.* (2023)
	Rapid prototyping and testing	Kenney (2003), Klippstein *et al.* (2018)
	Production of complex geometries	Goh *et al.* (2017), Hu and Qin (2020)
	Maintenance and repair operations	Huang *et al.* (2013), Cao and Shi (2023)
Automotive	Rapid prototyping and design	Ceretti *et al.* (2017), Yadav *et al.* (2020)
	Custom tooling and fixtures	Kumar *et al.* (2012), Lyu and Choi (2015)
	Production of end-use parts	Ning *et al.* (2015), Pati *et al.* (2015)
	Vehicle customization	Gray *et al.* (1998), Novakova-Marcincinova *et al.* (2012)
Scientific Industries	Custom laboratory equipment	Bollig *et al.* (2017), Górecka *et al.* (2020), Walters *et al.* (2020)
	Rapid prototyping and experimentation	Nabipour and Akhoundi (2020), Tosto *et al.* (2020), Naveed (2021)
	Educational models	Li *et al.* (2014), Fornells *et al.* (2020)
	Research and development	Mazurchevici *et al.* (2020), Wickramasinghe *et al.* (2020)

FDM is useful in improving tissue engineering through the construction of scaffolds that act as a growth support medium for cells and tissues to grow (Goyanes, Buanz, *et al.* 2015; Skowyra *et al.* 2015).

The aerospace sector relies on FDM as the technology produces lightweight yet highly performing components. Complex parts with integrated features like cooling channels in turbine blades for improved performance and lighter weights are manufactured using FDM. Rapid prototyping, functional testing of parts to iterate through the design, is also supported by this technology. On-demand spare parts production and the repairing of damaged parts through FDM reduce lead times and help in cutting down maintenance costs; hence, it enhances efficiency during operations.

FDM expedites the prototyping and design process in the automobile sector by the rapid fabrication of functional prototypes for testing and validation. Custom tooling and fixtures manufactured by FDM optimize assembly operations and decrease expenses. The method is also used for making end-use parts, especially those for custom and low-volume cars, where traditional manufacturing approaches may be impractical. Also, FDM can customize vehicles by making parts that fit specific customer preferences.

FDM is a technology that has significant advantages for both research and teaching. The technology can be used to create custom-made laboratory and experimental equipment that is engineered to the exact necessities. 3D printing of parts and tools is the main technique used in the creation of new experimental setups. 3D educational models and anatomical models have helped teachers develop hands-on teaching tools which have been successful in explaining difficult problems to both teachers and students.

FDM production technology has gained a predominant position in modern production and industrial branches and is, at the same time, a guarantee of good quality and the best use of production time for the consumer. Its fast and inexpensive production of intricate pieces has been pivotal in medical, aerospace, automotive, and scientific fields. With improvement of this technology, FDM will enable its applications in new sectors and the optimization of processes will be a bigger step toward its role as the leading state of contemporary design and manufacturing. Table 7.1 provides the application areas of FDM and its specific uses.

7.2 MEDICAL INDUSTRY

The medical industry has used FDM technology in numerous ways for medical applications. FDM technology is primarily focused on surgical guides, prosthetics and orthotics, medical equipment and implants, medicine, and tissue engineering. FDM has also provided a new and easy way for the manufacturing of respirator face shields and other personal protective equipment (PPE) during the COVID-19 crisis (Oladapo *et al.* 2021).

7.2.1 Prosthetics and Orthotics

FDM has substantially enhanced prosthetic limbs and orthotic devices. The production of prosthetics is labor-intensive and costly in the case of traditional methods, but FDM can produce prosthetic limbs and orthotics in very precise and cost-effective ways (Xu *et al.* 2023). FDM offers several features like comfort and mobility for the prosthesis that help to lead the quality of life of the persons (Teixeira *et al.* 2019). Moreover, braces (custom-designed orthotic devices) are very helpful for the users. Such braces that are made to perfectly fit the unique anatomy of a patient can render significantly better support and less of an injury or discomfort risk. To concentrate on a similar issue, there has been a significant boost to the attachable osteoid casts. In this case, functioning uses of low-intensity pulsed ultrasound (LIPUS) are used to promote the process of bone regeneration. Mechanical load applied to bone cells, which have broken into pieces due to stress, leads to a decrease in the time taken to heal by up to 38%. Two equal sections of the cast are produced and subsequently linked by a flexible pin. In the process of customizing the cast for the patient, a 3D scanner is used, which is then shaped up with computer-aided design (CAD) software (Awad *et al.* 2018).

7.2.2 Surgical Guides and Models

The accuracy as well as the flexibility of FDM technology has indeed made it indispensable in the introduction of surgical planning and the production of personal anatomical models. Thus, surgeons may now make meticulous surgical plans through innovative guides that are customized and take the patient's unique anatomy from imaging data as a basis (Rasoulianboroujeni *et al.* 2019). Subsequently, orthopedic surgery is an area where patient-specific guides are used to precisely locate screws and implants, which in turn is the instrument to avoid errors in the surgery and achieve better outcomes. 3D anatomical models, which are created with FDM technology, produce visual acute images of the patient's unique anatomy, helping surgeons to perform practice surgical procedures and plan their approach in a more methodical way (Scoutaris *et al.* 2018). Not only is the doctor's capability enhanced, but also the operational time is reduced and the patient's safety is increased. Some of the properties of FDM that make it stand out have been thoroughly studied for its potential to be used in the pharmaceutical industry. This technology has been used to make 3D printed models of different tablets that are also called 3D printlets. The word "printlet" refers to an oral dosage form made via 3D printing. As a manufacturing process, FDM is capable of engineering printlets with a wide range of geometries and sizes.

7.2.3 Medical Devices and Implants

FDM's capacity to manufacture instruments in the medical sector with biocompatible materials has emerged as an effective method for the development of medical devices and implants. Custom implants, such as cranial plates, dental implants, and hearing aids, can be created that are tailored to the patient's precise anatomical needs (Meaney and Goyen 2007). This uniqueness ensures an easier fit and better compatibility with the patient's body. Thus, the chances of complications are reduced and general outcomes are well-managed (Murphy and Atala 2014). In addition to that, FDM not only is a rapid prototyping technique for medical devices but also allows for iterative testing. Consequently, the development process is shortened, which in turn enables new medical technologies to be implemented more quickly. The execution of FDM for the manufacturing of 3D discs, beads, and catheters was proven by Weisman *et al.* (2015). With a view to the creation of the drug concentration profile and enhancement of mechanical strength, the model drugs gentamicin and methotrexate were combined for better results. The study of the drug-loaded and the 3D structures showed respective similarities of some forms against certain bacteria. Consequently, it was thus clearly proved that the pharmacological behavior was not changed by the 3D printing method. The research team that composed the study stated that their outcomes obviously revealed that this technology could make and deliver antimicrobial and chemotherapeutic drugs on request only, and thus increase the therapies to be more efficient and optimized (Weisman *et al.* 2015).

7.2.4 Tissue Engineering

Tissue engineering employs FDM to print out that could be biodegradable scaffolds with the function of cell support and tissue development. They are scaffold-printed as composite materials, in which the cells would interact and migrate in the network of pores created by the extracellular scaffold production. The researchers can fabricate the scaffolds with specific characteristics such as the elimination of inaccurate pore size from the structures and the limitation of the geometry, as well as the mechanical properties (Mandal, Hazra, *et al.* 2024). The scaffolds can be implemented in the treatment of various tissues to make them function in a good, stable, and effective way, as shown by the fact that Zein *et al.* (2002) have several examples concerning it. This clinical application in the field of regenerative medicine, in addition to its significance in the tissue repair and replacement sectors, has been one of the most promising aspects of the field (Winarso *et al.* 2022). FDM has truly been the limelight of the medical sector, which has revolutionized artificial limbs and body diagnostics, thus leading to the patient care impetus and the growth of medical research (Kumar 2023; Kumar *et al.* 2024). Enhanced FDM is capable of overcoming many difficult problems modern medicine has faced, namely the production of custom prostheses and orthoses, patient-specific surgical guides and models, personalized medical devices and implants, and quite complicated scaffolds for tissue engineering; therefore, it is a huge step forward in modern medicine (Wang *et al.* 2023). Its capability to manufacture intricate, custom-made products at a fast rate and a low cost is the main reason why the technology has become a vital instrument in the advancement of medical technologies all over the world (Upadhaya, Kumar, *et al.* 2019; Upadhaya, Talinungsang, *et al.* 2019).

7.3 AEROSPACE INDUSTRY

FDM is now the essential technology for the aerospace industry by providing creative solutions that help in manufacturing processes, reduce costs, and make better component performance in aviation. This section is devoted to the important applications of FDM in aerospace technology, the applications mainly being as lightweight component manufacturing, rapid prototyping and testing, production of complex geometries, and use in maintenance and repair operations (Vashishtha 2011; Klippstein *et al.* 2018).

7.3.1 Lightweight Component Manufacturing

Aviation engineering prefers a less amount of weight of components and minimizing the weight leads to several benefits like fuel efficiency, diminish pollutants, and good overall performance. FDM plays a major role to achieve the less amount of weight of components (Alami *et al.* 2023). These materials give the basic properties of mechanics while making the weight distinctly less than that of metal measurements (Najmon *et al.* 2019). Internal structures, like honeycomb structures, can be created with the help of precise lasers, which not only contribute to weight saving but also lead to an increase in the strength and durability of the final product. These are primarily important for non-core parts used in outer and common internal devices.

7.3.2 Rapid Prototyping and Testing

FDM is a revolution in the prototyping stage of aerospace development. Creating the prototypes of aerospace components has always been a time-consuming and expensive experience. FDM enables rapid prototyping, the ability to create many iterations of the design, which in turn allows the engineers to produce and assess them more efficiently (Kenney 2003). Thus, the design validation is significantly faster, as prototypes can be checked for form, fit, and function in the early stage of the development cycle. Engineers can detect and eliminate the possible issues before the final production, which also decreases the likelihood of any errors and delays (Klippstein *et al.* 2018). The capability to fabricate the functional prototypes at the time of aerodynamic testing and structural analysis would guarantee that the design requirements are strict enough to be achieved.

7.3.3 Production of Complex Geometries

The aerospace industry has not had an adequate number of complex geometry parts. Even so, the process is even more difficult when conventional manufacturing techniques are applied. FDM makes the production of complex shapes and internal features (Hu and Qin 2020). The ability to construct these kinds of parts is of great use not only in manufacturing, production, and faster processes but also in car making, machinery manufacturing, and so on. Simply using FDM, aerospace manufacturers are capable of small part customization that is related to only some applications, which support lean processes and reduce material waste (Goh *et al.* 2017). Additionally, FDM's layer-by-layer construction method enables the production of parts with internal cavities and undercuts, which are difficult or impossible to achieve with traditional subtractive methods. Through the measurement of extruder's bar mount vibrations, Tlegenov *et al.* (2018) demonstrated that nozzle clogging in fused filament fabrication 3D printers can be monitored using an accelerometer sensor. Utilizing the fundamental process modeling, the suggested technique is highly effective for monitoring nozzle clogging in fused filament fabrication 3D printers.

7.3.4 Maintenance and Repair Operations

One of the most important moves in the aerospace industry is the ability to make replacement parts and tools on the spot. The greater part of them is additive manufactured and applied in aerospace applications, making them tool-resistant. 3D printing is the most popular method of rapid prototyping because it can be used in a wide range of industries (Huang *et al.* 2013). The aerospace industry is one of the most interesting fields for its implementation in 3D printing technologies. In addition, for industries, it is highly beneficial because it also shortens the production and distribution chains. This technology decreases the need for large inventories of spare parts; thus, it helps to reduce long-term storage costs (Cao and Shi 2023). The best parts about FDM are that it reduces the cost and time (besides other aspects of automation) and saves the weight of the assemblies through the use of fewer raw materials, thus making the process eco-friendly. Therefore, it can function with minimum human labor and make the work more straightforward, faster, and safer for the person working. As long as the technology is getting better, more aspects of its usage will be involved, and the utilization of the newest and best materials will be crucial in safeguarding the changes in aerospace design, manufacturing, and maintenance. Being a multipurpose and high-end computer device, FDM has become a key step to the zero-waste engineering-for-the-future that is for quality aerospace engineering and excellence of execution (Zhang *et al.* 2023).

7.4 AUTOMOTIVE INDUSTRY

FDM has emerged as a reversible technology in the automotive industry, which enables both advancing the productivity of manufacturing and cutting down the production costs; meanwhile, new vehicle design development is brought to an increased velocity. This section explains the specific cases of FDM application in the automotive sector. In particular, it addresses rapid prototyping and design, custom tooling and fixtures, production of end-use parts, and the potential for vehicle customization in the automotive sector.

7.4.1 Rapid Prototyping and Design

Aviation engineering prefers a less amount of weight of components, and minimizing the weight leads to several benefits like fuel efficiency, reducing pollutants, and good overall performance. FDM plays a major role in achieving the least amount of weight of components. FDM is a technology that accelerates prototyping; thus now the designers and engineers can produce and evaluate various generations of the parts and assemblies in a short time by limiting time and cost inputs (Ceretti *et al.* 2017). Thus, the design cycle is accelerated, which leads to quick development and testing of new vehicles. Engineers can assess prototypes' form, fit, and function at the early stage of development, identify and resolve potential issues/threats that may occur in production (Yadav *et al.* 2020). The iterative approach to the design phase leads to a more accurate design and, in turn, reduces the risk of very expensive errors. Another fascinating production of FDM is the production of acoustic devices. The need to absorb or reflect the movement of sound waves through the air is at the core of several sectors. Consequently, the very economic production of porous, cellular, and metamaterials, which are to be used as acoustic liners in engines, turbofans, ventilation systems, and other applications, is required (Arjunan, Baroutaji, Robinson, *et al.* 2022; Arjunan, Baroutaji, and Robinson 2022).

7.4.2 Custom Tooling and Fixtures

The automotive manufacturing process is very reliant on specialized tooling and devices to help with accurate and effective delivery. FDM implemented the most cost-effective and most flexible tool fabrication process (Lyu and Choi 2015). Custom jig, fixture, and manufacturing assist design complete the task to improve daily production. Making time and money in the upcoming time is the best advantage of 3D-printed tooling over traditional methods, which are basically expensive, lengthy, and need skilled labor (Kumar *et al.* 2012).

7.4.3 Production of End-Use Parts

Although FDM is mainly a means of rapid prototyping, it is ever more often that it gets utilized for the production of final automotive parts. FDM materials move to a new level, for example, by means of high-performance thermoplastics, that make it possible to use them for the production of such components that will meet the high requirements of the car industry (Pati *et al.* 2015). Aircraft ducts and brackets are just some of the parts that can be manufactured via FDM. This technique is favored for individual and small series, in which production through traditional means is often too expensive. One of the great advantages of FDM is that it allows the production of parts only when they are needed, which decreases the inventory and lead times (Ning *et al.* 2015).

7.4.4 Vehicle Customization

FDM's primary strength in vehicle customization lies in the ability to rapidly deliver tailor-made parts. The drawing model produces the best choice of motors for individual car design. With the increasing demand for personal configuration, consumers want their cars to look special and unique, and also a customized interior to outdoor components (Novakova-Marcincinova *et al.* 2012; Anand *et al.* 2024). FDM stands as the technology that makes it possible for manufacturers to tailor the parts according to the specific customer's interests. This level of personalization concerns not only the appearance; powerful innovations as unique dashboard layouts, adjustable gear, and custom storage are also examples of possible customizations (Gray *et al.* 1998). The versatility feature of FDM promotes a more customer-focused and interactions toward the vehicle designs. This improvement is major because FDM processes have provided the automotive sector with methods such as rapid prototyping, custom tooling, and on-demand parts in line with customer design specifics. This appliance that enhances the production of intricate profiles of high-tensile material inevitably solves the challenge car makers are stuck with. Consequently, FDM increases the precision of the design, lowers the production costs of parts, and raises the process speed of product development, the quality of the technology used, as well as the producer's agility and efficiency.

7.5 SCIENTIFIC INDUSTRIES

FDM is known as a technology that is an essential component of different industries. It helps in developing research, sending research to other researchers, and upgrading educational programs. This section deals with diverse FDM applications in science fields, such as in custom laboratory equipment, rapid prototyping, activities, and experimentations, and design that is nonstandard, and also in research and development. The use of FDM in the printing of electrochemical cells to be used in energy storage devices arose from one of the other studies conducted and was part and direct printing for biomedical scaffolding and conductors for the electronics sector (Zhang *et al.* 2017; Manzanares Palenzuela *et al.* 2018).

7.5.1 Custom Laboratory Equipment

Scientific research generally requires highly complex tools and instruments that are either not readily available or cost too much to procure with traditional methods. FDM is the future of this technology and develops the rapid production of such specific lab tools and equipment that will be further suited to certain experiments (Walters *et al.* 2020). The researchers use computer-aided software to model a 3D or CAD image of components that they can print for their laboratory equipment. These parts may include the custom pipette holders, microscope mounts, reaction vessels, and experimental apparatus (Bollig *et al.* 2017). This capability permits the creation of customized equipment that exactly fulfils the individual experiment requirements, which, as a result, leads to the improvement of the lab's accomplish ability and precision. FDM as a means of devising laboratory equipment also brings to bear a very cost-effective and time-efficient method of cutting costs and lead times consequent upon traditional manufacturing processes (Górecka *et al.* 2020).

7.5.2 Rapid Prototyping and Experimentation

The use of FDM that has been used in serving scientists in research work in a very big way is prototyping. They can quickly make prototypes based on experimental apparatus and testing rig options, which helps in iteratively optimizing and improving quality by 3D printing (Nabipour and Akhoundi 2020; Tosto *et al.* 2020; Wang *et al.* 2021). This process thus accelerates the cycle of the development; researchers tend to have the ability to in a fraction of the time of the tests be conducted and the designs be improvement due to the agile nature of the methods used. Among the other supplies, the researcher has a casting machine to make his or her inventions produce the necessary building materials that will be used by him or her in the research process. The product was developed and submitted for use in the research for the casting process (Abeykoon *et al.* 2020; Naveed 2021; Ankita and Sengupta 2024).

7.5.3 Educational Models

FDM is a very essential tool in the field of education as it offers a higher level of teaching and learning experience. The technology enables teachers to create 3D detailed models of science concepts and structures that are usually very complex. For example, molecular models, anatomical structures, and geological formations can be printed to be used as hands-on learning tools to provide the students with physical materials to understand their contents (Li *et al.* 2014). These models enable students to visualize and understand the difficult scientific concepts that would otherwise be almost impossible to understand; therefore, the comprehension and retention are also improved. Moreover, the models made with FDM are often used in interactive demonstrations and laboratory exercises, which are in remission a more engaging educational environment and a highly effective one. The use of FDM can be made available at all educational institutions, and thus, teachers can make very many models that can help in their specific curricula (Fornells *et al.* 2020).

7.5.4 Research and Development

With the use of FDM, the research and development sector is able to perform actively across various scientific disciplines. Among them are chemistry, biology, physics, and engineering, where FDM is effective in making specialized equipment for the laboratory and the production of components that are used as part of cutting-edge research works. Thus, in the case of microfluidics, FDM can design unique channels and devices to control and manipulate fluids at the microlevel with the aid of the direct-write technologies (Wickramasinghe *et al.* 2020). Biology is another field where custom 3D prototypes and bioreactors are used for printing purposes to support experiments like cell culture and tissue engineering. The availability of the means to write and make replacement parts on demand greatly accelerates the innovation and provides new flexibility for the executives who want to take the route or approach that they have never taken before (Mazurchevici *et al.* 2020).

FDM has not only provided scientific industries with innovative solutions for lab equipment, but also as a useful production process of high-speed prototyping, educational models, and research and development. FDM can produce complex and high-precision parts fast and at lower costs that mostly attracts scientists and educators. FDM helps to increase productivity in research and teaching by lessening the time people spend on experiments, providing them with new experimental techniques, and providing a more enriching educational experience. With the technology constantly developing, scientific applications of the solution are expected to grow, which would, in turn, lead to new discoveries and make further progress possible. The multi-faceted and universal nature of the technology and its capacity make it one of the most essential components of the ongoing search for scientific knowledge and innovation.

7.6 TECHNICAL CHALLENGES OF FDM PROCESS

FDM is a popular additive manufacturing process that builds parts layer by layer using thermoplastic materials. Despite its widespread use, the FDM process presents several technical challenges shown in Table 7.2.

TABLE 7.2
Technical Challenges of FDM Process

Parameter	Challenge	Cause
Layer adhesion and delamination	Achieving strong interlayer bonding is crucial for part strength. Poor adhesion between layers can lead to delamination, where layers separate under stress	Inconsistent temperatures, material properties, or improper settings can weaken the bonding between layers
Warping	Warping occurs when parts cool unevenly, leading to internal stresses that cause the part to deform, especially at the corners or edges	Rapid cooling, especially in larger prints or when using materials like ABS, can cause differential shrinkage and warping
Surface finish quality	The surface finish of FDM parts often shows visible layer lines, requiring post-processing to achieve a smooth finish	The layer-by-layer nature of FDM inherently creates a stepped surface. Factors like layer height, print speed, and nozzle size affect surface quality
Dimensional accuracy	Achieving precise dimensions can be difficult due to material shrinkage, warping, and other factors that affect part accuracy	Variations in material flow, temperature control, and mechanical precision of the printer can lead to deviations from the desired dimensions
Material limitations	FDM is limited to thermoplastic materials, which may not meet the mechanical, thermal, or chemical resistance requirements for certain applications	The process involves melting and extruding thermoplastics, restricting it to materials that can be processed this way
Support structures	Complex geometries often require support structures, which can be difficult to remove and may leave marks on the part	Overhangs and bridges need temporary supports during printing, adding time and complexity to the process
Residual stresses	Residual stresses within the printed part can lead to cracking or failure under load.	Uneven cooling and thermal gradients during printing can induce internal stresses in the material
Overheating and heat management	Managing the heat generated during printing is essential to avoid issues like stringing, sagging, or overheating of the printed part	Inadequate cooling systems or incorrect temperature settings can cause problems with material extrusion and part quality
Environmental sensitivity	FDM printing can be sensitive to environmental factors like temperature and humidity, which can affect material performance and print quality	Some materials absorb moisture from the air, leading to issues like bubbling or weak prints. Ambient temperature fluctuations can also affect the print

7.7 CONCLUSION

The 3D printing industry has witnessed a growth over the last few years, and it is expected FDM's ability to create intricate and highly customized parts at low cost and short time has been the sole reason for its popularity and fostering of its usage in these industries. The FDM technology is expected to advance the making of new applications and opportunities, as it is expected to evolve even more in the future so that it becomes the primary tool in the present-day technology and design era. FDM is a proof that additive manufacturing works by positing its wide range of uses and major roles in various fields. The technology has shown how industries can use it to change the way they work while removing inefficiencies and giving consumers synchronized access to the best-fitting products. Thus, FDM is voicing the future of manufacturing and design, the main driver of progress and new directions across all sectors. FDM is a clear example of how additive manufacturing can really transform things through its effectiveness and innovation. Its multipurpose nature and outstanding achievements in different areas give evidence of the fact that it greatly advances the technology, changing business processes, and customizing solutions. Future-wise, there is no doubt FDM will be of immense benefit in the upcoming manufacturing and design era, thus contributing to progress and innovation in every field.

REFERENCES

Abeykoon, C., Sri-Amphorn, P., and Fernando, A., 2020. Optimization of fused deposition modeling parameters for improved PLA and ABS 3D printed structures. *International Journal of Lightweight Materials and Manufacture*, 3 (3), 284–297.

Alami, A.H., Ghani Olabi, A., Alashkar, A., Alasad, S., Aljaghoub, H., Rezk, H., and Abdelkareem, M.A., 2023. Additive manufacturing in the aerospace and automotive industries: recent trends and role in achieving sustainable development goals. *Ain Shams Engineering Journal*, 14 (11), 102516.

Anand, R., Meshram, G.S., and Kumar, P., 2024. Experimental analysis of high conductivity solid-state electrolyte $Li_7La_3Zr_2O_{12}$ (LLZO) for Li-ion Battery. *2024 IEEE 3rd International Conference on Electrical Power and Energy Systems (ICEPES)*, 12, 1–6.

Ankita, R. and Sengupta, S., 2024. Advancements in digital signal processing: design and implementation of IIR bandpass optical filter using optisystem. *2024 International Conference on Advances in Modern Age Technologies for Health and Engineering Science (AMATHE)*, (7), 1–4.

Arjunan, A., Baroutaji, A., and Robinson, J., 2022. *Advances in Acoustic Metamaterials. In:* A. Arjunan, A. Baroutaji, and J. Robinsoni, eds. Oxford: Elsevier, 1–10.

Arjunan, A., Baroutaji, A., Robinson, J., and Wang, C., 2022. *Characteristics of Acoustic Metamaterials. In:* A. Arjunan, A. Baroutaji, J. Robinson, and C. Wang, ed. Oxford: Elsevier, 35–45.

Awad, A., Gaisford, S., and Basit, A.W., 2018. Fused deposition modelling: advances in engineering and medicine. *In*: A.W. Basit and S. Gaisford, eds. *3D Printing of Pharmaceuticals*. Cham: Springer International Publishing, 107–132.

Belter, J.T. and Dollar, A.M., 2015. Strengthening of 3D printed fused deposition manufactured parts using the fill compositing technique. *PLoS One*, 10 (4), e0122915.

Bollig, L.M., Hilpisch, P.J., Mowry, G.S., and Nelson-Cheeseman, B.B., 2017. 3D printed magnetic polymer composite transformers. *Journal of Magnetism and Magnetic Materials*, 442, 97–101.

Cao, S. and Shi, Z., 2023. Application and prospect of 3D printing technology in aerospace. *In*: *2023 International Conference on Mechatronics, IoT and Industrial Informatics (ICMIII)*, Institute of Electrical and Electronics Engineers (IEEE), 66–69.

Ceretti, E., Ginestra, P., Neto, P.I., Fiorentino, A., and Da Silva, J.V.L., 2017. Multi-layered scaffolds production via fused deposition modeling (FDM) using an open source 3D printer: process parameters optimization for dimensional accuracy and design reproducibility. *Procedia CIRP*, 65, 13–18.

Daminabo, S.C., Goel, S., Grammatikos, S.A., Nezhad, H.Y., and Thakur, V.K., 2020. Fused deposition modeling-based additive manufacturing (3D printing): techniques for polymer material systems. *Materials Today Chemistry*, 16, 100248.

Fornells, E., Murray, E., Waheed, S., Morrin, A., Diamond, D., Paull, B., and Breadmore, M., 2020. Integrated 3D printed heaters for microfluidic applications: ammonium analysis within environmental water. *Analytica Chimica Acta*, 1098, 94–101.

Goh, G.D., Agarwala, S., Goh, G.L., Dikshit, V., Sing, S.L., and Yeong, W.Y., 2017. Additive manufacturing in unmanned aerial vehicles (UAVs): challenges and potential. *Aerospace Science and Technology*, 63, 140–151.

Górecka, Ż., Idaszek, J., Kołbuk, D., Choińska, E., Chlanda, A., and Święszkowski, W., 2020. The effect of diameter of fibre on formation of hydrogen bonds and mechanical properties of 3D-printed PCL. *Materials Science and Engineering: C*, 114, 111072.

Goyanes, A., Buanz, A.B.M., Hatton, G.B., Gaisford, S., and Basit, A.W., 2015. 3D printing of modified-release aminosalicylate (4-ASA and 5-ASA) tablets. *European Journal of Pharmaceutics and Biopharmaceutics*, 89, 157–162.

Goyanes, A., Robles Martinez, P., Buanz, A., Basit, A.W., and Gaisford, S., 2015. Effect of geometry on drug release from 3D printed tablets. *International Journal of Pharmaceutics*, 494 (2), 657–663.

Gray, R.W., Baird, D.G., and Helge Bøhn, J., 1998. Effects of processing conditions on short TLCP fiber reinforced FDM parts. *Rapid Prototyping Journal*, 4 (1), 14–25.

Hazrat Ali, M. and Abilgaziyev, A., 2021. Fused deposition modeling based 3D printing: design, ideas, simulations. *In*: H.K. Dave and J.P. Davim, eds. *Fused Deposition Modeling Based 3D Printing*. Cham: Springer International Publishing, 23–42.

Hu, C. and Qin, Q.-H., 2020. Advances in fused deposition modeling of discontinuous fiber/polymer composites. *Current Opinion in Solid State and Materials Science*, 24 (5), 100867.

Huang, S.H., Liu, P., Mokasdar, A., and Hou, L., 2013. Additive manufacturing and its societal impact: a literature review. *The International Journal of Advanced Manufacturing Technology*, 67 (5), 1191–1203.

Jin, Y., Li, H., He, Y., and Fu, J., 2015. Quantitative analysis of surface profile in fused deposition modelling. *Additive Manufacturing*, 8, 142–148.

Kenney, P., 2003. Rapid prototyping of an aircraft model in an object-oriented simulation. *In*: *AIAA Modeling and Simulation Technologies Conference and Exhibit*. American Institute of Aeronautics and Astronautics.

Khaled, S.A., Burley, J.C., Alexander, M.R., Yang, J., and Roberts, C.J., 2015. 3D printing of tablets containing multiple drugs with defined release profiles. *International Journal of Pharmaceutics*, 494 (2), 643–650.

Klippstein, H., Diaz De Cerio Sanchez, A., Hassanin, H., Zweiri, Y., and Seneviratne, L., 2018. Fused deposition modeling for unmanned aerial vehicles (UAVs): a review. *Advanced Engineering Materials*, 20 (2), 1–17.

Kovalcik, A., Sangroniz, L., Kalina, M., Skopalova, K., Humpolíček, P., Omastova, M., Mundigler, N., and Müller, A.J., 2020. Properties of scaffolds prepared by fused deposition modeling of poly(hydroxyalkanoates). *International Journal of Biological Macromolecules*, 161, 364–376.

Kumar, P., 2023. *Recent Progress in Computational*. Springer Nature Singapore.

Kumar, P., Ahuja, I.P.S., and Singh, R., 2012. Application of fusion deposition modelling for rapid investment casting – a review. *International Journal of Materials Engineering Innovation*, 3 (3–4), 204–227.

Kumar, P., Prasad, P.S., David, J., and Guerrero, M., 2024. *Different Curing Methods Used for Polymer Composites*. Springer Nature Singapore.

Lee, C.H., Cook, J.L., Mendelson, A., Moioli, E.K., Yao, H., and Mao, J.J., 2010. Regeneration of the articular surface of the rabbit synovial joint by cell homing: a proof of concept study. *The Lancet*, 376 (9739), 440–448.

Li, X., Cui, R., Sun, L., Aifantis, K.E., Fan, Y., Feng, Q., Cui, F., and Watari, F., 2014. 3D-printed biopolymers for tissue engineering application. *International Journal of Polymer Science*, 2014 (1), 829145.

Liao, G., Li, Z., Cheng, Y., Xu, D., Zhu, D., Jiang, S., Guo, J., Chen, X., Xu, G., and Zhu, Y., 2018. Properties of oriented carbon fiber/polyamide 12 composite parts fabricated by fused deposition modeling. *Materials & Design*, 139, 283–292.

Lyu, M.-Y. and Choi, T.G., 2015. Research trends in polymer materials for use in lightweight vehicles. *International Journal of Precision Engineering and Manufacturing*, 16 (1), 213–220.

Mandal, S., Basak, S., Kumar, P., Satpathy, B., Das, S., and Das, K., 2024. CTAB-facilitated ex-situ synthesis of chitosan-based ZnO, CuO, and ZnO/CuO nanocomposites for improved optical emission. *Optical Materials*, 155, 115813.

Mandal, S., Hazra, C., Jena, S., Kumar, P., Sen, R., Das, S., and Das, K., 2023. Lipopeptide and glycolipid assisted growth of ZnO micro-/nano-structures: Evaluating the role of chemical and microbial green surfactant. *Journal of Industrial and Engineering Chemistry*, 126, 352–364.

Mandal, S., Hazra, C., Joshi, N., Jain, N., and Kumar, P., 2024. A comparison of chemical and biogenic surfactant mediated synthesis of ZnO/CuO: highlighting the antimicrobial activity of the bio-functionalized metal oxide nanoparticles. *Colloids and Surfaces A: Physicochemical and Engineering Aspects*, 690 (December 2023), 133780.

Mandal, S., Kumar, P., Satpathy, B., Das, K., and Das, S., 2024. Nanostructured metal oxide based coating for the protection and conservation of cultural heritage: a comprehensive review. *Journal of Cultural Heritage*, 69, 94–112.

Manzanares Palenzuela, C.L., Novotný, F., Krupička, P., Sofer, Z., and Pumera, M., 2018. 3D-printed graphene/polylactic acid electrodes promise high sensitivity in electroanalysis. *Analytical Chemistry*, 90 (9), 5753–5757.

Mazurchevici, A.D., Nedelcu, D., and Popa, R., 2020. Additive manufacturing of composite materials by FDM technology: a review. *Indian Journal of Engineering and Materials Sciences*, 27 (2), 179–192.

Meaney, J.F.M. and Goyen, M., 2007. Recent advances in contrast-enhanced magnetic resonance angiography. *European Radiology*, 17 (Suppl 2), B2–6.

Murphy, S.V. and Atala, A., 2014. 3D bioprinting of tissues and organs. *Nature Biotechnology*, 32 (8), 773–785.

Nabipour, M. and Akhoundi, B., 2020. An experimental study of FDM parameters effects on tensile strength, density, and production time of ABS/Cu composites. *Journal of Elastomers & Plastics*, 53 (2), 146–164.

Najmon, J.C., Raeisi, S., and Tovar, A., 2019. 2 – Review of additive manufacturing technologies and applications in the aerospace industry. *In*: F. Froes and R.B.T.-A.M. for the A.I. Boyer, eds. *Additive Manufacturing for the Aerospace Industry*, Elsevier, 7–31.

Naveed, N., 2021. Investigate the effects of process parameters on material properties and microstructural changes of 3D-printed specimens using fused deposition modelling (FDM). *Materials Technology*, 36 (5), 317–330.

Ning, F., Cong, W., Qiu, J., Wei, J., and Wang, S., 2015. Additive manufacturing of carbon fiber reinforced thermoplastic composites using fused deposition modeling. *Composites Part B: Engineering*, 80, 369–378.

Novakova-Marcincinova, L., Novak-Marcincin, J., Barna, J., and Torok, J., 2012. Special materials used in FDM rapid prototyping technology application. *In*: *2012 IEEE 16th International Conference on Intelligent Engineering Systems (INES)*. Institute of Electrical and Electronics Engineers (IEEE), 73–76.

Oladapo, B.I., Ismail, S.O., Afolalu, T.D., Olawade, D.B., and Zahedi, M., 2021. Review on 3D printing: fight against COVID-19. *Materials Chemistry and Physics*, 258, 123943.

Pati, F., Jang, J., Lee, J.W., and Cho, D.-W., 2015. Chapter 7 – Extrusion bioprinting. *In*: F. Pati, J. Jang, J.W. Lee, and D-W. Cho, eds. *Essentials of 3D Biofabrication and Translation*, Boston: Academic Press, 123–152.

Prasad, P.S., David, J., Guerrero, M., and Kumar, P., 2024. *Natural and Synthetic Fiber-Filled Polymer Composites Used as Anticorrosive Materials*. Springer Nature Singapore.

Rasoulianboroujeni, M., Fahimipour, F., Shah, P., Khoshroo, K., Tahriri, M., Eslami, H., Yadegari, A., Dashtimoghadam, E., and Tayebi, L., 2019. Development of 3D-printed PLGA/TiO_2 nanocomposite scaffolds for bone tissue engineering applications. *Materials Science and Engineering: C*, 96, 105–113.

Scoutaris, N., Ross, S.A., and Douroumis, D., 2018. 3D printed "Starmix" drug loaded dosage forms for paediatric applications. *Pharmaceutical Research*, 35 (2), 34.

Singh, D., Singh, R., Boparai, K.S., Farina, I., Feo, L., and Verma, A.K., 2018. In-vitro studies of SS 316 L biomedical implants prepared by FDM, vapor smoothing and investment casting. *Composites Part B: Engineering*, 132, 107–114.

Skowyra, J., Pietrzak, K., and Alhnan, M.A., 2015. Fabrication of extended-release patient-tailored prednisolone tablets via fused deposition modelling (FDM) 3D printing. *European Journal of Pharmaceutical Sciences*, 68, 11–17.

Talinungsang, U. D., Kumar, P., and Purkayastha, D.D., 2019. Superhydrophilicity of photocatalytic ZnO/SnO_2 heterostructure for self-cleaning applications. *Journal of Sol–Gel Science and Technology*, 92 (3), 575–584.

Teixeira, B.N., Aprile, P., Mendonça, R.H., Kelly, D.J., and Thiré, R.M. da S.M., 2019. Evaluation of bone marrow stem cell response to PLA scaffolds manufactured by 3D printing and coated with polydopamine and type I collagen. *Journal of Biomedical Materials Research Part B: Applied Biomaterials*, 107 (1), 37–49.

Tlegenov, Y., Hong, G.S., and Lu, W.F., 2018. Nozzle condition monitoring in 3D printing. *Robotics and Computer-Integrated Manufacturing*, 54, 45–55.

Tosto, C., Saitta, L., Pergolizzi, E., Blanco, I., Celano, G., and Cicala, G., 2020. Methods for the characterization of polyetherimide based materials processed by fused deposition modelling. *Applied Sciences (Switzerland)*, 10 (9), 3195.

Upadhaya, D., Kumar, P., and Dhar Purkayastha, D., 2019. Superhydrophobic ZnO/TiO_2 heterostructure with significantly enhanced photocatalytic activity. *Journal of Materials Science: Materials in Electronics*, 30 (11), 10399–10407.

Upadhaya, D., Talinungsang, K. P., and Purkayastha, D.D., 2019. Tuning the wettability and photocatalytic efficiency of heterostructure ZnO–SnO_2 composite films with annealing temperature. *Materials Science in Semiconductor Processing*, 95 (February), 28–34.

Vashishtha, V.K., 2011. Advancement of rapid prototyping in aerospace industry – a review. *Science and Technology*, 3 (3), 2486–2493.

Walters, J.G., Ahmed, S., Terrero Rodríguez, I.M., and O'Neil, G.D., 2020. Trace analysis of heavy metals (Cd, Pb, Hg) using native and modified 3D printed graphene/poly(lactic acid) composite electrodes. *Electroanalysis*, 32 (4), 859–866.

Wang, P., Zou, B., Ding, S., Li, L., and Huang, C., 2021. Effects of FDM-3D printing parameters on mechanical properties and microstructure of CF/PEEK and GF/PEEK. *Chinese Journal of Aeronautics*, 34 (9), 236–246.

Wang, W., Liu, P., Zhang, B., Gui, X., Pei, X., Song, P., Yu, X., Zhang, Z., and Zhou, C., 2023. Fused deposition modeling printed PLA/nano β-TCP composite bone tissue engineering scaffolds for promoting osteogenic induction function. *International Journal of Nanomedicine*, 18, 5815–5830.

Weisman, J.A., Nicholson, J.C., Tappa, K., Jammalamadaka, U., Wilson, C.G., and Mills, D., 2015. Antibiotic and chemotherapeutic enhanced three-dimensional printer filaments and constructs for biomedical applications. *International Journal of Nanomedicine*, 10, 357–370.

Wickramasinghe, S., Do, T., and Tran, P., 2020. FDM-based 3D printing of polymer and associated composite: a review on mechanical properties, defects and treatments. *Polymers*, 12 (7), 1–42.

Winarso, R., Anggoro, P.W., Ismail, R., Jamari, J., and Bayuseno, A.P., 2022. Application of fused deposition modeling (FDM) on bone scaffold manufacturing process: a review. *Heliyon*, 8 (11), e11701.

Xu, X., Wang, H., Shen, L., Yang, Q., and Yang, Y., 2023. Application and evaluation of fused deposition modeling technique in customized medical products. *International Journal of Pharmaceutics*, 640, 122999.

Yadav, D.K., Srivastava, R., and Dev, S., 2020. Design & fabrication of ABS part by FDM for automobile application. *Materials Today: Proceedings*, 26, 2089–2093.

Zein, I., Hutmacher, D.W., Tan, K.C., and Teoh, S.H., 2002. Fused deposition modeling of novel scaffold architectures for tissue engineering applications. *Biomaterials*, 23 (4), 1169–1185.

Zhang, F., Wei, M., Viswanathan, V.V, Swart, B., Shao, Y., Wu, G., and Zhou, C., 2017. 3D printing technologies for electrochemical energy storage. *Nano Energy*, 40, 418–431.

Zhang, H., Hui, J., Jingxiang L.V., Lee, C.-H., Yan, Z., Wang, J. J., Guo, L., and Xu, Z., 2023. A novel method to combine fused deposition modelling and inkjet printing in manufacturing multifunctional parts for aerospace application. *Journal of Materials Research and Technology*, 24, 4405–4426.

Zhao, H., He, Y., Fu, J., and Qiu, J., 2018. Inclined layer printing for fused deposition modeling without assisted supporting structure. *Robotics and Computer-Integrated Manufacturing*, 51, 1–13.

Zindani, D. and Kumar, K., 2019. An insight into additive manufacturing of fiber reinforced polymer composite. *International Journal of Lightweight Materials and Manufacture*, 2 (4), 267–278.

8 A Rapid Tool Fabrication Using Fused Deposition Modeling

Rohit Anand, Pankaj Kumar, and Ganesh Sahadeo Meshram

8.1 INTRODUCTION

Fused deposition modeling (FDM) is one of the additive manufacturing (AM) technologies that has found large acceptance due to its use to make durable, functional parts in a cost-effective manner (Masood 1996). The three-dimensional (3D) printing process works by extruding through a nozzle a heated thermoplastic filament layer by layer. The nozzle moves by following the digital design of the object (Gibson *et al.* 2021; Berman 2012). Once it has come out of the nozzle, it cools and becomes hard for forming the required shape. The FDM method is compatible with a variety of materials, such as polyacrylic acid (PLA), acrylonitrile butadiene styrene (ABS), polyethylene terephthalate glycol (PETG), and nylon, which are selected depending on the application's requirements for durability, flexibility, or heat resistance (Tymrak *et al.* 2014; Nikzad *et al.* 2011; Ning *et al.* 2015; Shofner *et al.* 2003). One of the major advantages of FDM is that it is possible to make complex geometries, which are very difficult or impossible to manufacture using conventional subtractive manufacturing methods. While methods such as computer numerical control (CNC) machining cut away material from a block, FDM adds material only where it is needed and yields less waste with the possibility of creating overhangs, hollow structures, and internal channels. In addition, FDM 3D printers are quite cheaper than other additive technologies, such as selective laser sintering (SLS) or stereolithography (SLA), thus making them even more feasible for companies with a tight budget or individuals who are relatively new to 3D printing (Golubovic *et al.* 2024; Mandal, Kumar, *et al.* 2024).

The rapid tooling fabrication (RTF) process in the industrial application area is a major application of FDM. RTF comprises rapid manufacturing of tooling parts, including molds, dies, jigs, and fixtures. Traditionally, such tools have been machined, cast, or forged and mostly are labor-intensive, time-consuming, and costly, and also bounded by design constraints (Agarwala *et al.* 1996). With the use of FDM, tooling parts can now be produced faster and for a lower cost with very minimal waste and short lead times. This process is, therefore, of great value to industries requiring customization or low-volume production in aerospace, automotive, as well as medical device manufacturing (Iftikhar *et al.* 2013). In addition to cost, there are other advantages of using FDM in RTF. It is an additive process and, thus, gives designers the opportunity to design more complex geometries in tools than more difficult

DOI: 10.1201/9781003686941-8

processes or impossible geometries compared to the traditional techniques of production. For example, molds containing complex internal cooling channels, lightweight jigs, or ergonomic fixtures can all be produced with FDM in a fraction of the time it would take with traditional techniques. This accelerates the manufacturing time while also improving the output and performance of the delivered tool (Ebrahimzadeh *et al.* 2016; Anand *et al.* 2024).

Another major advantage of FDM-based RTF is high-speed iteration and customization. If a mold or fixture needs to be changed, engineers can just modify the attached digital design and print the tool without needing to overhaul an entire production process. This is flexibility that can be appreciated in fast industries where designs change that frequently. FDM and RTF offer a number of benefits to modern manufacturing when integrated. FDM is affordable, easy to use, and can make strong, complex parts. This makes it a top choice for 3D printing. It is also used in RTF, which helps companies save money, speed up production times, and be more creative.

8.1.1 Benefits of Using FDM for Rapid Tool Fabrication

1. *Speed and Efficiency:* The most significant advantage of using FDM in rapid tooling (RT) is that design and production time will be saved. If traditional tooling such as CNC machining or injection molding is used, it may take weeks or even months to have the working tool. On the other hand, one can obtain a tool within days through FDM. This is because the prototyping cycle and the design iteration cycles become reduced. Therefore, designs are verified much faster, and a new product hits the market much earlier.
2. *Cost-Effectiveness:* FDM saves on expensive and time-consuming machining. Traditional tooling always brings along considerable costs in terms of material wastage, tool wear, and use of highly skilled labor. The reverse is the case with FDM, which has very minimal material wastage, as only the amount needed to build the part is utilized. Compared with other materials and metals used for tooling processes, the thermoplastic used in FDM is cost-effective in most cases.
3. *Customization:* FDM offers excellent flexibility and customization in the tooling process. Design changes can be easily incorporated with no need for a retooling step in the production of customized tools. This is very useful for applications in which bespoke tools need to be produced for special purposes or where design changes are relatively frequent.
4. *Complexity:* FDM excels in creating complex internal geometries, undercuts, and intricate features that are often difficult to produce using more traditional methods. This has become an important capability when designing tools with complex designs, for example, conformal cooling channels in injection molds; these improve the efficiency of the molding process and, importantly, its quality. The layer-by-layer deposition of FDM allows making these intricate features without demanding special machining setups or assembly of parts.

The uses of FDM-based RT include automotive, aerospace, medical, and consumer goods. In the world of automobiles, FDM makes jigs, fixtures, and assembly aids to create the most efficient and accurate production process. Likewise, the use of FDM for complex composite parts by tools made from lightweight material characterizes the aerospace industry. The medical industry also utilizes FDM for manufacturing custom surgical guides and prosthetics that are prepared uniquely to suit patients. Consumer goods manufacturers apply FDM for short-run molding and prototyping in a way that is quite speedy for market testing and developing products.

8.2 CASE STUDY AND EXAMPLES

According to Golubovic *et al.* (2024), mechanical properties are the key concern of materials produced from ABS by using AM technologies, especially FDM, SLA, and digital light processing (DLP; Golubovic *et al.* 2024). The article focuses on the comparison of traditional thermoplastic ABS and an innovative ABS resin material when discussing the tensile properties. The configuration of the specimens was according to the ISO 527-2 standard for tensile testing. To obtain reliable results, five specimens were made for each AM technology. FDM specimens were produced using the Creality CR-10 Smart Pro machine, whereas the Kings 600 Pro and Creality LD-002R machines were applied for SLA and DLP specimens, respectively. The tensile tests were carried out both under room temperature and at a humidity of 55% using the Shimadzu AGS-X machine. From 2D and 3D microscopy studies, the topography and morphology of the images were visualized. The report concluded that ABS resin material is much better than the traditional ABS based on AM technology. SLA technology seems to be more favorable for improved tensile strength, whereas DLP offers higher elongation and energy absorption. The research has outlined the potentials of ABS resin in receiving AM parts that have tailored mechanical properties for a specific application.

According to Kuo and Li (2016a), the paper focuses on the demand for cost-effective and efficient design and manufacturing of sheet metal forming dies, especially in luxury car and aerospace areas where low-volume mass customization occurs. Traditional methods are characterized by high expenses and time consumption. This work involves the use of silicone rubber molds and epoxy resins in the production of sheet metal forming dies. The type of die-making material chosen includes silicone rubber (KE-1310ST) and epoxy resins (174 AB). The master model is designed in 3D CAD software. The punch and die prototypes are fabricated using an FDM system with the use of thermoplastic ABS filament. Support removal and surface finishing are done as post-processing. Prototype-based silicone rubber molds were prepared. Punch and die pieces were made using epoxy resins. Aluminum–magnesium alloy sheets with a thickness of 0.6 mm were shaped by making use of sheet metal forming dies through the hydraulic cylinder machine. Optical measurement instruments were used to assess the dimensional accuracy of the components that were produced. The recommended technology significantly reduced costs of production (to 87.2%) and time spent on manufacturing compared to the traditional methods. The products had quite good dimensional accuracy with errors within the limits from 0.2 to 0.6 mm.

The punches made from epoxy resins proved more resistant to wear as opposed to those created directly from ABS plastics through a 3D printer. The fabricated dies seemed strong enough to support small to medium levels of production (Figure 8.1).

Kuo *et al.* (2017a) discusses a cost-effective and efficient method for manufacturing wax injection molds with complex geometrical cooling channels using RT and 3D printing techniques. In other words, classic approaches such as straight-line channels and copper ducts that are commonly used to create cooling channels are very limited in their efficiency and design flexibility. Even though laser sintering and melting are possible techniques for creating intricate channels, they are expensive and long-lasting. The authors propose a new method, which is the use of aluminum-filled epoxy resin molds, which are actually less thermally conductive compared to steel molds but are a perfect solution for short production runs. The process is comprised of the following steps. (1) The preparation stage, which is accomplished by using Pro/ENGINEER software to design the cooling channels and then creating hollow channels with a 3D printer using ABS material. (2) Mold fabrication stage: Here the vacuum machine is used to remove the air bubbles while the molds are being cured in a convection oven. (3) Channel removal stage: The authors explained that they tried three methods for removing the ABS channels from silicone rubber molds: ultrasonic vibrations with acetone liquid, flushing with acetone liquid, and acetone vapor. The most efficient method is using ultrasonic vibrations with acetone liquid and flushing with acetone liquid. (d) The application, which helps in the demonstration of the technique on a wax injection mold for a car's shell, confirming the incorporation of cooling channels in both core and cavity halves. It is shown that the proposed technique is practicable for the creation of molds with conformal cooling channels. The

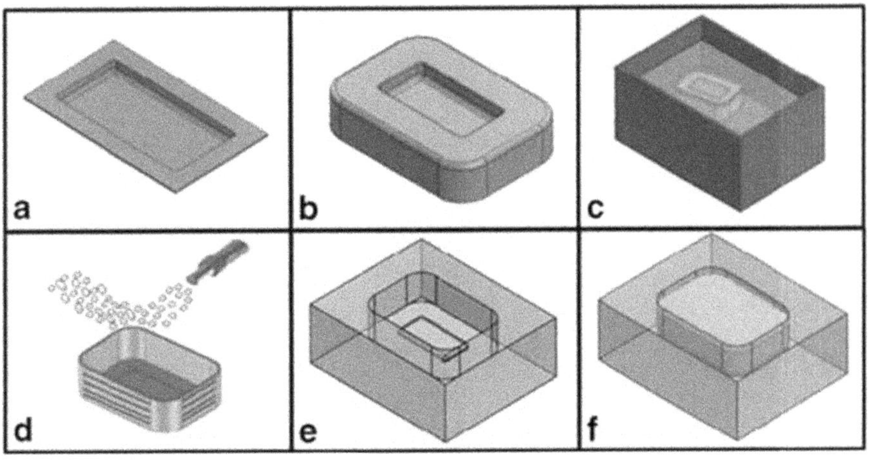

FIGURE 8.1 Schematic diagram of manufacturing sheet metal forming dies. (a) Designing the master model, (b) fabrication of prototype of master model, (c) removing the support materials of prototype, (d) surface treatment of prototype by sandblasting, (e) fabrication of the silicone rubber mold, and (f) fabrication of the sheet metal forming dies. (From Kuo and Li (2016a).)

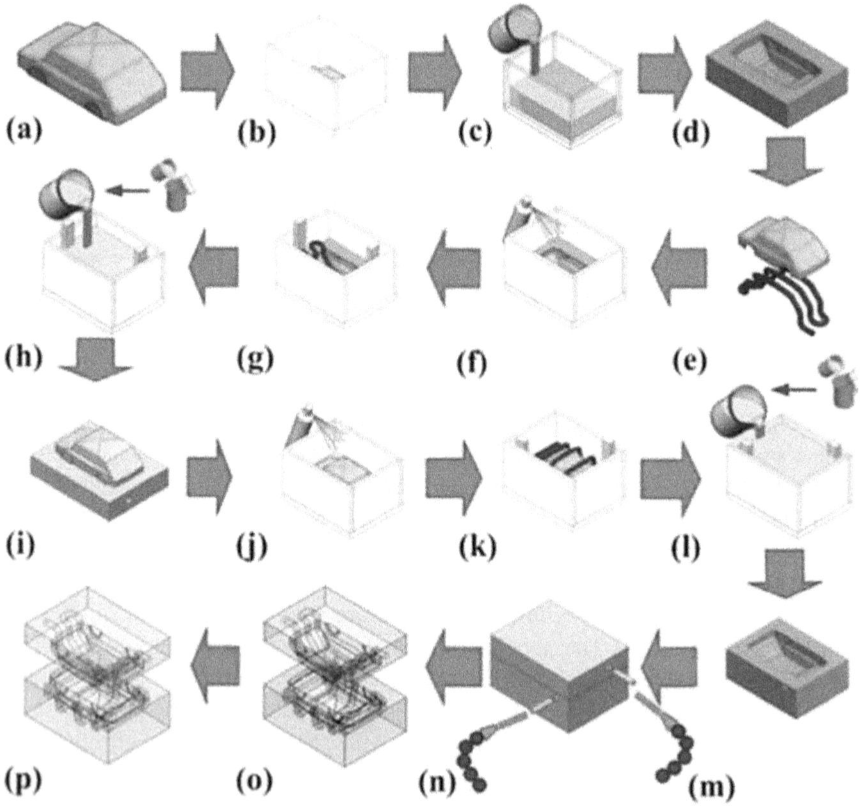

FIGURE 8.2 Process layouts for fabricating a wax injection mold with conformal cooling channels. (From Kuo *et al.* (2017a).)

advantageous side of this method is that it is quick, simple, and cheap compared to old methods. The technical paper acknowledges that acetone flushing is the most effectual way to get rid of the channels. Furthermore, molding the recycled aluminum-filled epoxy resin powder is a factor that facilitates the raw materials cost, and, at the same time, the mold surface quality is acceptable. Constructing thin shells of ABS channel with support materials is suggested, as it is inexpensive and the preparation time is short (Figure 8.2).

The Masood and Song (2004) research focuses on the development of a metal-based composite material for FDM for injection molding dies and inserts. In this direction, the research is in developing a composite of iron particles and nylon (P301) that can be extruded into filaments that can be processed using FDM. The composite material is an iron particle-reinforced nylon (P301) matrix (Table 8.1; Kuo *et al.* 2017a).

The composition includes plasticizers and surfactants that enhance flow, dispersion, and toughness. Dried composite materials were melted in a Brabender Plasti-Corder single-screw extruder and drawn out into filaments. An iron/nylon composite material was successfully developed for FDM during the course of this research—one

TABLE 8.1
Three Samples Were Prepared with Varying Iron Content and Particle Sizes

Sample Name	Composite Material	Particle Size (μm)
Sample 1	70% nylon, 30% iron	50–80 μm particles
Sample 2	60% nylon, 40% iron	50–80 μm particles
Sample 3	60% nylon, 40% iron	<30 μm particles

that can be used to make directly injectable molding inserts and save considerable time and expense associated with RT. Future work will concern the optimization of a composite formulation and FDM processing parameters with a perspective of enhancement in mechanical properties and performances for the developed parts.

According to Kuo and You (2018a), the study focuses on developing wax filaments for fabricating wax patterns using AM technology and identifying optimal process parameters using the Taguchi method. The goal is to produce low-cost, high-accuracy wax patterns suitable for precision casting, specifically for creating injection molding tooling with conformal cooling channels.

Key points:

- *Formulation of Wax Filament:* A low melting point wax filament below 70°C was formulated. The newly developed wax filament cost is significantly lower than the cost of commercially available ones at 0.3 NTD/g as compared to 6 NTD/g.
- *Experimental Design:* Cylindrical rod (10 mm × 10 mm). Key process parameters: Nozzle temperature, print speed, bed temperature, and nozzle flow rate. Optimal values for these parameters were obtained by using the Taguchi method's L9 orthogonal array, with a focus on the reduction of peak-to-valley (P-V) values for better surface quality.
- *Optimal Process Parameters:* Nozzle temperature: 64°C, print speed: 60 mm/s, bed temperature: 40°C, and nozzle flow rate: 100%.
- *Results and Analysis:* The nozzle flow showed the most important effect on the forming accuracy, followed by the print speed on the wax pattern (Figure 8.3). The analysis of variance (ANOVA) result indicated extensive significant damage due to these factors by the wax mold.
- *Surface Treatment:* The dipping of wax patterns in hot water at 65°C for 10 seconds improved the surface quality without affecting the form.
- *Application:* The designed wax filament can demonstrate the feasibility of building conformal cooling channels for the tooling used during the process of injection molding. It is efficient, cost-effective, and also supports the principles of green manufacturing. This concludes that the developed wax filament is practical for the AM and investment casting industries. Optimal process parameters enhance the form accuracy of wax patterns. This method enables rapid fabrication of high-quality, complex geometries in wax patterns, which improves the efficiency and cost-effectiveness of producing injection molding tooling.

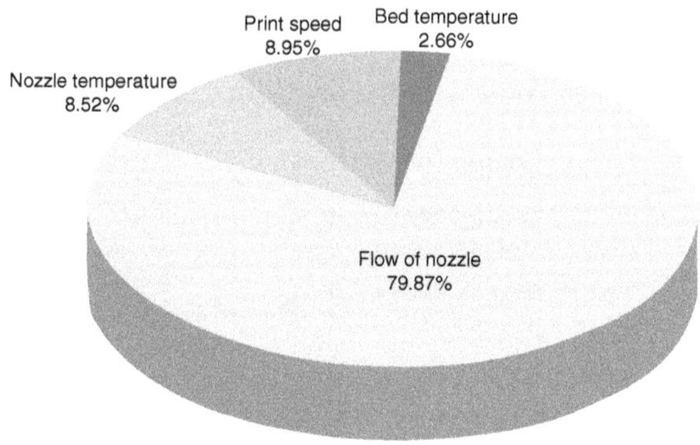

FIGURE 8.3 Schematic illustration of the percentage of contribution. (From Kuo and You (2018a).)

8.3 APPLICATIONS OF FDM

8.3.1 Sensing and Shielding Application

The development of composite filament has greatly enhanced the application capabilities of FDM. The FDM filament becomes conductive when conductive particles like graphene, carbon fiber, carbon nanotubes, and conductive carbon black are added. This allows for the creation of low-voltage functional components. Matrix and reinforcement are mixed to start the process of making conductive filament. Researchers frequently use the solvent casting approach to ensure that the reinforcement and matrix are properly mixed. The addition of the polymer matrix and reinforcement into the appropriate solvent constitutes the fundamental process in solvent casting: mixing the mixture for the necessary amount of time and removing the solvent by drying the agitated mixture in a regulated environment. Using the hot-melt extrusion (HME) method, the composite film obtained from solvent casting could be transformed into a filament with the necessary diameter. Modern HME systems may mix the reinforcement and matrix independently, doing away with the requirement for the solvent casting technique. 3D-printed sensors work based on measuring changes in resistance in response to external stimuli. It is discovered that the poly(vinylidene fluoride)/carbon nanotube composite sample created by FDM can detect distinct solvent vapors. The 3D-printed composite swells, and its resistance rises when it comes into contact with vapor. The solvent can be identified by measuring the change in resistance. The kind of polymer and solvent, the type of reinforcement, and the loading percent all affect how sensitive composite parts are. Better sensitivity to acetone is demonstrated by 3D-printed PVDF/CNT composite (Kennedy *et al.* 2017). Forces applied can be detected via a multiaxial force sensor made of a TPU/CNT nanocomposite. A sensor's resistance varies in response to an outside force (Kim *et al.* 2017). The PLA/graphene composite temperature sensor can detect variations in air and water temperatures up to 70°C. As the temperature rises, composite material becomes more resistant. According to Sajid *et al.* (2018), the 3D-printed PLA/graphene temperature sensor had a response and recovery time of 6 and 14 seconds. Similar to this, temperature change was measured using a smart cap that was three-dimensionally printed using conductive PLA filament (Marasso *et al.* 2018). Utilizing recycled ABS and reinforcement materials such as graphite, $ZnCl_2$, MnO_2, and NH_4Cl, composite filament was effectively utilized in the creation of energy storage devices (ESDs). The ESD suggested by the aforementioned material combination can function at temperatures up to 95°C. When acetone was applied, the voltage of the ESD increased from 0.04 to 0.8 V (Singh *et al.* 2019). Jayanth and Senthil (2019) constructed a capacitive-type void fraction measurement sensor using ABS/CB conductive filament. The created system's realized capacitance value is directly proportional to the fluid's volume. There is a noticeable variation in the obtainable capacitance value across different void percentages. The neutron radiation shielding potential of 3D-printed ABS was reduced from 50% to 72% by adding 20% boron nitride (Woosley *et al.* 2018). Conductive polymer composite is taken into consideration as a substitute material for efficient electromagnetic interference (EMI) shielding because metal-type EMI shielding is heavy and oxidative. Different proportions of carbon nanotubes, carbon black, and a hybrid CNT/CB combination added to ABS show varying levels of shielding effectiveness. CNT exhibits the strongest shielding effect among the different types of reinforcement. Additionally, the printing direction affects shielding potential (Schmid *et al.* 2006). In 8–12 GHz, FDM-printed PLA components containing 10% graphene were able to shield 16 dB, but unreinforced PLA could only shield 2 dB. By absorbing or reflecting the incident energy, shielding takes place. Increasing the loading of reinforcement could raise the possible shielding percentage (Prashantha and Roger 2017). It is important to address problems with nozzle blockage, brittleness, and warping of the printed component while printing conductive filament. For composite filament printing to work properly, the reinforcement should be distributed uniformly. The utilization of 3D-printed conductive polymer composite results in a shielding effect, but at a relatively lower level than that of metal-based shielding. While shielding efficiency is increased by increasing the filler fraction, printability suffers as a result. Printability and shielding effectiveness must therefore be balanced. A 3D-printed sensor is capable of sensing external stimuli; however, its reproducibility and repeatability are restricted to a limited number of cycles. The type of polymer and reinforcement used determines how well an FDM chemical sensor detects chemicals. Consequently, it is challenging to detect many compounds using a single polymer composite. In addition, the printed 3D sensors are influenced by the surrounding atmosphere. Once more, depending on the composite utilized, different results are received from the force sensor. This taught it that FDM parts can function as sensors and shields, but further research is needed to get the best outcome. The focus of future research could be on creating a high-performance polymer composite application that can tolerate high working temperatures.

8.3.2 Scaffold Development

A scaffold is a porous structure used in tissue engineering that promotes cell adhesion, proliferation, and differentiation by functioning as a synthetic extracellular matrix. The following qualities are essential in scaffold materials:

- A surface chemistry that promotes cell adhesion and development;
- Needed bioresorbency or biodegradability;
- Adequate mechanical strength; moreover
- Good processing capacity (Hutmacher 2001).

The scaffold's mechanical characteristics, porosity, pore size, and exterior geometry all affect how well it works. Melt electrospinning, solution electrospinning, emulsion freeze-drying, thermally induced phase separation, melt molding, solvent casting, and particle leaching are examples of conventional methods used to build scaffold structures. However, using conventional methods makes it impossible to produce repeatably highly interwoven systems. Rapid scaffold design that satisfies patient criteria can be produced using AM techniques. Additionally, AM offers improved control over the scaffold structure. The benefits of adopting FDM for scaffold construction include low manufacturing costs, a variety of material availability, and the capacity to create intricate designs. Thus, the current research being done on scaffold construction using FDM is reviewed. Poly(*e*-caprolactone) (PCL) is frequently used to build scaffolds because it is biocompatible and allows for a slower rate of hydrolysis-based biodegradation. FDM of scaffolds from L-lactide/e-caprolactone 75/25 mol percent copolymer poly(L-lactide-co-ε-caprolactone) (PLC) and PCL/bioactive glass (BAG) composite has been demonstrated by Korpela *et al.* (2013). Better cell proliferation was obtained with PLC as opposed to PCL because PLC is more hydrophobic and allows proteins to connect more quickly before cells adhere. The scaffold that was created by AM has pores that are bigger than the size of the cells. Thus, utilizing FDM and electrospinning techniques, Naghieh *et al.* (2017) created a compound scaffold of macro-sized PLA fiber and nano-sized PCL gelatin fiber (Naghieh *et al.* 2017). According to Lam *et al.* (2009), the FDM-printed PCL/TCP composite scaffold degraded more quickly than a standard PCL scaffold. Similar to pure PCL scaffolds, scaffolds made with PCL/calcium phosphate composite show improved cell adhesion, proliferation, and a greater rate of disintegration (Schantz *et al.* 2005). In 2003, Kalita *et al.* created scaffolds using composite filament made of polypropylene and polypropylene/tricalcium phosphate, featuring pore sizes of 160 mm (Kalita *et al.* 2003). Two weeks of in vitro testing show that the scaffold made using the suggested material combination is safe and can support reasonable cell development. The rigidity of scaffolds made from PLA/HA microspheres is comparable to that of trabecular bone. According to Corcione *et al.* (2019), the addition of HA increases surface roughness while decreasing mechanical properties. Type 1 collagen over PLA scaffolds is rendered immobile by the application of polydopamine (PDA) coating. Better cell response and extracellular matrix deposition were thus accomplished, and PLA scaffolds' osteoinductivity was increased (Teixeira *et al.* 2019). When compared to structures made of pure polylactic-co-glycolic acid (PLGA)/TiO_2, scaffolds printed from this combination exhibit greater wettability, glass transition temperature, and compression modulus (Rasoulianboroujeni *et al.* 2019). With the aforementioned efforts, it is evident that FDM may be used to construct the scaffold with the necessary geometry. On the other hand, the FDM-printed scaffold has a somewhat lesser capacity for cell proliferation and differentiation and lacks cell adherence. The inclusion of reinforcement such as tricalcium phosphate, HA, TiO_2, etc. alters the functionality of a polymer-based scaffold, although more research is necessary (Talinungsang *et al.* 2019; Upadhaya, Kumar, *et al.* 2019; Upadhaya, Talinungsang, *et al.* 2019; Kumar 2023; Kumar *et al.* 2024; Mandal, Basak, *et al.* 2024; Mandal, Hazra, Jena, *et al.* 2024; Mandal, Hazra, Joshi, *et al.* 2024; Prasad *et al.* 2024). Additional disadvantages of using FDM to create scaffolds include the limited selection of polymer materials available, the achievable maximum resolution, the potential need for support structures when creating overhanging features, and the occasional need for post-processing to enhance functionality. Additionally, methods for enhancing the attainable osseointegration and strength are required.

8.3.3 Rapid Tooling

RT, a development of rapid prototyping (RP), is the technique of creating tools more quickly. It can be classified as either direct or indirect. Indirect RT methods build tooling of sufficient strength using conventional techniques indirectly utilizing RP master designs, while direct RT methods use additively generated tooling to fabricate prototypes and functional parts from a variety of materials. While RT requires far less tooling time and expense than traditional technologies, it can only create a restricted number of components. Spray metal, epoxy resin, and silicone rubber tooling are a few examples of indirect RT techniques. The cost of producing Al–Mg alloy sheets can be lowered by up to 87% with epoxy resin punch and dies made using an indirect quick tooling process (Kuo and Li 2016b). Kuo and You (2018b) used FDM to create a conformal cooling channel and indirect tooling to create wax filament. Forty ABS and 70 low-density polyethylene pieces may be produced using injection molding inserts made from nylon/iron composite filament. High costs are associated with the photolithography wax stamping method used to develop paper microdevices. Thus, the idea of a PDMS stamp for creating paper microdevices was put out, and it has been questioned whether or not FDM-printed molds may be used to create PDMS stamps. Failures were observed below the minimal barrier and channel width of the FDM mold that produced optimal flow, which were 1 and 1.1 mm, respectively (Montgomery *et al.* 2018). Forming DC04 and S355MC material allowed for the study of the stamping capacity of FDM-produced dies. One hundred pieces of DC04 material were successfully molded within tolerance using 3D-printed dies. After shaping a few components, the die deforms due to the greater hardness of S355MC, which makes it difficult to shape within tolerance standards (Durgun 2015). Electrode for electric discharge machining made using the FDM process and covered in copper-machine-titanium alloy. The 3D-printed electrode's material removal rate and tool wear rate are marginally lower than those of the traditional copper electrode (Danade *et al.* 2019).

The only reason the investment casting method makes sense for large-scale production is the increased expense of creating wax patterns. RP technology is used in the process of creating sacrificial patterns, which is known as "rapid investment casting." The time and expense associated with creating metallic tooling for creating wax patterns are eliminated when RP is incorporated into IC (Lee *et al.* 2004).

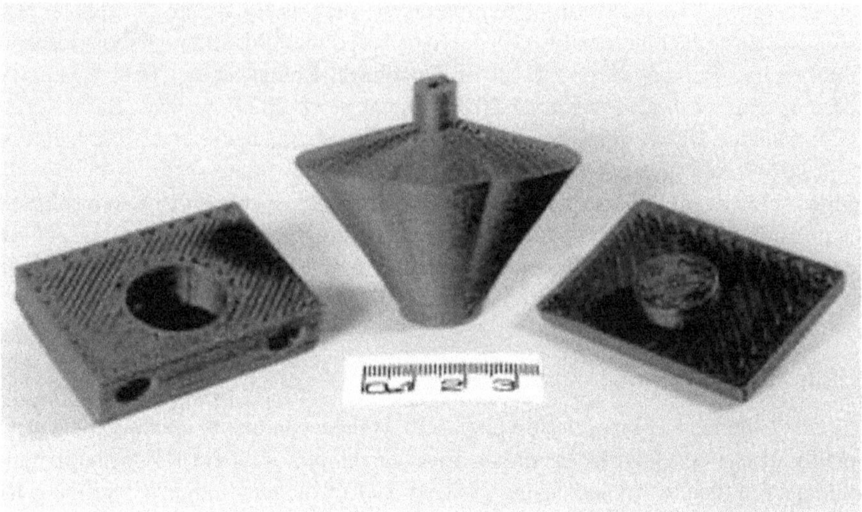

FIGURE 8.4 Examples of parts and inserts produced on the FDM system. (From Masood and Song (2004).)

Researchers have noted problems with incomplete pattern burning, ash present after burning, and ceramic shell breaking when utilizing RP patterns directly for IC. Because aluminum has a lesser heat conductivity than other materials, aluminum-filled epoxy resin molds used to generate investment casting wax patterns require longer cycle times. As a result, Kuo *et al.* (2017b) produced an aluminum-filled epoxy resin mold with conformal cooling channels created using FDM. During the construction of the aluminum-filled epoxy resin mold, thin shell cooling channels made of ABS material were added. The integrated conformal circuit was then flushed out of the finished mold using acetone. The suggested technique facilitates effective heat transfer. Its usage as direct tooling is impeded by its poor surface smoothness and lack of high-performance thermoplastic filament for FDM. While indirect quick tooling can be effectively completed using FDM, the associated time and expense are significantly greater. Low-volume production of bespoke components is possible using quick tooling based on FDM. Higher resolution can only be achieved to a certain extent with quick tooling based on FDM. Even when reinforcements like iron and aluminum powder are added to FDM-based tooling to increase its toughness, the performance still falls short of that of traditional tooling methods (Figures 8.4 and 8.5).

8.4 CHALLENGES AND CONSIDERATIONS

Although RTF with FDM presents great advantages, such as shorter lead times, economies of scale, and design freedom, it simultaneously introduces a large number of challenges that can affect the quality and operational efficiency of the tooling fabricated. The following list describes some of the most important challenges associated with the use of FDM in RTF.

FIGURE 8.5 A complex-shaped injection molding insert produced on FDM. (From Masood and Song (2004).)

8.4.1 Surface Finish and Dimensional Accuracy

In general, visible layer lines may appear in FDM parts and can contribute to poor surface finishes. Tooling applications may be used, for example, in the form of molds or fixtures, where smooth surface qualities are sometimes required so that these can work correctly when injection molding. This may impact the phenomenon insofar as the materials do not flow well or the precision in molded parts is compromised. This involves additional post-production, which might involve chemical smoothing or sanding to create a better surface finish of the product, therefore needing more time and money for the entire process. Moreover, the dimensional accuracy is also a problem. Compared to SLA or SLS, FDM is not as precise as the other types of AM. The precision of the final product depends on some parameters, including the resolution of the printer, material shrinkage, and the printing orientation. Variations in temperature in printing will cause warping of the prints, especially with bigger or more complex parts.

8.4.2 Material Limitations

FDM works primarily with thermoplastic materials, such as PLA, ABS, and PETG, which have shown to be very useful in the manufacturing world. Nevertheless, these plastics are not always the most ideal materials for use in tooling applications. They exhibit poor heat resistance, mechanical strength, and wear resistance. For example, molds that are to be used in injection molding or fixtures that are exposed to high temperatures will warp or degrade very quickly when printed from the more conventional FDM materials. However, while there are certainly specialty thermoplastics available with significantly higher thermal resistance and life than FDM materials, they tend to be expensive and a good deal more difficult to print. They often need specific printers that can print at higher temperatures and require strict environmental control, which can be expensive as well as complicated.

8.4.3 Print Time and Scaling Limitations

FDM can save lead times significantly compared to more traditional manufacturing techniques. However, it is relatively slow at printing large or complex parts. The printing time increases with the size and complexity of tooling. And for parts of larger size, prints will take hours or days. This imposes a scaling limit on using RTF with FDM, especially in high-volume manufacturing where time is a constraint. Sometimes, big tooling parts must be printed in sections and then assembled, which introduces some added complexities with regard to alignment, bonding, and structural integrity.

8.4.4 Mechanical Strength and Wear Resistance

The mechanical strength of the printed tools, although FDM is known to produce strong parts, is usually anisotropic, meaning the strength is directionally dependent on the layer buildup of the printing. The FDM parts typically are weak in the Z-axis as the layers bond rather than melt together. That is the main weakness in this material for tooling that would require strength or must last long. Another would be wear resistance; FDM tools generally degrade faster if they are placed under repeated usage or abrasive or high-impact conditions. In some cases, the tool will need additional support by coatings or material reinforcements that raise the complexity and the cost of the process.

8.4.5 Post-Processing Needs

In general, FDM tools demand significant post-processing to attain industrial usage standards. As has already been discussed, smoothing surfaces, correcting dimensions, and assembling parts may all be needed after printing. Furthermore, other post-processing operations, such as heat treatment or infiltration with other materials, for instance, epoxy, may be required to enhance strength, hardness, or thermal properties in printed tools.

8.4.6 Limited Material Compatibility for Tooling Applications

FDM does not support all materials, especially those used in high-performance tooling for aerospace, automotive, and medical applications. Standard FDM technology is incapable of printing high-performance alloys, ceramics, or other advanced composites.

8.4.7 Environmental Sensitivity

Environmental sensitivities of FDM printing depend on temperature and humidity as factors that may cause warping, cracking, or unevenness during printing. This results in issues with the strength and accuracy of the size of the printed tool for applications where precision or high strength is important. Controlled environments need to be maintained during production to produce such tools and add costs and complexity in manufacturing. FDM allows tools to be built rapidly by making cheaper production costs, higher flexibility designs, and shorter production cycles. However, there are many challenges, such as surface finish and dimensional accuracy, material, mechanical strength, and post-process requirements, that need to be overcome to ensure these produced tools meet the general standards for industrial use. Often, the challenge is overcome by balancing the selection of materials, print settings, and post-processing steps that can at least partially counterbalance these drawbacks while still benefiting from unique FDM capabilities.

8.5 FUTURE TRENDS AND INNOVATION

Due to its ease of operation, low cost, and ability to process most thermoplastics and composites, research in FDM has sped up. Composite filaments could be printed by FDM and thus have been applied to 4D printing, RT, scaffolding, sensors, and drug delivery devices. Indeed, it has immense potential, with a considerable amount of work still going on for improvement. The primary drawback of manufacturing microfluidic devices through FDM is the low resolution and lack of transparency. Therefore, there is a need for methodologies that enhance the resolution of FDM. Moreover, the medical applications of polymers necessitate sterilization, which raises experimental research in the field concerning the sterilization impact on the mechanical properties of the FDM-printed product. Post-processing techniques have been proposed to enhance FDM part features, yet many approaches are material-specific, without a standardized methodology applicable to all materials. This post-processing research can be extended to advanced materials like HIPS, PPSF, PEEK, and composites. The applications of developed composite filaments are limited and need further exploration. Most of the sensors are prepared using low melting point filaments, including conductive PLA, ABS, and TPU. They cannot be used for sensing at temperatures higher than 250°C. Therefore, future studies could consider high-performance conductive composites such as PEEK/CNT, PEEK/graphene, and PEKK/graphene for sensing purposes. The strength of sensing ability in 3D printing depends on the type of filler and percentage loading. Therefore, future work is expected on how the concentration of fillers influences different mechanisms of sensing. The brittleness of conductive filaments hinders effective printing, calling for studies on the impact of adding various plasticizers to improve mechanical characteristics. Electroplating can enhance polymer conductivity; thus, the effect of electroplating on the mechanical and sensing properties of FDM-printed samples should be investigated. Efficient EMI shielding can be achieved with metamaterials, and the EMI shielding potential of 3D-printed metamaterial structures needs exploration. Recycling often degrades polymer properties; therefore, analyzing the FDM printability of recycled polymer and composite filaments is necessary.

FDM can produce scaffolds with intricate designs, but their osseointegration ability is limited. Adding reinforcements such as TiO_2, TCP, PCL, HA, and collagen is suggested to improve this. Further methods for enhancing the functionality of FDM-printed scaffolds are needed, along with a detailed analysis of their inflammatory response. Controlling the drug release rate of 3D-printed drug delivery systems (DDSs) is challenging, and suitable techniques for this are in demand. Identifying parameters for developing and printing various drug-loaded filaments is also crucial, alongside efforts to develop filaments that meet regulatory standards. FDM's application in microfluidics lowers costs and enables complex profile fabrication, but voids and high roughness affect fluid flow. Solutions to these issues are needed. With the limited dimensions of most 3D printers, various weldability studies on FDM-printed components are necessary. Further investigation into FDM's role in topology optimization is also warranted.

Modifying the printing head allows FDM to print continuous fiber-reinforced composites, but reported applications are limited, indicating a need for further research. FDM's introduction to the fashion industry allows for innovative designs, yet issues with filament materials and polymer adhesion to fabrics limit its use. Addressing these challenges requires research attention. In summary, FDM as an AM technique is growing in application development, necessitating further contributions to achieve its full potential.

8.6 CONCLUSION

Extensive research has been done on the use of FDM, which includes its growth and limitations. Starting from its development from prototyping to the fabrication of functional parts, FDM still faces major challenges due to high surface roughness, anisotropy, and strength reduction, among others. The existence of conductive filaments opened the route for FDM to fabricate chemical sensors, force sensors, void fraction measurement sensors, electrodes for electrochemical devices, and even radiation shielding. However, more research needs to be conducted to establish the correct filler percentages for each type of sensing application. With RT, FDM is faster and less expensive than the conventional method; however, the strength of tooling made from FDM is weaker, which produces fewer parts. FDM also enables the development of different DDS, such as controlled-release, sustained-release, and immediate-release dosage forms; drug-loaded implants; multi-medication dosage forms; complex-shaped medicines; and taste-masked drugs. Understanding the interaction of drugs and drugs during filament production and printing is necessary to reach the desired goal. The nature of the drug, the amount of loading capacity, the matrix characteristics, and the DDS design are relevant factors for the drug release rates that are chemically bound, as summaries are chemically bound. FDM-fabricated biocompatible and biodegradable scaffolds have shown enough porosity and supported cell attachment and growth. However, improvements in functionalities such as better osseointegration, reduced inflammation, and desirable biodegradability are still in demand. The ability of FDM to create lightweight structures with significant geometric design freedom makes it promising for automotive applications. Regarding the results in the present research case history, it is confirmed that the conductive polymer composites made by FDM are effective to sense solvents. The research on what effect the type of reinforcing agent and its volume of loading would have on solvent sensing is suggested to be carried out. Altogether, FDM holds the potential to manufacture functioning items, but the right choice of process parameters and printing materials is a prerequisite to fabricating success.

REFERENCES

Agarwala, M.K., van Weeren, R., Bandyopadhyay, A., Whalen, P.J., Safari, A., and Danforth, S.C., 1996. Fused deposition of ceramics and metals: an overview. *1996 International Solid Freeform Fabrication Symposium*, 385–392.

Anand, R., Meshram, G.S., and Kumar, P., 2024. Experimental analysis of high conductivity solid-state electrolyte $Li_7La_3Zr_2O_{12}$ (LLZO) for Li-ion battery. In: *2024 IEEE 3rd International Conference on Electrical Power and Energy Systems (ICEPES)*. IEEE, 1–6.

Berman, B., 2012. 3-D printing: the new industrial revolution. *Business Horizons*, 55 (2), 155–162.

Corcione, C.E., Gervaso, F., Scalera, F., Padmanabhan, S.K., Madaghiele, M., Montagna, F., Sannino, A., Licciulli, A., and Maffezzoli, A., 2019. Highly loaded hydroxyapatite microsphere/PLA porous scaffolds obtained by fused deposition modelling. *Ceramics International*, 45 (2), 2803–2810.

Danade, U.A., Londhe, S.D., and Metkar, R.M., 2019. Machining performance of 3D-printed ABS electrode coated with copper in EDM. *Rapid Prototyping Journal*, 25 (7), 1224–1231.

Durgun, I., 2015. Sheet metal forming using FDM rapid prototype tool. *Rapid Prototyping Journal*, 21 (4), 412–422.

Ebrahimzadeh, A., Azimifar, F., and Nosouhi, R., 2016. Design and manufacturing of integrated drilling and cutting orthopedic bone-specific surgical guide. *Materials and Manufacturing Processes*, 31 (5), 608–611.

Gibson, I., Rosen, D.W., Stucker, B., Khorasani, M., Rosen, D., Stucker, B., and Khorasani, M., 2021. *Additive Manufacturing Technologies*. Springer.

Golubovic, Z., Bojovic, B., Petrov, L., Sedmak, A., Milovanović, A., Mišković, Ž., and Milošević, M., 2024. Comparative analysis of ABS materials mechanical properties. *Procedia Structural Integrity*, 56, 153–159.

Hutmacher, D.W., 2001. Scaffold design and fabrication technologies for engineering tissues—state of the art and future perspectives. *Journal of Biomaterials Science, Polymer Edition*, 12 (1), 107–124.

Iftikhar, A., Khan, M., Alam, K., Imran Jaffery, S.H., Ali, L., Ayaz, Y., and Khan, A., 2013. Turbine blade manufacturing through rapid tooling (RT) process and its quality inspection. *Materials and Manufacturing Processes*, 28 (5), 534–538.

Jayanth, N. and Senthil, P., 2019. Application of 3D printed ABS based conductive carbon black composite sensor in void fraction measurement. *Composites Part B: Engineering*, 159, 224–230.

Kalita, S.J., Bose, S., Hosick, H.L., and Bandyopadhyay, A., 2003. Development of controlled porosity polymer-ceramic composite scaffolds via fused deposition modeling. *Materials Science and Engineering: C*, 23 (5), 611–620.

Kennedy, Z.C., Christ, J.F., Evans, K.A., Arey, B.W., Sweet, L.E., Warner, M.G., Erikson, R.L., and Barrett, C.A., 2017. 3D-printed poly (vinylidene fluoride)/carbon nanotube composites as a tunable, low-cost chemical vapour sensing platform. *Nanoscale*, 9 (17), 5458–5466.

Kim, K., Park, J., Suh, J., Kim, M., Jeong, Y., and Park, I., 2017. 3D printing of multiaxial force sensors using carbon nanotube (CNT)/thermoplastic polyurethane (TPU) filaments. *Sensors and Actuators A: Physical*, 263, 493–500.

Korpela, J., Kokkari, A., Korhonen, H., Malin, M., Närhi, T., and Seppälä, J., 2013. Biodegradable and bioactive porous scaffold structures prepared using fused deposition modeling. *Journal of Biomedical Materials Research Part B: Applied Biomaterials*, 101 (4), 610–619.

Kumar, P., 2023. Recent progress in computational techniques in various coating materials. In: *Materials Horizons: From Nature to Nanomaterials*. Springer Nature, 217–228.

Kumar, P., Prasad, P.S., and Guerrero, J.D.M., 2024. Different curing methods used for polymer composites. In: *Materials Horizons: From Nature to Nanomaterials*. Springer Nature, 43–53.

Kuo, C.-C., Chen, W.-H., and Xu, W.-C., 2017a. A cost-effective approach for rapid manufacturing wax injection molds with complex geometrical shapes of cooling channels. *The International Journal of Advanced Manufacturing Technology*, 91, 1689–1695.

Kuo, C.-C., Chen, W.-H., and Xu, W.-C., 2017b. A cost-effective approach for rapid manufacturing wax injection molds with complex geometrical shapes of cooling channels. *The International Journal of Advanced Manufacturing Technology*, 91, 1689–1695.

Kuo, C.-C. and Li, M.-R., 2016a. A cost-effective method for rapid manufacturing sheet metal forming dies. *The International Journal of Advanced Manufacturing Technology*, 85 (9), 2651–2656.

Kuo, C.-C. and Li, M.-R., 2016b. A cost-effective method for rapid manufacturing sheet metal forming dies. *The International Journal of Advanced Manufacturing Technology*, 85 (9), 2651–2656.

Kuo, C.-C. and You, Z.-Y., 2018a. Development of injection molding tooling with conformal cooling channels fabricated by optimal process parameters. *The International Journal of Advanced Manufacturing Technology*, 96, 1003–1013.

Kuo, C.-C. and You, Z.-Y., 2018b. Development of injection molding tooling with conformal cooling channels fabricated by optimal process parameters. *The International Journal of Advanced Manufacturing Technology*, 96, 1003–1013.

Lam, C.X.F., Hutmacher, D.W., Schantz, J., Woodruff, M.A., and Teoh, S.H., 2009. Evaluation of polycaprolactone scaffold degradation for 6 months in vitro and in vivo. *Journal of Biomedical Materials Research Part A: An Official Journal of The Society for Biomaterials, The Japanese Society for Biomaterials, and The Australian Society for Biomaterials and the Korean Society for Biomaterials*, 90 (3), 906–919.

Lee, C.W., Chua, C.K., Cheah, C.M., Tan, L.H., and Feng, C., 2004. Rapid investment casting: direct and indirect approaches via fused deposition modelling. *The International Journal of Advanced Manufacturing Technology*, 23, 93–101.

Mandal, S., Basak, S., Kumar, P., Satpathy, B., Das, S., and Das, K., 2024. CTAB-facilitated ex-situ synthesis of chitosan-based ZnO, CuO, and ZnO/CuO nanocomposites for improved optical emission. *Optical Materials*, 155, 115813.

Mandal, S., Hazra, C., Jena, S., Kumar, P., Sen, R., Das, S., and Das, K., 2024. Lipopeptide and glycolipid assisted growth of ZnO micro-/nano-structures: evaluating the role of chemical and microbial green surfactant. *Journal of Industrial and Engineering Chemistry*, 129, 352–364.

Mandal, S., Hazra, C., Joshi, N., Jain, N., Kumar, P., Sen, R., Das, S., and Das, K., 2024. A comparison of chemical and biogenic surfactant mediated synthesis of ZnO/CuO: highlighting the antimicrobial activity of the bio-functionalized metal oxide nanoparticles. *Colloids and Surfaces A: Physicochemical and Engineering Aspects*, 690, 1–15.

Mandal, S., Kumar, P., Satpathy, B., Das, K., and Das, S., 2024. Nanostructured metal oxide based coating for the protection and conservation of cultural heritage: a comprehensive review. *Journal of Cultural Heritage*, 69, 94–112.

Marasso, S.L., Cocuzza, M., Bertana, V., Perrucci, F., Tommasi, A., Ferrero, S., Scaltrito, L., and Pirri, C.F., 2018. PLA conductive filament for 3D printed smart sensing applications. *Rapid Prototyping Journal*, 24 (4), 739–743.

Masood, S.H., 1996. Intelligent rapid prototyping with fused deposition modelling. *Rapid Prototyping Journal*, 2 (1), 24–33.

Masood, S.H. and Song, W.Q., 2004. Development of new metal/polymer materials for rapid tooling using fused deposition modelling. *Materials & Design*, 25 (7), 587–594.

Montgomery, R.H., Phelan, K., Stone, S.D., Decuir, F., and Hollins, B.C., 2018. Photolithography-free PDMS stamps for paper microdevice fabrication. *Rapid Prototyping Journal*, 24 (2), 361–367.

Naghieh, S., Badrossamay, M., Foroozmehr, E., and Kharaziha, M., 2017. Combination of PLA micro-fibers and PCL-gelatin nano-fibers for development of bone tissue engineering scaffolds. *International Journal of Swarm Intelligence and Evolutionary Computation*, 6 (1), 1–4.

Nikzad, M., Masood, S.H., and Sbarski, I., 2011. Thermo-mechanical properties of a highly filled polymeric composites for fused deposition modeling. *Materials & Design*, 32 (6), 3448–3456.

Ning, F., Cong, W., Qiu, J., Wei, J., and Wang, S., 2015. Additive manufacturing of carbon fiber reinforced thermoplastic composites using fused deposition modeling. *Composites Part B: Engineering*, 80, 369–378.

Prasad, P.S., Guerrero, J.D.M., and Kumar, P., 2024. Natural and synthetic fiber-filled polymer composites used as anticorrosive materials. In: *Materials Horizons: From Nature to Nanomaterials*. Springer Nature, 181–202.

Prashantha, K. and Roger, F., 2017. Multifunctional properties of 3D printed poly (lactic acid)/graphene nanocomposites by fused deposition modeling. *Journal of Macromolecular Science, Part A*, 54 (1), 24–29.

Rasoulianboroujeni, M., Fahimipour, F., Shah, P., Khoshroo, K., Tahriri, M., Eslami, H., Yadegari, A., Dashtimoghadam, E., and Tayebi, L., 2019. Development of 3D-printed PLGA/TiO_2 nanocomposite scaffolds for bone tissue engineering applications. *Materials Science and Engineering: C*, 96, 105–113.

Sajid, M., Gul, J.Z., Kim, S.W., Kim, H.B., Na, K.H., and Choi, K.H., 2018. Development of 3D-printed embedded temperature sensor for both terrestrial and aquatic environmental monitoring robots. *3D Printing and Additive Manufacturing*, 5 (2), 160–169.

Schantz, J.-T., Brandwood, A., Hutmacher, D.W., Khor, H.L., and Bittner, K., 2005. Osteogenic differentiation of mesenchymal progenitor cells in computer designed fibrin-polymer-ceramic scaffolds manufactured by fused deposition modeling. *Journal of Materials Science: Materials in Medicine*, 16, 807–819.

Schmid, G., Eidenschink, U., and Regensburg, B.M.W., 2006. Rapid manufacturing with FDM in jig and fixture construction. Available at: https://scholar.google.com/scholar?hl=en&as_sdt=0%2C5&q=Schmid%2C+G.%2C+Eidenschink%2C+U.%2C+and+Regensburg%2C+B.M.W.%2C+2006.+Rapid+manufacturing+with+FDM+in+jig+and+fixture+construction.+&btnG=

Shofner, M.L., Lozano, K., Rodríguez-Macías, F.J., and Barrera, E.V, 2003. Nanofiber-reinforced polymers prepared by fused deposition modeling. *Journal of Applied Polymer Science*, 89 (11), 3081–3090.

Singh, R., Singh, H., Farina, I., Colangelo, F., and Fraternali, F., 2019. On the additive manufacturing of an energy storage device from recycled material. *Composites Part B: Engineering*, 156, 259–265.

Talinungsang, Upadhaya, D., Kumar, P., and Purkayastha, D.D., 2019. Superhydrophilicity of photocatalytic ZnO/SnO_2 heterostructure for self-cleaning applications. *Journal of Sol–Gel Science and Technology*, 92 (3), 575–584.

Teixeira, B.N., Aprile, P., Mendonca, R.H., Kelly, D.J., and Thiré, R.M. da S.M., 2019. Evaluation of bone marrow stem cell response to PLA scaffolds manufactured by 3D printing and coated with polydopamine and type I collagen. *Journal of Biomedical Materials Research Part B: Applied Biomaterials*, 107 (1), 37–49.

Tymrak, B.M., Kreiger, M., and Pearce, J.M., 2014. Mechanical properties of components fabricated with open-source 3-D printers under realistic environmental conditions. *Materials & Design*, 58, 242–246.

Upadhaya, D., Kumar, P., and Purkayastha, D.D., 2019. Superhydrophobic ZnO/TiO_2 heterostructure with significantly enhanced photocatalytic activity. *Journal of Materials Science: Materials in Electronics*, 30 (11), 10399–10407.

Upadhaya, D., Talinungsang, Kumar, P., and Purkayastha, D.D., 2019. Tuning the wettability and photocatalytic efficiency of heterostructure ZnO–SnO_2 composite films with annealing temperature. *Materials Science in Semiconductor Processing*, 95, 28–34.

Woosley, S., Galehdari, N.A., Kelkar, A., and Aravamudhan, S., 2018. Fused deposition modeling 3D printing of boron nitride composites for neutron radiation shielding. *Journal of Materials Research*, 33 (22), 3657–3664.

9 Study on the Effect of Process Parameters on Mechanical Properties of Nylon 12 Carbon Fiber Composite Using Fused Deposition Modeling

Pratap Chandra Padhi, Ravindra Reddy N., Karthi S., and Arpan Paul

9.1 INTRODUCTION

In the modern era, additive manufacturing (AM) is extensively used in mold-making to produce complex components at a competitive cost. It is a process of three-dimensional (3D) printing where the physical object is manufactured from a digital computer-aided design (CAD) file by laying down successive layers of material to produce complex shapes without the need for tooling [1]. The study focuses on development and process optimization related to different aspects of fused deposition modeling (FDM). This technology is highly preferred for new product development instead of conventional machining methods, and soft data alone is sufficient to generate an object [2]. The CAD data is generated based on the shape and size of the drawing and is converted into an STL file for 3D printing. Manufacturing industries use various AM processes for prototype development, which is used in different engineering applications [3]. Researchers have developed various polymer composites that are 3D-printed in FDM to study their physical and mechanical properties. Many new-generation materials are used in medical implants, spacecraft components, and electronic parts developed by 3D printing, whose properties have been studied based on functional requirements [4]. Conventionally, the products are manufactured by converting raw materials into finished ones through the material removal process, such as milling, drilling, and grinding operations. These are expensive and time-consuming, and are simplified by using the FDM process, where the

DOI: 10.1201/9781003686941-9

components are produced using the material addition method. The products are built much faster using a layer-by-layer material deposition process.

The study explores the effects of variation in FDM process parameters. Nylon 12CF filament has been used to evaluate the print quality and mechanical properties. Nylon 12CF is a composite material known for its excellent tensile strength, heat resistance, specific strength, fatigue properties, and durability. It is useful in various industries, including automotive, aerospace, sports equipment, oil and gas, and is used for parts that require high performance [5]. Its strength and lightweight characteristics make it ideal for producing functional prototypes, end-use parts, and load-bearing components. The strength and rigidity of Nylon 12CF replace metal in many applications and validate designs as faster functional prototypes instead of costly and time-consuming metal prototypes.

Substantial research has been reported on FDM process parameter optimization on product quality, mechanical properties, surface finish, dimensional accuracy, and build time for different plastic materials. Galetto et al. [6] have evaluated the effect of independent variables on fill density, print speed, extruder temperature, layer height, build orientation, and retraction speed on surface roughness and dimensional precision of polylactic acid (PLA) components using the FDM process. The experiment was done using the Design of Experiments (DoE) approach, and input factors were optimized using mathematical models. These results found that variable process parameter settings show different output responses in terms of the quality of the product. Mohamed et al. [7] have optimized the print properties of input variables to improve the quality of build parts using acrylonitrile butadiene styrene (ABS) filament in FDM using the statistical design method. It was revealed that input variables in FDM are crucial factors and should be selected as suitable for the improvement of physical and mechanical properties. Sharma et al. [8] have investigated various process parameters like infill density, deposition speed, and layer height, and optimized the surface finish of the part in FDM using polyvinyl chloride (PVC) with polypropylene (PP) and hydroxyapatite (HAP) composite filament applied in bio-sensing work. The study was carried out using the L9 Taguchi design and found that the infill density is a significant variable affecting the surface properties. Dey and Yodo [9] reviewed the literature on FDM research, focusing on optimizing various input variables to enhance output responses. Their study utilized different plastic filaments and applied a statistical design of experiments along with optimization tools. Sood et al. [10] have analyzed the influence of control factors like part build direction, layer height, raster angle, raster width, and air gap in the FDM process of ABS P400 plastic parts. The experimental investigation was done using the response surface methodology (RSM) FCC design in MINITAB software, and the data was analyzed. The optimal process parameter level was determined to find the best compressive strength. Nagendra et al. [11] have studied the surface quality of the 3D-printed product using aramid filament in the FDM process by optimizing the process variables. The experimental design was planned by using Taguchi's orthogonal design. Patil et al. [12] optimized the input variables in the FDM process using control factors such as infill pattern (IP), infill density, print speed, and layer height on surface finish. The statistical tool was modeled to analyze the effect of various input variables. It was observed that the layer height is the most significant variable that controls the surface phenomenon.

In another study, Eswaran et al. [13] investigated the error of circularity of the FDM-printed parts, and the 3D printing was done using Taguchi's design to optimize the control factors. The input variables considered are orientation angle and infill density to reduce the circularity error. Rao and Rai [14] studied the optimal time required in the FDM process, dimensional deviation in printed parts, and their mechanical properties by using single and multi-objective optimization teaching–learning-based optimization algorithms, and the Pareto analysis solutions were obtained. Zhang et al. [15] have analyzed the wear and thermal study of FDM parts using different grades of Nylon filaments, like Nylon composites, and found that the Nylon 618 grade has shown the best wear properties. Ahn et al. [16] have found the optimal setting of the input variables of the ABS components printed in FDM. The mechanical strength of FDM components was compared with injection-molded parts. Alafaghani et al. [17] investigated the effect of ABS filament input variables on FDM build parts. The input variables considered are infill density, layer height, IP, and extrusion temperature using the Taguchi design of experiment. Vishwas et al. [18] have studied the mechanical properties in the optimization of process variables in FDM parts using ABS and Nylon filament. Bedi et al. [19] studied the correlation of process variables without putting responses in the FDM process using SiC/Al_2O_3-blended with low-density polyethylene (LDPE), and experiments were conducted as per Taguchi's L9 design to optimize the responses. Christiyan et al. [20] have analyzed the importance of printing speed and layer height on the flexural strength of components according to ASTM D790. The composite parts were printed in the FDM process using ABS and hydrous magnesium silicate filaments, and maximum flexural value was found at minimal print speed and layer thickness. Chacón et al. [21] have studied the outcomes of process characteristics on flexural strength and tensile properties by developing a quadratic equation for each part orientation angle in the FDM process. They found maximum flexural strength in the build direction, which was influenced by thickness of layer in each pass and speed during printing. Mahmood et al. [22] have observed the mechanical strength of ABS parts printed in FDM, which depends on part thickness, shell number, and infill density. The part strength increases with infill density and the number of shells but is inversely related to the cross-sectional area. Raju et al. [23] found that the print layer's thickness and the part's orientation angle are the most effective parameters in the FDM process to optimize the tensile properties.

Aw et al. [24] have analyzed the effect of IP and infill density of the FDM part, which is highly significant on the tensile modulus (TM) of ABS/ZnO and conductive ABS/ZnO build parts. It was observed that 100% infill density contributed to optimum tensile strength in the build part. Sun et al. [25] have investigated the print speed and the tool path, which have to be decided based on the part geometry. Also, the bond strength of the layer and product characteristics are influenced by the build platform temperature, rate of cooling, and extrusion temperature. Paul and Das [26] analyzed the influence of layer height, print speed, and build orientation on printing time and cost using Taguchi's DoE. Their study demonstrated that these three parameters significantly affect both response factors. Peng et al. [27] used response surface design for experimentation, considering the input variables such as layer thickness (LT), filling velocity, and extrusion velocity to study the dimensional deviation, build time, and deformation due to warpage. Further, fuzzy logic and genetic algorithm (GA) were used to effectively enhance the build part properties in FDM. Srivastava et al. [28] have used the central composite design (CCD) method in RSM and conducted 86 experiments in FDM on Fortus 250mc modeler, considering the input variables as slice height, contour width, air gap, raster width, raster angle, and orientation angle to optimize the responses, print time, and volume of support material. Rinanto et al. [29] used the Taguchi design to conduct the FDM experiments, considering the variables infill density, extrusion temperature, and raster angle to optimize the tensile strength, build time, and energy consumption. Subsequently, Process Capability Ratio (PCR) with Technique for Order Preference by Similarity to the Ideal Solution (TOPSIS) multi-objective optimization algorithms were adopted to get the optimal setting of input parameters. Phatak and Pande [30] analyzed the design and orientation angle of build parts in FDM based on the material used, part quality, and build time. The weighted average GA was applied to optimize the process variables in the FDM build part. Paul and Anand [31] have corelated cylindrical tolerance and part orientation angle in the FDM process. Optimum build orientation was found by developing mathematical models and graphical techniques. Barua et al. [32] have studied the process parameter optimization of ABS M30 parts built in FDM for the response surface roughness using the multi-objective optimization on the basis of the ratio analysis (MOORA) technique and desirability function analysis. Gajdoš and Slota [33] conducted an experimental study by changing the process temperature and base plate layout and found that it affected the build part. The FDM build parts were studied by tomography to find the variation in layer, dimensions, and volume of unfilled parts, and it was observed that the homogeneity of the structure was affected by the part geometry.

A thorough review of the literature indicates that minimal research has been conducted on analyzing process parameters and their impact on parts fabricated using Nylon–CF engineering polymer composite material in the FDM process, which is widely utilized in the automotive, aerospace, and other load-bearing applications. Considering the importance of the material, the study aims to optimize the variable printing parameters on the mechanical properties of Nylon 12CF-printed parts in the FDM process.

9.2 EXPERIMENTAL PROCEDURE

9.2.1 Materials and Methods

The experimental investigation begins with the selection of appropriate print parameters for 3D printing of Nylon 12CF polymer composite filament. Different combinations of process variables were chosen to analyze the responses in the FDM process based on work experience by referring to the machine manufacturer's data book and literature review. The experiments were conducted in a FDM machine

(Make: MakerBot; Model: MethodX, USA), as shown in Figure 9.1, having the capacity to print parts up to a size of 200 × 170 × 150 mm^3 with an accuracy of 0.2 mm. It works on the principle of extrusion of semi-molten polymeric material deposited layer over layer as per the standard STL-CAD file, and the FDM machine is capable of using multiple build filament materials in a build and support association. Quick Slice software is used to prepare the incoming STL data for the preparation of the model. The present study uses Nylon 12CF filament of standard diameter 1.75 mm for printing tensile and flexural specimens.

Nylon 12CF is an emerging engineering polymer with excellent tensile strength and stiffness. It is moisture-resistant, easy to process, and provides a smooth final surface finish without the need for postprocessing. The melting temperature of Nylon 12 carbon is relatively high compared to general-purpose plastics. The recommended extrusion temperature for FDM 3D printing with this material is typically between 240°C and 270°C. Setting the temperature on the 3D printer is crucial to achieving optimal printing results. A heated build platform is often used for FDM printing with

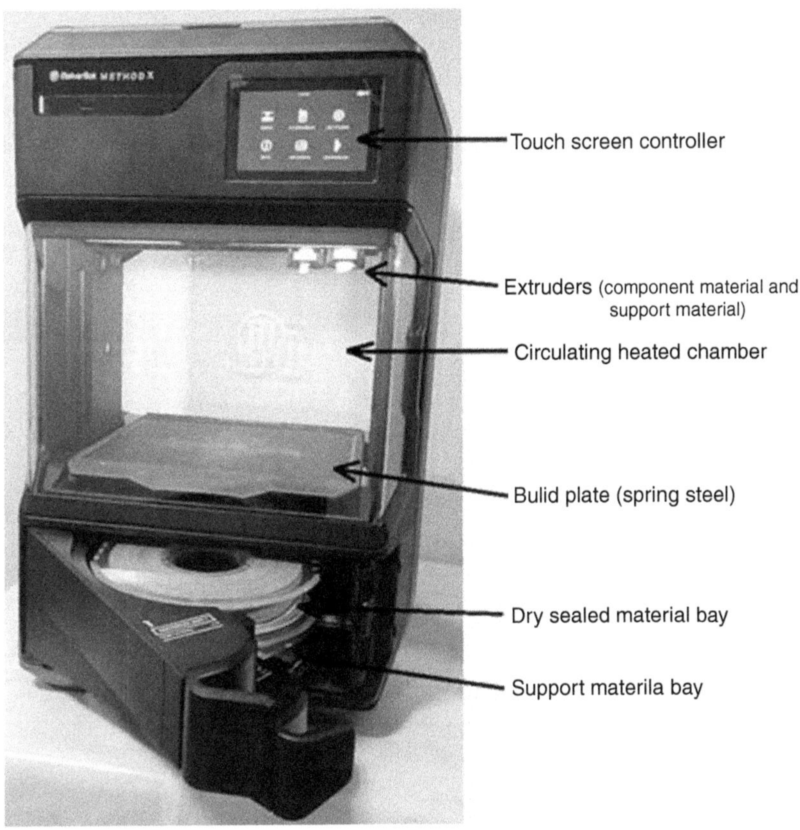

FIGURE 9.1 FDM-3D printing machine (Make: MakerBot; Model: MethodX).

Nylon 12CF to ensure good adhesion and minimize warping. The recommended bed temperature is typically around 60–80°C.

In the present experimentation, the 3D printing in FDM was carried out at an extrusion melt temperature of 250°C with 100% infill density. To have better adhesion properties of the build part and to minimize warping, a heated build platform maintained at 80°C has been used for FDM printing of Nylon 12CF. The print speed was kept uniform at 35 mm/sec for better print quality and dimensional accuracy and to reduce the risk of nozzle clogging. The CAD models of the tensile test specimens were developed using CATIA software, as shown in Figure 9.2, and exported as STL (stereolithography) files in FDM software. The FDM in-built software divides the 3D model into slices and generates the tool path for printing. The Nylon 12CF filament is supplied to the machine head via the spool of the machine. The specimens have been printed using the input process parameters of the FDM machine.

The tensile properties of Nylon 12CF specimens change depending on the selection of operational variables in FDM. The input factors considered in this study are print temperature, IP, LT, part orientation angle, print speed, infill density, and part geometry. Therefore, the mechanical properties of Nylon 12CF specimens were tested to assess the precise tensile properties of the build tensile specimen 3D printed as per Taguchi's design of experiments. When building tensile specimens in FDM, it is essential to consider the appropriate design and printing parameters to ensure accurate and reliable results. The tensile stress at break and TM for all nine numbers of 3D-printed specimens shown in Figure 9.3 were tested in a universal testing

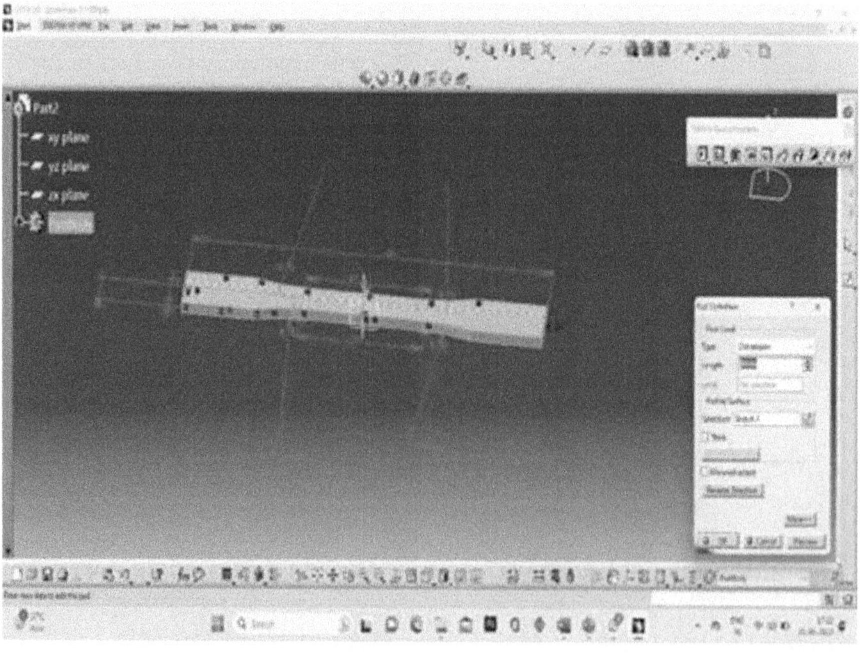

FIGURE 9.2 CATIA 3D model of tensile test specimen.

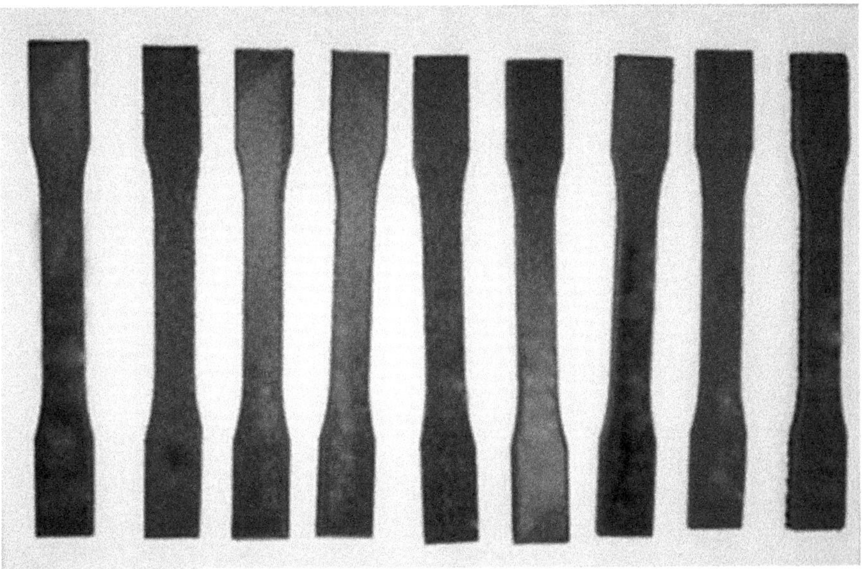

FIGURE 9.3 FDM-printed specimens for tensile test.

machine (UTM), Make: Instron 3382, as per ASTM D638 standard for the specimen dimensions type I (dog bone-shaped) at a speed of 5 mm/min, as shown in Figure 9.4. The dimensions of the tensile 3D-printed specimens are shown in Figure 9.5.

The flexural test is a standard method used to assess the mechanical properties of materials. In the present study, nine flexural test specimens, as per Taguchi's experimental design to determine the flexural properties, are 3D printed in FDM with a cross-section of 3.2 mm × 12.7 and a length of 127 mm, as shown in Figure 9.6.

The flexural test was done in a UTM machine (Make: Instron 3382) in the specimens subjected to a three-point bending test as shown in Figure 9.7, and the bending forces were applied to measure their strength, flexural modulus (FM), maximum flexural load at break (MFB), and other mechanical properties.

The LT of the 3D-printed specimen has been selected based on the desired resolution and the capabilities of the FDM machine. Thinner layers generally provide better resolution but increase the printing time. A common choice is a LT of 0.1–0.5 mm, which balances resolution and print time. Similarly, the infill density determines the internal structure of the specimen. To achieve better mechanical properties, it is recommended to use a high infill density; hence, in this study, the infill density is considered 100% to achieve consistent and reliable results. A solid infill helps ensure uniform mechanical properties throughout the specimen. As the print orientation angle also affects the part's strength in FDM, 0° and 90° orientation angle print directions have been considered for the study. Similarly, the IP plays a major role in evaluating the mechanical behavior of the 3D-printed specimen. So printing was done using various recommended IP shapes: linear, hexagonal, and diamond.

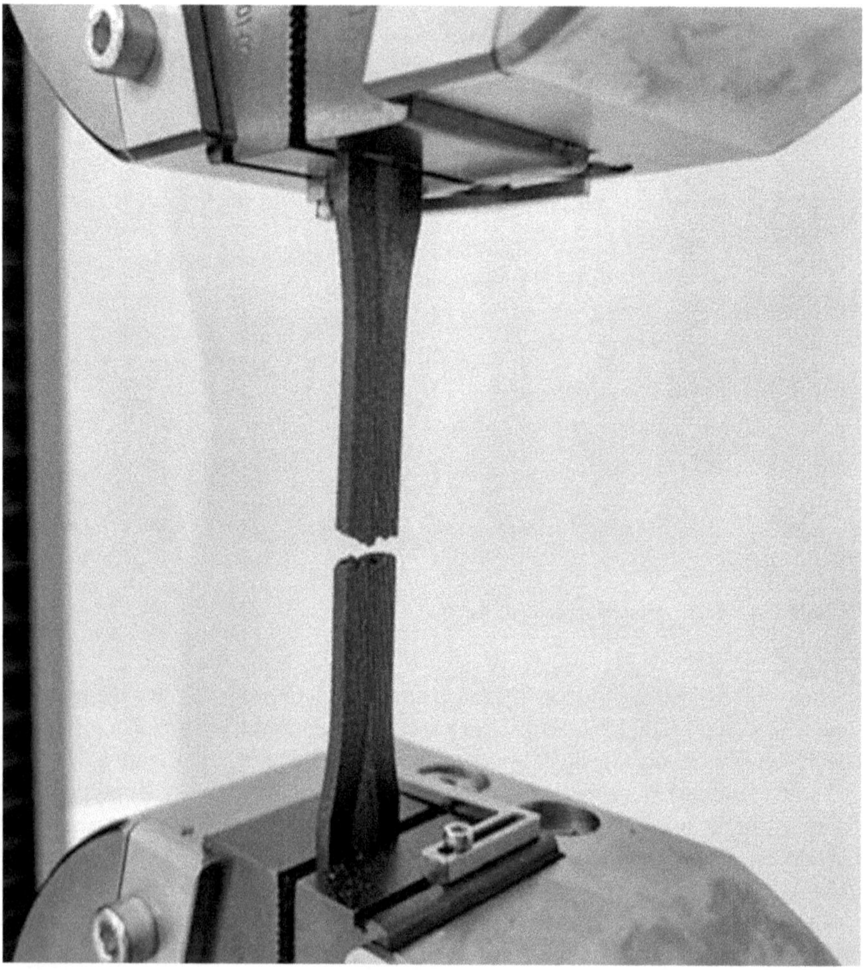

FIGURE 9.4 Fracture specimen after tensile test in universal testing machine (Make: Instron 3382).

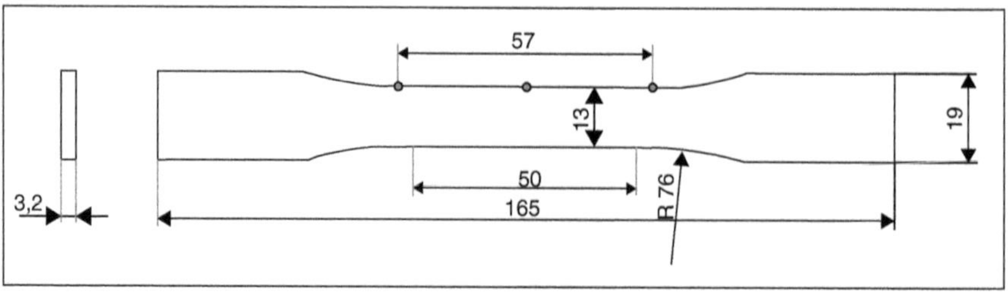

FIGURE 9.5 Specimen dimensions as per ASTM D638.

FIGURE 9.6 FDM-printed specimens as per ASTM standard for flexural test.

9.2.2 Design of Experiment and Experimentation

In this study, the selected process parameters are IP, LT, and part orientation angle in Z direction (POZ) for experiential study of the mechanical properties of Nylon 12CF in FDM. The independent variables and fixed parameters were chosen based on the review of the literature and work experience in FDM – 3D printing of various engineering polymers.

Taguchi's orthogonal array design was proposed to minimize the number of experiments needed to determine the optimal test parameters. The orthogonal matrix provides the shortest possible combinations of process parameters at different levels and evaluates their significance for optimization. In this study, all designs, plots, and analyses were performed using Minitab 17 statistical software. The process parameters, levels, and fixed parameters considered in FDM are detailed below and shown in Tables 9.1 and 9.2.

In this experimental investigation, the samples were 3D printed in FDM using Nylon 12CF as filament, considering the process variables at three different levels to measure the mechanical properties, namely tensile stress at break (TB), TM, FM, and MFB in the universal testing machine. The experiments were conducted following Taguchi's L9 orthogonal design, and the output responses of each variable at different levels were obtained from UTM experimental data, which are shown in Table 9.3.

The Taguchi design of experiment method is used for solving single-response optimization problems but cannot optimize the multi responses. Meanwhile, the accomplishment of the manufactured products is qualified based on several response

FIGURE 9.7 Sample during flexural testing in UTM (Make: Instron 3382).

TABLE 9.1
Process Parameters and Their Levels Considered for the Experimentations

Sl. No	Process Parameters	Units	Symbol	Level 1	Level 2	Level 3
1	IP		A	Linear	Hexagonal	Diamond
2	LT	mm	B	0.1	0.3	0.5
3	POZ	Degree	C	0	45	90

TABLE 9.2
Fixed Parameters for the Selected FDM Machine

Control Factors	Symbols	Fixed Parameters
Part orientation X (POX)	Degree	0
Part orientation Y (POY)	Degree	0
Infill density	%	100
FDM chamber temperature	°C	93
Extrusion temperature	°C	250
Filament thickness	mm	1.75
Print speed	mm/sec	35

TABLE 9.3
Taguchi's L9 Orthogonal Array with Experimental Results

	Control Factors			Response Variables			
Exp. No.	IP (A)	LT (B)	POZ (C)	TB	TM	FM	MFB
1	L	0.1	0	45.975	2361.77	2810.82	124.096
2	L	0.3	45	36.386	1623.94	2405.26	99.231
3	L	0.5	90	23.197	911.06	1244.06	61.542
4	H	0.1	45	49.143	3045.12	3192.30	151.628
5	H	0.3	90	43.877	1918.60	2490.82	118.034
6	H	0.5	0	24.234	1599.22	1615.25	78.695
7	D	0.1	90	36.084	1615.60	2176.43	94.756
8	D	0.3	0	46.508	2416.15	3109.23	127.820
9	D	0.5	45	29.716	1792.64	2093.65	86.070

characteristics. Hence, in this study, multi-objective optimization using the MOORA technique was used to convert multiple response values into a single overall assessment value for optimization using Taguchi's method.

9.3 RESULTS AND DISCUSSION

9.3.1 Analysis of the Response TB

The effect of input variables IP, LT, and POZ on TB on the response tensile stress at break was evaluated based on the experiments conducted in UTM described in Table 9.3 using Minitab17 software. The graphical representation of the main effect of the three input variables at three different levels, 1, 2, and 3, for the mean data on the response tensile stress at break, is shown in Figure 9.8. The main effect plot and 3D surface plot reveal that the TB is optimal at level (A2, B1, C1).

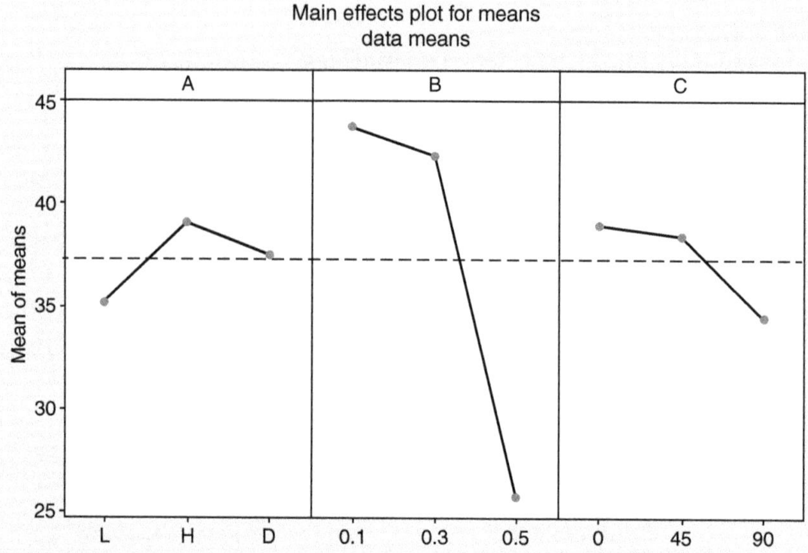

FIGURE 9.8 Main effect plot for means of TB.

TABLE 9.4
ANOVA of TB

Symbol	Machining Factor	Degrees of Freedom	Sum of Squares	Mean Squares	*F*	*P*	Percentage Contribution
A	IP	2	22.98	11.49	0.20	0.831	2.97
B	LT	2	600.46	300.23	5.33	0.158	77.69
C	OZ	2	36.90	18.45	0.33	0.753	4.78
Error		2	112.63	56.31			14.56
Total		8	772.97				100.00

It is evident from the main effect plot that process parameter setting with a hexagonal IP at a LT of 0.1 mm and part orientation angle at 0 degrees in X, Y, and Z direction yields maximum tensile stress at the break for the 3D-printed tensile specimen in FDM. It was noticed that the lower LT provides better strength in the part. The tensile stress at break decreases as the layer height and part orientation angle increase in Nylon 12CF specimens fabricated using FDM.

Further, the effect of input variables with respect to tensile stress at break was analyzed to find out more accurately the optimal combination of print parameters in FDM by using analysis of variance (ANOVA). The results of ANOVA for the performance characteristics of tensile stress at break are shown in Table 9.4.

It was observed that the process parameters, LT, and POZ, followed by IP, significantly affect both the mean and the variation in the tensile stress values at break. LT is the most dominant factor, with a percentage contribution of 77.69% followed by POZ of 14.56% and IP of 2.97%, respectively. The total variation in the response coefficient of determination (R^2) is 85.43%.

9.3.2 Analysis of the Response TM

The TM of the 3D-printed specimens was recorded in a UTM tensile testing machine to investigate the ratio of stress versus elastic strain under tension. The high TM represents the rigidity of the material, and more stress is needed to generate a specified strain. This study evaluated the effect of process parameters on the TM of the 3D-printed FDM specimens, as shown in Table 9.3.

The average values of TM for each process parameter at levels 1, 2, and 3 for data means are plotted in Figure 9.9 using MINITAB software. It is observed that the TM is highest at the process variable setting in FDM at level (A2, B1, C2). The hexagonal IP with 0.1 mm LT at 45° part orientation reveals the best result of the tensile specimens. The TM decreases with an increase in layer height, and it first increases and then decreases for part orientation angle in 3D printing of Nylon 12CF composite.

The effect of the control variables on TM recorded values of the 3D-printed nine numbers of specimens in UTM has been statistically analyzed using ANOVA, shown in Table 9.5.

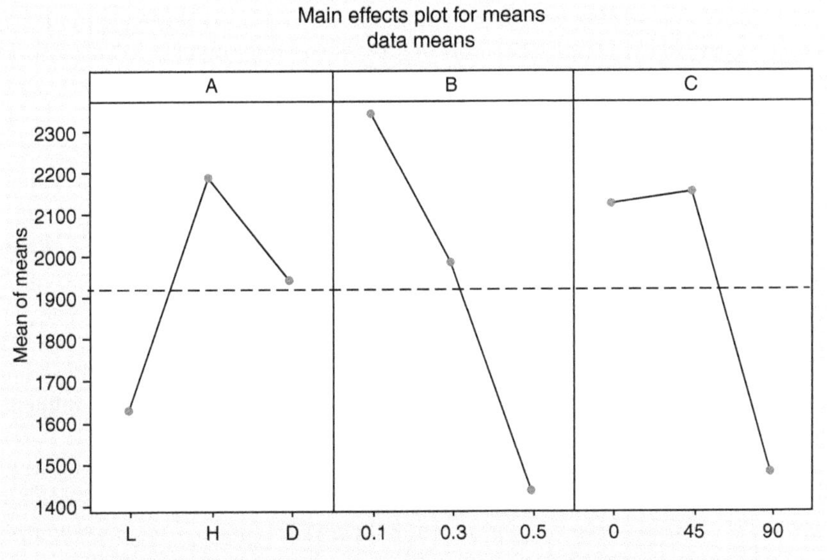

FIGURE 9.9 The main effect plots for means TM.

TABLE 9.5
ANOVA of TM

Symbol	Machining Factor	Degrees of Freedom	Sum of Squares	Mean Squares	*F*	*P*	Percentage Contribution
A	IP	2	464673	232337	1.05	0.487	15.36
B	LT	2	1252145	626073	2.84	0.260	41.41
C	OZ	2	867260	433630	1.97	0.337	28.67
Error		2	440534	220267			14.56
Total		8	3024612				100.00

It was found that the FDM process parameter variable layer height has been found to be the most significant factor affecting the TM, having a percentage contribution of 41.41%, followed by part orientation angle in the Z axis of 28.67% and IP of 15.36%, respectively. The coefficient of determination (R^2), representing the proportion of total variation explained in the response, is 85.44%.

9.3.3 Analysis of the Response FM

Flexural strength shows the optimal stress the specimen can withstand prior to bending and breaking, whereas the FM expresses the stiffness resistance to bend under applied force perpendicular to the sample under the three-point bend test.

The bending equation is given by $\sigma = My/I$, where σ represents stress, M denotes the bending moment, y is the distance from the neutral axis, and I refers to the moment of inertia. In the present experimental study, the flexural properties of all nine samples made out of Nylon 12CF were 3D printed in the FDM process and tested in a UTM machine as per ASTM D790. The effect of control variables on the FM was tested by conducting experiments as described in Table 9.3. Minitab17 software was used to evaluate the correlation between the control factors and FM.

The average values of FM for each process parameter at levels 1, 2, and 3 for raw data (data means) are plotted in Figure 9.10. The FM was found to be maximum at process parameter setting A3 B1 C2. The diamond IP with a 0.1-mm LT of printing at a 45° part orientation angle shows the best result among all the input variable combinations of 3D-printed samples in FDM. FM decreases with an increase in layer height, first increasing and then decreasing with print orientation angle.

The data was analyzed to study the influence of different print variables in FDM on the response FM using the statistical formula ANOVA in Minitab software. The input variables were compared across all levels' average data, and the influential factors affecting the output were determined.

The significance of each process variable on FDM-printed specimens on FM was determined by performing an ANOVA, as shown in Table 9.6. It is observed that layer height is the most important factor affecting the FM, having a percentage contribution of 64.95%, followed by part orientation angle around the Z axis, 19.17% and IP,

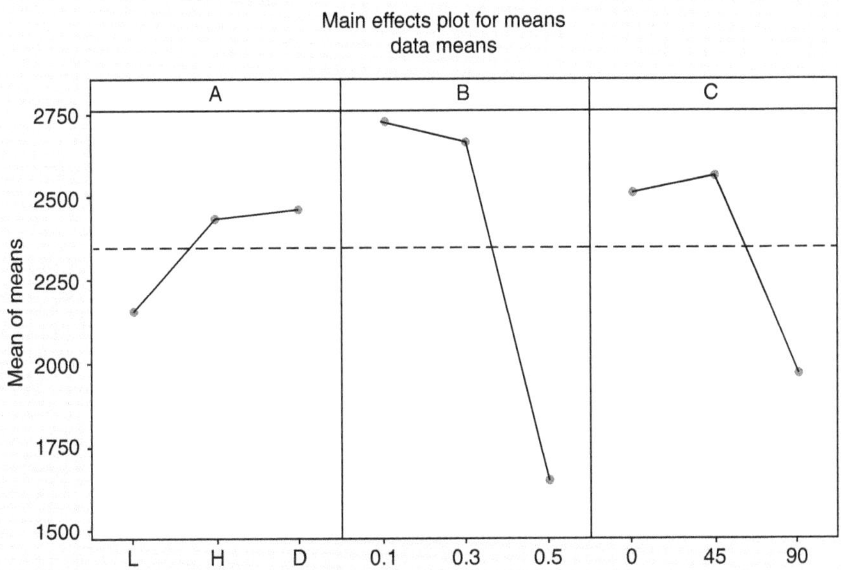

FIGURE 9.10 The main effect plot for means FM.

TABLE 9.6
ANOVA for FM

Symbol	Machining Factor	Degrees of Freedom	Sum of Squares	Mean Squares	F	P	Percentage Contribution
A	IP	2	172673	86336	0.47	0.678	5.11
B	LT	2	2195343	1097671	6.03	0.142	64.95
C	OZ	2	647744	323872	1.78	0.360	19.17
Error		2	364166	182083			10.77
Total		8	3379925				100.00

5.11%, respectively. The coefficient of determination (R^2), representing the proportion of total variation explained in the response, is 89.23%.

9.3.4 Maximum Flexural Load at Break

The flexural test has been conducted using a three-point bending test in UTM as per ASTM D790 with crosshead speeds of 2 mm/min to study the MFB. The test results of all nine FDM-printed specimens were recorded and shown in Table 9.3, and the data analysis was carried out to study the influence of each process parameter on MFB using Minitab 15 software.

The UTM testing data of all nine samples built in FDM are the average values of MFB for each process variable at levels 1, 2, and 3 for raw data (data means), as plotted in Figure 9.11. It is seen that the FM is maximum at the process variable level (A2, B1, C2). The figure shows that the process parameter setting of a hexagonal IP with 0.1-mm LT and part print orientation angle at 45° reveals the best result among all the input variable combinations in the FDM process. The flexural load at break decreases with increasing LT and exhibits a nonlinear trend with orientation angle, initially increasing and then decreasing in the FDM process.

Based on the flexural test carried out in UTM, the experimental values of flexural load at the break for all nine samples were analyzed using MINITAB software.

The 3D-printed samples in FDM were conducted using the flexural test in UTM, and the effect of the process variable on the flexural load at the break of all the samples was studied by performing ANOVA shown in Table 9.7. It was observed that LT was

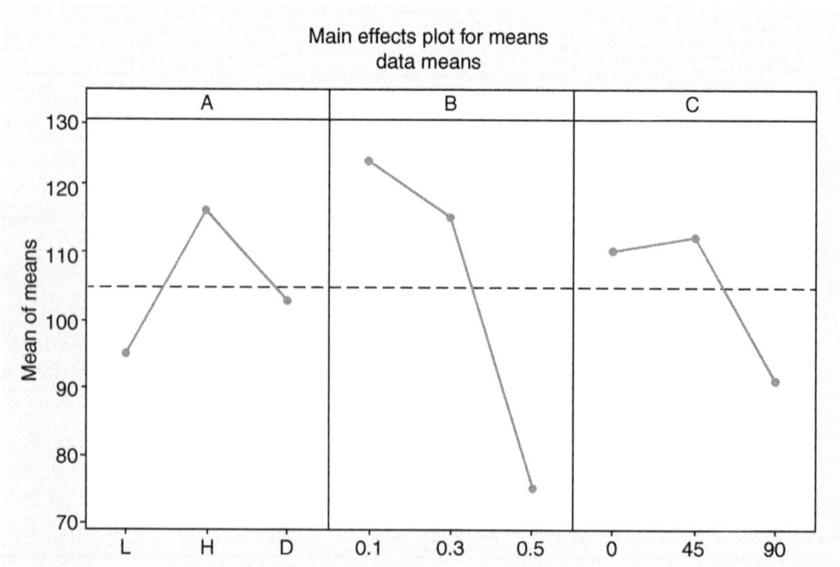

FIGURE 9.11 Main effect plot for means flexural load at break.

TABLE 9.7
ANOVA of Flexural Load at Break

Symbol	Machining Factor	Degrees of Freedom	Sum of Squares	Mean Squares	*F*	*P*	Percentage Contribution
A	IP	2	685.9	342.9	0.78	0.562	10.88
B	LT	2	3948.8	1974.4	4.49	0.182	62.62
C	OZ	2	791.7	395.9	0.90	0.526	12.56
Error		2	879.1	439.5			13.94
Total		8	6305.5				100.00

found to be the most vital factor affecting the FM, with a percentage contribution of 62.62%, followed by part orientation angle around the Z axis 12.56% and IP 10.88%, respectively. The R^2, which indicates the percentage of total variation in the response, is 86.10%.

9.4 MULTI-OBJECTIVE OPTIMIZATION ON THE BASIS OF THE RATIO ANALYSIS

The MOORA method is adopted to simultaneously optimize multiple objective functions to evaluate the optimal parametric combination of input factors and improve the print quality in the present experimental work in FDM. MOORA is a decision-making technique where the objective functions are controlled for each output statement from a set of existing alternatives [34]. This method is efficient in solving numerous complex multi-objective optimization problems. The performance of the varied output responses has been organized in a decision matrix, as shown in Equation (9.1):

$$Y = \begin{bmatrix} y_{11} & y_{12} & .. & .. & y_1 \\ y_{21} & y_{22} & .. & .. & y_2 \\ .. & .. & .. & .. & .. \\ .. & .. & .. & .. & .. \\ .. & .. & .. & .. & .. \\ y_1 & y_2 & .. & .. & y \end{bmatrix} \tag{9.1}$$

where y_{ij} represents the response values of the ith alternative on jth quality characteristics, m is the number of input variables, and n is the number of responses. The ratio system is formed by data normalization of the decision matrix, which is calculated by applying Equation (9.2):

$$y_{ij}^* = y_{ij} / [\sum_{i=1}^{m} y_{ij}^2]^{1/2} \tag{9.2}$$

where y^* indicates the normalized value y, a dimensionless number between 0 and 1 for ith option on jth characteristic. Further, the normalized value is added to resolve the maximization problem or subtracted for minimization problems. Some attributes are more important than other responses in certain cases. Hence, different weights will be given to the respective attributes and multiplied by their equivalent weight using Equation (9.3):

$$x_i = \sum_{j=1}^{g} w_j y_{ij}^* - \sum_{j=g+1}^{n} w_j y_{ij}^* \tag{9.3}$$

where $j = 1, 2, 3, \ldots, g$, with g representing the number of criteria to be maximized, while $(n - g)$ denotes the number of criteria to be minimized; w_j is the weight of jth criterion; and x_i is the assessment value of ith alternative. After computing the normalized assessment value, ranking is performed based on x_i, arranged from highest to lowest, to determine the best alternative among all attributes. The alternative with the highest x_i value is considered the best, as the ranking of x_i represents the final preference.

9.5 OPTIMIZATION OF PROCESS PARAMETERS USING THE MOORA METHOD

In this study, the multi-optimization MOORA method is applied to the responses of TB, TM, FM, and MFB to find out the optimal process parameter setting in 3D printing of FDM samples. The normalization of the performance measures was carried out by applying Equation (9.2), and the normalized assessment values were evaluated. An equal weight of 25% was taken for all four responses, and the summation of all the weights became 1. The overall computation value has been evaluated by applying Equation (9.3). The normalized assessment values of the performance measures and the overall assessment value are presented in Table 9.8 and are shown in Figure 9.12.

MINITAB software was used to evaluate the optimal parameter setting of assessment values based on the mean data at three different levels of all the process variables for the nine experimental samples, 3D printed in FDM, are shown in Table 9.9. Based on the mean response data and main effect plot, overall assessment values using the MOORA method reveal that the optimal level process

TABLE 9.8
Normalization of Performance Measures (y^*_{ij}) and Overall Assessment Value (x_i)

Exp. No.	y^*_{ij} (TB)	y^*_{ij} (TM)	y^*_{ij} (FM)	y^*_{ij} (MFB)	x_i
1	0.3994	0.3924	0.3860	0.3832	0.3903
2	0.3161	0.2698	0.3303	0.3064	0.3057
3	0.2015	0.1514	0.1708	0.1900	0.1784
4	0.4269	0.5060	0.4384	0.4682	0.4599
5	0.3812	0.3188	0.3421	0.3645	0.3516
6	0.2105	0.2657	0.2218	0.2430	0.2353
7	0.3135	0.2685	0.2989	0.2926	0.2933
8	0.4040	0.4015	0.4270	0.3947	0.4068
9	0.2581	0.2979	0.2875	0.2658	0.2773

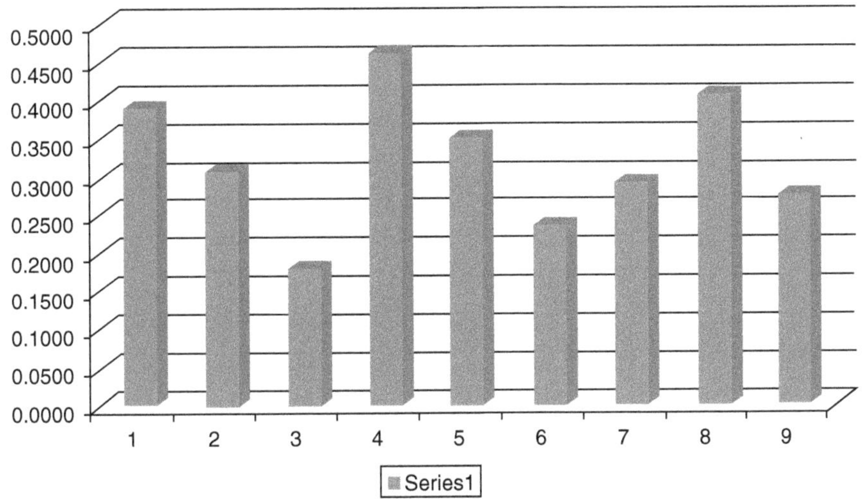

FIGURE 9.12 Overall assessment values of FDM experimental samples using MOORA method.

TABLE 9.9
Mean of the Overall Assessment Value

Level	A	B	C
1	0.2915	0.3812	0.3441
2	0.3489	0.3547	0.3476
3	0.3258	0.2303	0.2745
Delta	0.0575	0.1508	0.0731
Rank	3	1	2

parameter combination at (A2, B1, C2) yields the maximum value, as shown in Figure 9.13.

The process parameter setting of the FDM-tested sample of hexagonal IP with a layer height of 0.1 mm and part orientation angle 45° in Z direction proves the optimal value of tensile and flexural strength, as evident from multi-objective optimization using the MOORA method. The response value at optimal setting was found as follows: TB is 49.508 MPa, TM is 3045.12 MPa, FM is 3192.30 MPa, and MFB is 151.628 MPa.

The significance of each control factor on the performance measures—TB, TM, FM, and flexural load at break—was evaluated using ANOVA, as presented in Table 9.10. The analysis reveals that layer thickness (factor B) is the major influencing factor, contributing 61.81%, followed by POZ (factor C) at 16.22%, while IP (factor A) contributes only 7.97% in the FDM 3D printing of specimens.

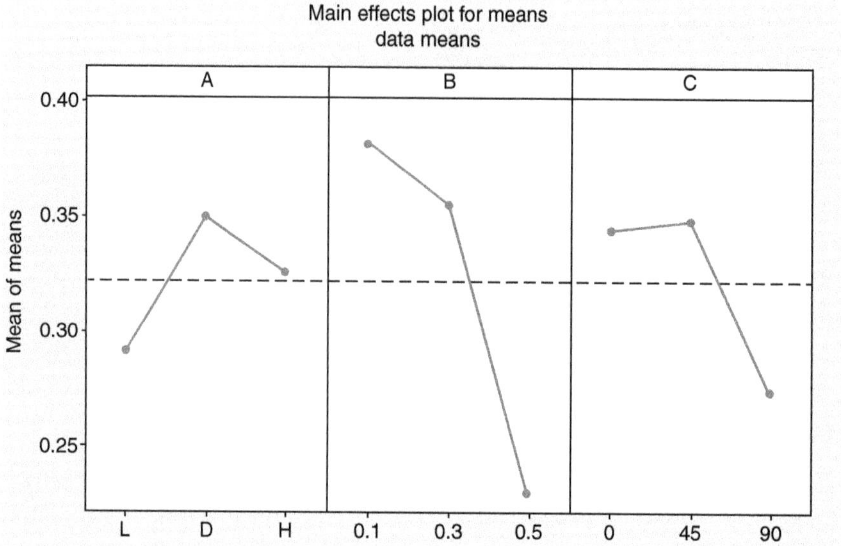

FIGURE 9.13 Response graph of overall assessment value.

TABLE 9.10
ANOVA of Overall Assessment Value

Symbol	Process Parameter	Degrees of Freedom	Sum of Squares	Mean Squares	*F*	*P*	Percentage Contribution
A	IP	2	0.005017	0.002509	0.57	0.637	7.97
B	LT	2	0.038907	0.019454	4.41	0.185	61.81
C	OZ	2	0.010212	0.005106	1.16	0.463	16.22
Error		2	0.008813	0.004407			14.00
Total		8	0.062950				100.00

The coefficient of determination (R^2), which quantifies the proportion of total variation in the response accounted for by the model's terms, is 86.01%.

9.6 CONFIRMATION EXPERIMENT

For validation of the results obtained, the confirmation study was done in FDM considering the optimal setting of the process variables at (A2, B1, C2), and the results were compared with the initial process parameter setting (A1, B1, C1) as shown in Table 9.11. The response values are improved at optimal levels compared with the initial setting of FDM variables. The performance characteristics are greatly enhanced through this analysis.

TABLE 9.11
Results of the Confirmation Experiment

	Optimal Machining Parameter		
	Initial Printing Parameters	Prediction	Experimental
Level	A1, B1, C1	A2, B1, C2	A2, B1, C2
TB (MPa)	45.975		49.143
TM (MPa)	2361.77		3045.12
FM (MPa)	2810.82		3192.30
Flexural load at break (MPa)	124.096		151.628
MOORA	0.3903	0.4336	0.4599

9.7 CONCLUSIONS

This experimental investigation aims to optimize the process variables: IP, LT, and POZ in FDM to maximize the responses: TB, TM, FM, and MFB. The experiments were carried out following Taguchi's L9 orthogonal array design, and the MOORA was used for multi-optimization. The optimal process parameter combination (A2, B1, C2)—hexagonal IP, 0.1 mm print LT, and a 45° orientation angle in the Z direction—yields the highest tensile and flexural strength in 3D-printed parts. The variables significantly affecting FDM-printed specimens were identified using the ANOVA technique. The validity of test results proves that there is a remarkable improvement in TB, TM, FM, and MFB concerning the initial setting of parameters. The study confirms the validity of Taguchi's principle combined with the MOORA method to improve the performance of the FDM build parts and optimize the responses in FDM operations.

REFERENCES

1. Lalegani Dezaki, M., Mohd Ariffin, M.K.A. and Hatami, S., 2021, "An overview of fused deposition modelling (FDM): research, development and process optimisation." *Rapid Prototyping Journal*, **27**(3), pp. 562–582.
2. Durgun, I. and Ertan, R., 2014, "Experimental investigation of FDM process for improvement of mechanical properties and production cost." *Rapid Prototyping Journal*, **20**(3), pp. 228–235.
3. Gibson, I., Rosen, D.W., Stucker, B., Khorasani, M., Rosen, D., Stucker, B. and Khorasani, M., 2021, *Additive Manufacturing Technologies*, vol. 17, pp. 160–186. Cham, Switzerland: Springer.
4. Paek, S.W., Balasubramanian, S. and Stupples, D., 2022, "Composites additive manufacturing for space applications: a review." *Materials*, **15**(13), p. 4709.
5. Wu, H., Fahy, W.P., Kim, S., Kim, H., Zhao, N., Pilato, L., Kafi, A., Bateman, S. and Koo, J.H., 2020, "Recent developments in polymers/polymer nanocomposites for additive manufacturing." *Progress in Materials Science*, **111**, p. 100638.

6. Galetto, M., Verna, E. and Genta, G., 2021, "Effect of process parameters on parts quality and process efficiency of fused deposition modeling." *Computers & Industrial Engineering*, **156**, p. 107238.
7. Mohamed, O.A., Masood, S.H. and Bhowmik, J.L., 2015, "Optimization of fused deposition modeling process parameters: a review of current research and future prospects." *Advances in Manufacturing*, **3**, pp. 42–53.
8. Sharma, R., Singh, R. and Batish, A., 2020, "On multi response optimization and process capability analysis for surface properties of 3D printed functional prototypes of PVC reinforced with PP and HAp." *Materials Today: Proceedings*, **28**, pp. 1115–1122.
9. Dey, A. and Yodo, N., 2019, "A systematic survey of FDM process parameter optimization and their influence on part characteristics." *Journal of Manufacturing and Materials Processing*, **3**(3), p. 64.
10. Sood, A.K., Ohdar, R.K. and Mahapatra, S.S., 2012, "Experimental investigation and empirical modelling of FDM process for compressive strength improvement." *Journal of Advanced Research*, **3**(1), pp. 81–90.
11. Nagendra, J., Srinath, M.K., Sujeeth, S., Naresh, K.S. and Prasad, M.G., 2021, "Optimization of process parameters and evaluation of surface roughness for 3D printed nylon-aramid composite." *Materials Today: Proceedings*, **44**, pp. 674–682.
12. Patil, P., Raykar, S.J., Bhamu, J. and Singh, D., 2022, "Modeling and analysis of surface roughness in fused deposition modeling based on infill patterns." *Indian Journal of Engineering and Materials Sciences*, **29**(1), pp. 92–99.
13. Eswaran, P., Sivakumar, K. and Subramaniyan, M., 2018, "Minimizing error on circularity of FDM manufactured part." *Materials Today: Proceedings*, **5**(2), pp. 6675–6683.
14. Rao, R.V. and Rai, D.P., 2016, "Optimization of fused deposition modeling process using teaching-learning-based optimization algorithm." *Engineering Science and Technology, an International Journal*, **19**(1), pp. 587–603.
15. Zhang, Y., Purssell, C., Mao, K. and Leigh, S., 2020, "A physical investigation of wear and thermal characteristics of 3D printed nylon spur gears." *Tribology International*, **141**, p. 105953.
16. Ahn, S.H., Montero, M., Odell, D., Roundy, S. and Wright, P.K., 2002, "Anisotropic material properties of fused deposition modeling ABS." *Rapid Prototyping Journal*, **8**(4), pp. 248–257.
17. Alafaghani, A., Ablat, M.A., Abedi, H. and Qattawi, A., 2021, "Modeling the influence of fused filament fabrication processing parameters on the mechanical properties of ABS parts." *Journal of Manufacturing Processes*, **71**, pp. 711–723.
18. Vishwas, M., Basavaraj, C.K. and Vinyas, M., 2018, "Experimental investigation using Taguchi method to optimize process parameters of fused deposition modeling for ABS and nylon materials." *Materials Today: Proceedings*, **5**(2), pp. 7106–7114.
19. Bedi, P., Singh, R. and Ahuja, I.P.S., 2020, "Multifactor optimization of FDM process parameters for development of rapid tooling using SiC/Al_2O_3-reinforced LDPE filament." *Journal of Thermoplastic Composite Materials*, **33**(5), pp. 581–598.
20. Christiyan, K.J., Chandrasekhar, U. and Venkateswarlu, K., 2016, "A study on the influence of process parameters on the mechanical properties of 3D printed ABS composite." *IOP Conference Series: Materials Science and Engineering*, **114**(1), p. 012109.
21. Chacón, J.M., Caminero, M.A., García-Plaza, E. and Núnez, P.J., 2017, "Additive manufacturing of PLA structures using fused deposition modelling: effect of process

parameters on mechanical properties and their optimal selection." *Materials & Design*, **124**, pp. 143–157.
22. Mahmood, S., Qureshi, A.J., Goh, K.L. and Talamona, D., 2017, "Tensile strength of partially filled FFF printed parts: meta modelling." *Rapid Prototyping Journal*, **23**(3), pp. 524–533.
23. Raju, M., Gupta, M.K., Bhanot, N. and Sharma, V.S., 2019, "A hybrid PSO–BFO evolutionary algorithm for optimization of fused deposition modelling process parameters." *Journal of Intelligent Manufacturing*, **30**, pp. 2743–2758.
24. Aw, Y.Y., Yeoh, C.K., Idris, M.A., Teh, P.L., Hamzah, K.A. and Sazali, S.A., 2018, "Effect of printing parameters on tensile, dynamic mechanical, and thermoelectric properties of FDM 3D printed CABS/ZnO composites." *Materials*, **11**(4), p. 466.
25. Sun, Q., Rizvi, G.M., Bellehumeur, C.T. and Gu, P., 2008, "Effect of processing conditions on the bonding quality of FDM polymer filaments." *Rapid Prototyping Journal*, **14**(2), pp. 72–80.
26. Paul, A., and Das, M., 2024, "A decision support system for the selection of FDM process parameters using MOORA." *Management Science Letters*, **14**(3), pp. 181–188.
27. Peng, A., Xiao, X. and Yue, R., 2014, "Process parameter optimization for fused deposition modeling using response surface methodology combined with fuzzy inference system." *The International Journal of Advanced Manufacturing Technology*, **73**, pp. 87–100.
28. Srivastava, M., Rathee, S., Maheshwari, S. and Kundra, T.K., 2018, "Multi-objective optimisation of fused deposition modelling process parameters using RSM and fuzzy logic for build time and support material." *International Journal of Rapid Manufacturing*, **7**(1), pp. 25–42.
29. Rinanto, A., Nugroho, A., Prasetyo, H. and Pujiyanto, E., 2018, "Simultaneous optimization of tensile strength, energy consumption and processing time on FDM process using Taguchi and PCR-TOPSIS." In *2018 4th International Conference on Science and Technology (ICST)*, pp. 1–5.
30. Phatak, A.M. and Pande, S.S., 2012, "Optimum part orientation in rapid prototyping using genetic algorithm." *Journal of Manufacturing Systems*, **31**(4), pp. 395–402.
31. Paul, R. and Anand, S., 2011, "Optimal part orientation in rapid manufacturing process for achieving geometric tolerances." *Journal of Manufacturing Systems*, **30**(4), pp. 214–222.
32. Barua, A., Jeet, S., Cherkia, H., Bagal, D.K. and Sahoo, B.B., 2019, "Parametric optimization of FDM processed part for improving surface finish using MOORA technique and desirability function analysis." *International Journal of Applied Engineering Research*, **14**(13), pp. 1–7.
33. Gajdoš, I. and Slota, J., 2013, "Influence of printing conditions on structure in FDM prototypes." *Tehnički Vjesnik*, **20**(2), pp. 231–236.
34. Barua, A., Jeet, S., Parida, B., Samantray, A. and Bagal, D.K., 2018, "Comparative evaluation and optimization of 4-cylinder CI engine camshaft material using finite element analysis: a hybrid MOORA technique and Taguchi based desirability function analysis approach." *International Journal of Technical Innovation in Modern Engineering & Science*, **4**(11), pp. 105–114.

10 Effect of Build Orientation and Infill Density on Mechanical Properties of FDM Build ABS Parts

Kumar Aviseck Jati and Pradeep Kumar Mishra

10.1 INTRODUCTION

Additive manufacturing (AM) is generally referred to as three-dimensional (3D) printing process. It is a process that creates 3D physical models layer by layer by building them directly from computer-aided design (CAD) files. The process began in the 1980s as a rapid prototyping technique (RPT) that allowed engineers and designers to realize their designs quickly. One of the key benefits of AM is that it can create complex designs that would otherwise be difficult to produce using traditional machining techniques. The process starts with a CAD model that is cut into thin layers. Then, the 3D printer builds the 3D object layer by layer using different materials such as plastic, metal, ceramics, or composite materials. This layer-by-layer fabrication approach allows for high design freedom and flexibility as complex geometries and inner structures can be easily created [1–3].

The creation of useful prototypes, design tools, and small-scale manufacturing are where AM has found success. However, more research is still needed to advance this process. This process needs to know how accurate the products are in terms of dimensions, how they perform mechanically, how easy it is to repeat the process, and how consistent they are. To be accepted as a reliable manufacturing process, AM needs to provide consistent and repeatable results, ensuring that each part produced meets the required specifications [4, 5]. This is particularly important in industries where product quality and reliability are paramount. Without reliable and consistent outcomes, it is challenging to use AM for mass production or in applications where precise tolerances are essential [6]. Fused deposition modeling (FDM) is one of the most popular AM techniques, but it is important to consider how reliable the process is, how it affects the mechanical properties of the material, and how accurate the measurements are. In FDM, the layers of the object are constructed by extruding molten thermoplastic material through a nozzle. FDM parts may vary in mechanical strength, surface finish, and dimensional accuracy because of the nature of this process [7, 8]. To overcome these challenges and enhance the use of AM in end

DOI: 10.1201/9781003686941-10

manufacturing, ongoing research and development efforts are focused on improving process control, material properties, and post-processing techniques. By gaining a deeper understanding of the underlying physics and optimizing process parameters, manufacturers can achieve greater reliability and consistency in AM processes. In addition, advances in material science, as well as the development of AM-specific materials, improve mechanical properties and product quality. Furthermore, the integration of 3D scanning with AM has the potential to enhance the efficiency and flexibility of design and development. 3D scanning allows for accurate and precise digital modeling of existing parts or objects, which can then be modified or reproduced using AM techniques [9–11].

Overall, even though AM has demonstrated its value in a few applications; more development is required to address issues with process repeatability, mechanical properties, and dimensional accuracy. AM has the potential to play a bigger role in end-use manufacturing environments as these areas continue to advance, providing benefits like shortened product cycle times, customized production, and the capacity to efficiently create complex geometries [9, 10]. With the help of this technology, 3D objects can be constructed layer by layer without the use of the specialized cutters, tools, or fixtures required in conventional manufacturing processes [12, 13].

One of the main benefits of AM is that it can produce high-level designs that are hard to fabricate using conventional techniques. It offers greater design flexibility and the potential to reduce product cycle times significantly. By utilizing 3D scanning alongside AM, manufacturers can accurately and precisely model parts digitally, enhancing the efficiency and flexibility of design and development processes. While AM has found extensive use in the production of functional prototypes, design aids, testing, and customizable parts in small-scale manufacturing environments, it is still not widely employed in end manufacturing environments [14, 15].

The lack of vital information about the process repeatability and consistency of the manufactured products is one of the main obstacles preventing the adoption of AM as a true manufacturing process. AM must exhibit dependable and consistent results to be accepted in actual production environments. Furthermore, it is important to have a good knowledge of the mechanical and geometrical characteristics of printed parts [16–19]. When it comes to AM processes, one of the biggest questions is how reliable the process is, how good the mechanical properties are, and how accurate the dimensions are. This is especially true for FDM processes. FDM involves the extrusion of molten thermoplastic material to build the layers of the object. Achieving consistent mechanical properties and dimensional accuracy can be more challenging with the FDM process compared to other processes [20, 21]. The quality and dependability of AM processes for final manufacturing applications are still being worked on. These issues are being focused on improvements in material development, process control, and post-processing techniques, which are bridging the gap between AM and conventional manufacturing processes [22, 23].

In view of the above, an attempt has been made to prepare the samples according to the ASTM standard and study the effect of build orientation and infill density on the tensile strength and compressive strength of FDM-built ABS plastic parts.

10.2 MATERIALS AND METHODS

For this work, ABS plastic was used as the working filament material for producing specimens by the FDM process. The monomer of ABS plastic consists of three main components: acrylonitrile, butadiene, and styrene, as shown in Figure 10.1. These monomers combine through a process called polymerization to form the ABS polymer.

Acrylonitrile provides chemical resistance and heat stability to the ABS polymer. It helps improve the overall mechanical properties and makes the material more rigid. Butadiene is responsible for enhancing the impact resistance and toughness of ABS. It increases the material's ability to withstand impacts and absorb energy without breaking or fracture. Styrene contributes to the ABS polymer's ease of processing and its ability to maintain dimensional stability. It helps lower the viscosity of the material, making it more flowable during the manufacturing process. The combination of these monomers in ABS results in a thermoplastic material with desirable properties for a range of applications. ABS is super strong in terms of tensile strength, impact strength, and flexibility, so it is extremely useful for making prototypes, production tools, and parts for other uses. It offers good temperature resistance and can be filled sparsely or solidly to meet specific requirements. However, ABS is also more susceptible to warping compared to some other thermoplastics. Care must be taken during the printing or manufacturing process to minimize warping effects and achieve accurate dimensional stability. The mechanical properties of ABS are presented in Table 10.1.

Using the SolidWorks modeling program, all specimens that were to be made were designed and exported as STL files. The STL file was imported into the software catalyst EX for FDM Insight. The ASTM D638 (type IV) and ASTM D695 (type IV) standard dimensions were used to generate a 3D CAD geometric model of the tensile test. Compressive test specimens are shown in Figures 10.2 and 10.3, respectively.

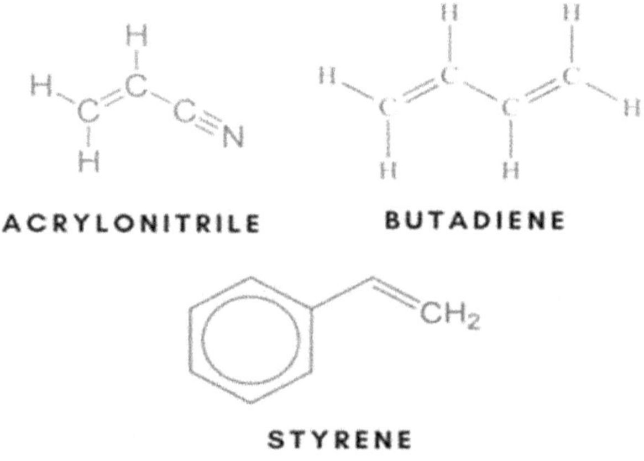

FIGURE 10.1 Monomers of ABS.

TABLE 10.1
Mechanical Properties of ABS

Property	Value
Tensile strength	46 MPa
Flexural strength	74 MPa
Melting temperature	200^0–280^0C
Glass transition	97^0C
Shrink rate	0.5–0.7%
Compressive yield strength	65 MPa
Compressive modulus	2.5 GPa

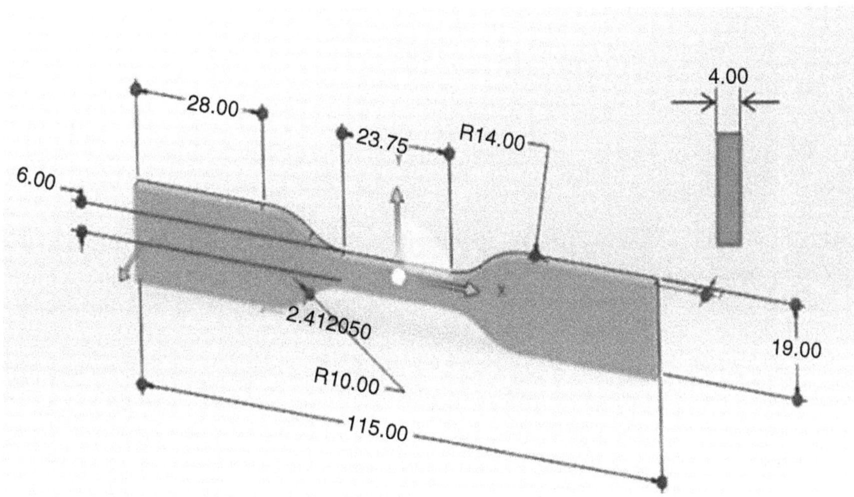

FIGURE 10.2 Tensile test specimen (standard ASTM D638).

Plastic filament used in FDM is melted and laid down on a build platform. The build platform is lowered by the distance equal to the layer thickness of each layer and a subsequent layer is added on top till completion of the specimen. To alter the mechanical properties, the height and layer orientation are adjusted. Infill density and build orientation are used in this process. The material's response to a mechanical stress determines how a part will behave mechanically. Depending on the force's direction, mechanical characteristics, and size of the component's geometry, the applied force can cause a component to deform.

The aim of this study is to investigate the impact of building orientation and infill density on mechanical properties. Further applications of this information include manufacturing for 3D printing procedures and design processes. The experiment employed two sample groups. The initial sample group has solid infill with build

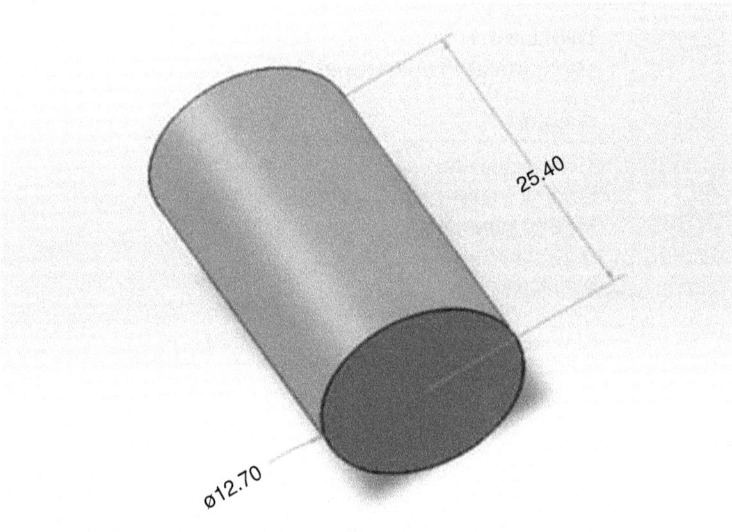

FIGURE 10.3 Compressive test specimen (standard ASTM D695).

orientations of 0^0, 15^0, and 30^0 with low infill density. The second sample group will be solidly produced with the same build orientation and high infill density.

Stratasys Mojo 3D printer was used to create components. ABS was the filament material utilized for part production. Utilizing a universal testing machine (UTM) model INSTRON 3382 with a capacity of 100KN, the compressive and tensile tests were carried out. All test specimens were tested at room temperature with a 0.5 mm/min crosshead speed for tensile and compressive tests. Here, specimens for tensile and compression testing were designed using SolidWorks software and exported as STL files. These files were then imported into a specific FDM machine (Stratasys Mojo 3D printer) for fabrication. The specimens were printed with ABS filament material, utilizing varying process parameters.

For the tensile test, flat specimens were prepared according to ASTM-D638 (type IV) standard, which specifies the dimensions and geometry of the specimens. The specimens were printed with alternating build orientations (0^0, 15^0, and 30^0) and infill densities, using low and high settings of infill density. A single printer nozzle with a diameter of 1 mm was used for fabrication. The tensile test specimens are shown in Figure 10.4.

Similarly, for the compression test, samples were prepared according to ASTM-D695 standard, which provides guidelines for compression testing of plastics. The specimens were printed with the same alternating build orientations and infill densities as in the tensile test. The compressive test specimens are shown in Figure 10.5. The material used for the FDM parts in this study is ABS M30, which is a specific grade of acrylonitrile butadiene styrene. ABS is known for its good impact resistance, rigidity, and heat resistance, making it widely adopted in the plastic manufacturing industry.

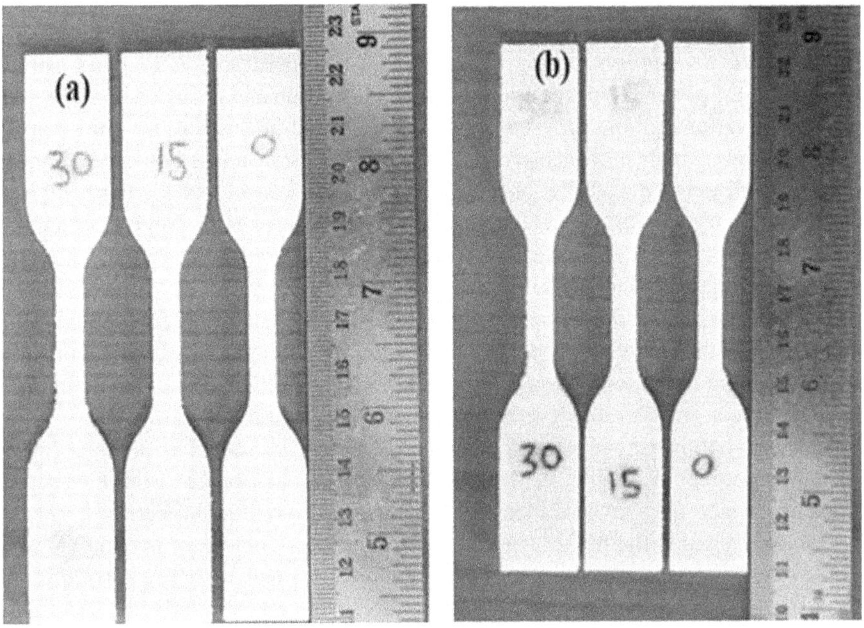

FIGURE 10.4 (a) Tensile test specimens of high infill density. (b) Tensile test specimens of low infill density.

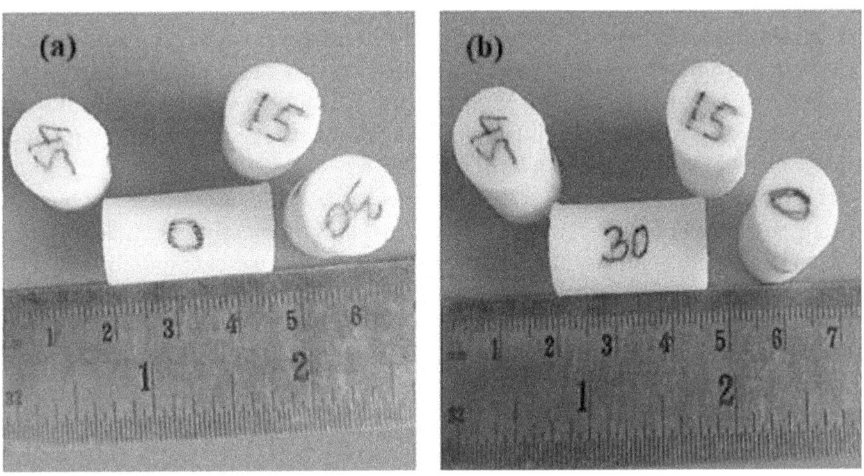

FIGURE 10.5 (a) Compressive test specimens of low infill density. (b) Compressive test specimens of high infill density.

The study aimed to evaluate the mechanical properties of the ABS specimens fabricated using FDM with different process parameters. By testing the tensile and compression performance, the quality and strength of the printed parts were assessed.

10.3 RESULTS AND DISCUSSION

This work focuses on presenting the mechanical properties of ABS parts printed using the FDM process, considering their built orientation and infill density. The built orientation varies at three different angles 0^0, 15^0, and 30^0, while the infill density is categorized as high infill density and low infill density. In this work, build direction and infill density were the only factors that were considered for fabrication of specimens for the tensile test and compression test. Other potential control parameters were maintained constant in this experiment as the difficulty of a test rises as the number of parameters increases.

The main aspect of the present study focuses on evaluation of tensile and compressive strength. The tensile strength is the amount of weight a material can handle before breaking under tension, while the compressive strength is the maximum load the material can withstand before failing under compression.

The results of the tensile test and compressive test are presented in Table 10.2, with the data organized according to the different built orientations and infill densities. This comparison may involve identifying which built orientation and infill density combination yielded the highest mechanical properties. The experimental results of tensile strength and compressive strength are shown in Table 10.2. The parts after the tensile test are shown in Figures 10.6 and 10.7, respectively. Similarly, the specimens after the compression test are shown in Figures 10.8 and 10.9, respectively.

The effect of build orientation and infill density on tensile strength is shown in Figure 10.10. From Figures 10.10, it can be found that the ultimate tensile strength decreases with an increase in the build orientation. It can be found that as the build orientation increases from 0^0 to 30^0, the ultimate tensile strength decreases for both high infill density and low infill density samples, because of the mode of fracture and the direction of loading [22, 24]. With increase in infill density the tensile strength increases, because, at higher infill density the specimens fabricated were with more density that withstand higher tensile load.

Figure 10.11 depicts the effect of infill density and build orientation on compressive strength. From Figure 10.11, it is found that the compressive strength decreases with an increase in build orientation and increases with an increase in infill density. It is found that as the build orientation increases from 0^0 to 30^0, the compressive strength

TABLE 10.2
Experimental Data of Tensile and Compressive Strength

Sl. No.	Infill Density	Built Orientation	Tensile Strength (MPa)	Compressive Strength (MPa)
1	High	0^0	26.66	89.42
2	High	15^0	24.82	75.72
3	High	30^0	22.35	75.52
4	Low	0^0	25.18	32.17
5	Low	15^0	24.22	23.91
6	Low	30^0	19.81	23.19

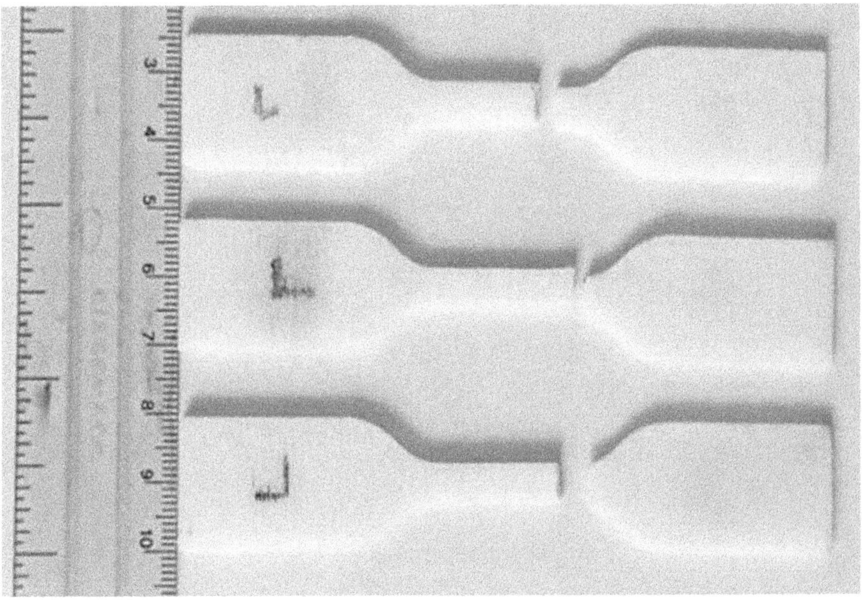

FIGURE 10.6 Specimens after tensile test for low infill density.

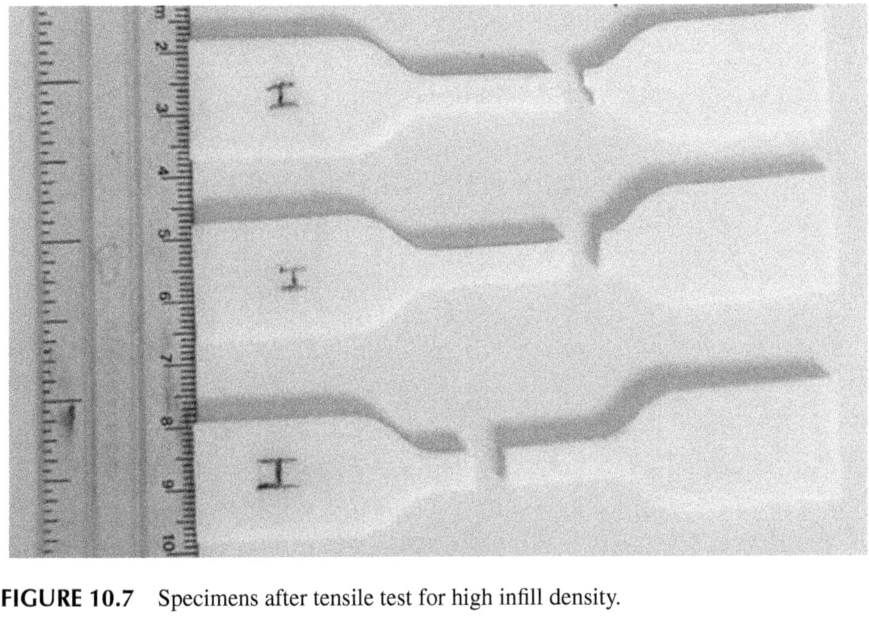

FIGURE 10.7 Specimens after tensile test for high infill density.

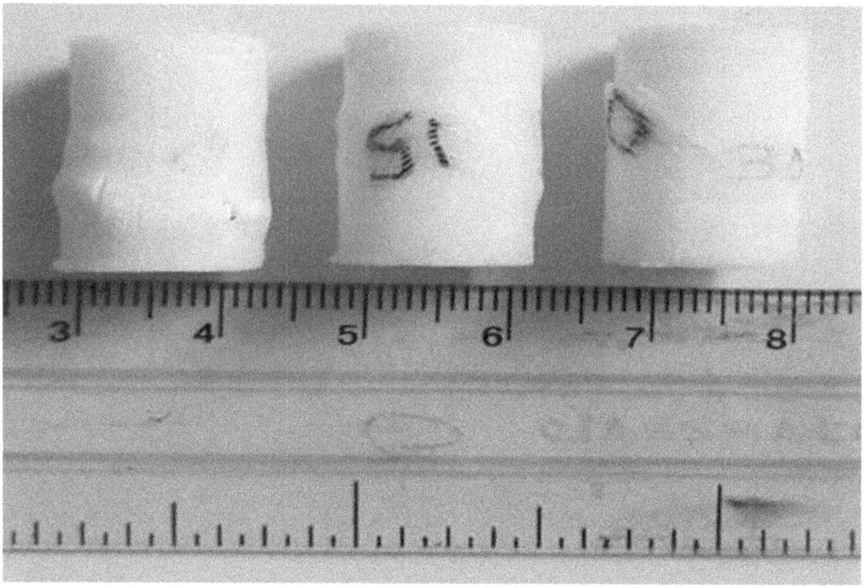

FIGURE 10.8 Specimens after compressive test for low infill density.

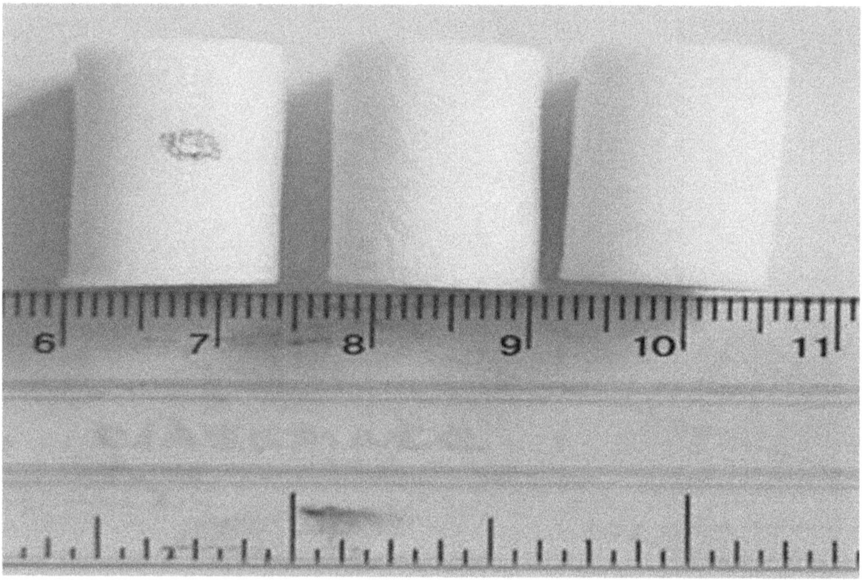

FIGURE 10.9 Specimens after tensile test high infill density.

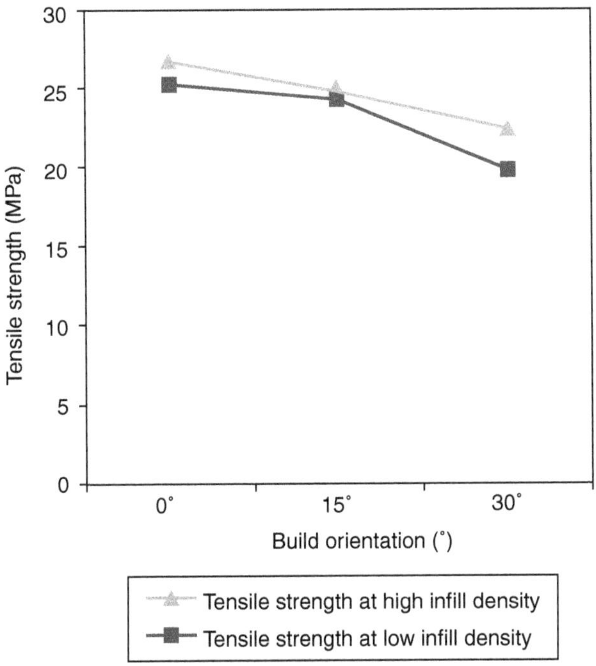

FIGURE 10.10 Effect of infill density and build orientation on tensile strength.

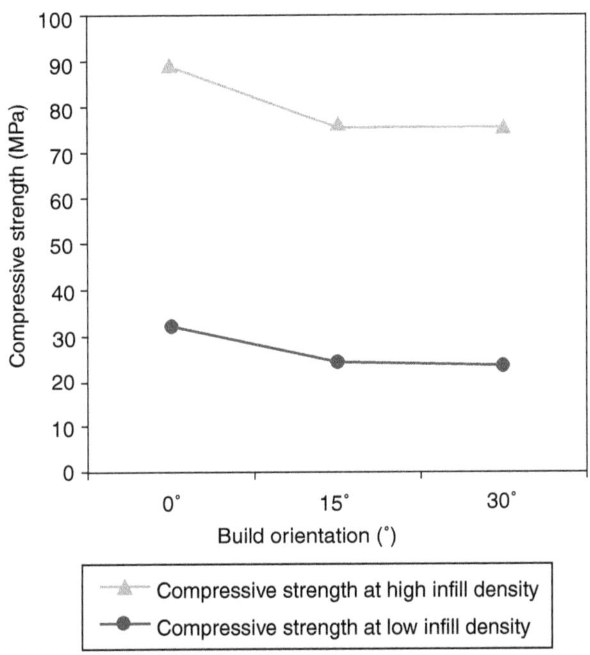

FIGURE 10.11 Effect of infill density and build orientation on compressive strength.

decreases for both high infill density and low infill density samples. This is due to the type of fracture and the manner in which the load is applied. With increase in infill density the compressive strength increases, because, at higher infill density, the specimens fabricated were with more density that withstand higher compressive load.

10.4 CONCLUSIONS

In this work, ABS plastic was used as the working filament material for producing specimens by FDM process. The parameters were varied at different levels to fabricate specimens were build orientation (0^0, 15^0, and 30^0) and infill density (high and low). In this study, experimental results were used to investigate the relationship between process variables and the strength of FDM specimens. Here, compressive strength and tensile strength of FDM-fabricated parts were evaluated. The conclusions from this work are discussed as follows:

- Strength is clearly impacted by infill density and build orientation.
- Tensile strength decreased with an increase in built orientation, and it increased with an increase in infill density. It is maximum at 0^0 build orientation, that is, 26.66 MPa and minimum at 30^0 orientation, that is, 22.35 MPa for high infill density. Similarly, tensile strength is maximum for the low infill density at 0^0 build orientation which is 25.18 MPa. At 30^0 build orientation and low infill density, tensile strength is 19.81 MPa.
- Compressive strength also increased with increase in infill density and decreased with increased build orientation. It is maximum at 0^0 build orientation, that is, 89.42 MPa and minimum at 30^0 build orientation, that is, 75.52 MPa for high infill density. Similarly, for low infill density, it is 32.17 MPa at 0^0 build orientation and 23.19 MPa at 30^0 build orientation.
- Thus, it is demonstrated that the mechanical properties of the FDM-fabricated ABS parts significantly affected their build orientation and infill density, and higher tensile strength and compressive strength can be achieved at low build orientation (0^0) and high infill density.

REFERENCES

1. K M Agarwal, P Shubham, D Bhatia, P Sharma, H Vaid, R Vajpeyi, Analyzing the impact of print parameters on dimensional variation of ABS specimens printed using fused deposition modelling (FDM), *Sensor International* 3 (2022): 100149.
2. E Monaldo, S Marfia, Multiscale technique for the analysis of 3D-printed materials, *International Journal of Solids and Structures* 232 (2021): 111173.
3. S Mahmood,S D Talamona, K L Goh, A J Qureshi, Fast deviation simulation for 'fused deposition modeling' process, *Procedia CIRP* 43 (2016): 327–332.
4. C Casavola, A Cazzato, V Moramarco, G Pappalettera, Residual stress measurement in fused deposition modelling parts, *Polymer Testing* 58 (2017): 249–255.
5. T Yao, Z Deng, K Zhang, S Li, A method to predict the ultimate tensile strength of 3D printing polylactic acid (PLA) materials with different printing orientations, *Composites Part B* 163 (2019): 393–402.

6. A Equbal, A K Sood, M I Equbal, I A Badruddin, Z A Khan, RSM based investigation of compressive properties of FDM fabricated part, *CIRP Journal of Manufacturing Science and Technology* 35 (2021): 701–714.
7. N Vidakis, M Petousis, A Vairis, K Savvakis, A Maniadi, On the compressive behavior of an FDM Steward Platform part, *Journal of Computational Design and Engineering* 4 (2017): 339–346.
8. M N Sudin, F R Ramli, M R Alkahari, M A Abdullah, Comparison of wear behavior of ABS and ABS composite parts fabricated via fused deposition modelling, *International Journal of Advanced and Applied Sciences* 5(1) (2018): 164–169.
9. P Rezaeian, M R Ayatollahi, A N Kivi, S M Javad, Effect of printing speed on tensile and fracture behavior of ABS specimens produced by fused deposition modeling, *Engineering Fracture Mechanics* 266 (2022): 108393.
10. V Tambrallimath, R Keshavamurthy, D Saravanabavan, A Badari, M C Jeevan, Numerical and experimental analysis of thermal conductivity of PC-ABS nanocomposite reinforced with graphene developed by fused deposition modeling, *Materials Today: Proceedings* 46 (2021): 8964–8967.
11. H B Mamo, A D Tura, A J Santhosh, N Ashok, D K Rao, Modeling and analysis of flexural strength with fuzzy logic technique for a fused deposition modeling ABS components, *Materials Today: Proceedings* 57 (2022): 768–774.
12. S Warke, V S Puranik, Comparison of energy consumption of ABS and PLA while 3 D printing with fused deposition modeling process, *Materials Today: Proceedings* 66 (2022): 2098–2103.
13. M T Mastura, S Najwa, A Kudus, R Jumaidin, R Nadlene, Development of eco-friendly kenaf fiber reinforced ABS (KRABS) composite filament of fused deposition modeling, *Materials Today: Proceedings* 70 (2022): 363–367.
14. D Foltut, E Valean, V Dzitac, L Marsavina, The influence of temperature on the mechanical properties of 3D printed and injection molded ABS, *Materials Today: Proceedings* 78 (2023): 210–213.
15. R R Mohan, R Venkatraman, S Raghuraman, Experimental analysis on density, micro-hardness, surface roughness and processing time of acrylonitrile butadiene styrene (ABS) through fused deposition modeling (FDM) using Box Behnken Design (BBD), *Materials Today Communications* 27 (2021): 102353.
16. P Diana, B Florin, V Daniel, C M Cosmin, M Rodica, Effects of multiple sterilizations and natural aging on the mechanical behavior of 3D-printed ABS, *Mechanics of Materials* 148 (2020): 103423.
17. R Behzad, J Sina, D Fereidoon, L Todd, Failure analysis and mechanical characterization of 3D printed ABS with respect to layer thickness and orientation, *Journal of Failure Analysis and Prevention* 3 (2016): 467–481.
18. P K Nayak, A K Sahu, S S Mahapatra, Effect of process parameters on the mechanical behavior of FDM and DMLS build parts, *Materials Today: Proceedings* 22 (2020): 1443–1451.
19. P K Nayak, A K Sahu, S S Mahapatra, Influence of layer thickness and orientation on tensile strength and surface roughness of FDM-built parts, in: C Prakash, S Singh, J P Davim (eds), *Characterization, Testing, Measurement, and Metrology*. CRC Press (2020).
20. J Christiyan, U Chandrasekhar, K Venkateswarlu, A study on the influence of process parameters on the mechanical properties of 3D printed ABS composite, *Materials Science and Engineering* 114 (2016): 012109.
21. D Cole, J Riddick, I Jaim, K Strawhecker, N Zander, Interfacial mechanical behavior of 3D printed ABS, *Journal of Applied Polymer Science* 30 (2016): 43671 .

22. G Cwikla, C Grabowik, K Kalinowski, I Paprocka, P Ociepka, The influence of printing parameters on selected mechanical properties of FDM/FFF 3D-printed parts, *Materials Science and Engineering* 227 (2017): 012033.
23. M Jaroslav, H Eva, Z Milan, K Petr, Effect of processing parameters on mechanical properties of 3D printed samples, *Materials Science Forum* 919 (2018): 230–235.
24. M Eryildiz, Effect of build orientation on mechanical behaviour and build time of FDM 3D-printed PLA parts: an experimental investigation, *European Mechanical Science* 5(3) (2021), 116–120.

11 Study on Effect of Process Parameters on Mechanical Properties of Build Parts Using Fused Deposition Modeling

Borra N. Dhanunjayarao, N. V. Swamy Naidu, Sateesh Kumar Adapa, Uzwalkiran Rokkala, Roop Sandeep Bommidi, Gill Santosh Kumar, Ch. Bharath Kumar, Usha Kiran Sanivada, and P. V. N. Sai Kumar

11.1 INTRODUCTION

Fused deposition modeling/fused filament fabrication (FDM/FFF) is one of the more important additive manufacturing (AM) technologies, which have found applications in many areas characterized by complex and custom-made parts. Since its development in the late 1980s, FDM has undergone massive evolution in materials, hardware, and software as shown in Figure 11.1. This evolution has made it more accessible for both industrial and consumer applications and has turned FDM into the method of choice in fabricating intricate geometries using thermoplastic materials [1]. Among these materials, polylactic acid (PLA) is most commonly used in FDM due to its biodegradability, ease of processing, and satisfactory mechanical properties. However, the mechanical performance of parts produced via FDM is strongly dependent on the optimization of process parameters [2].

It is critical to understand how process parameters influence the enhancement in part quality, improvement in mechanical properties, and expansion in application scope of FDM [3]. The mechanical behavior of FDM-built parts is significantly influenced by variables like layer thickness, print speed, extrusion temperature, infill density, and build orientation [4]. These parameters directly influence the strength, durability, and overall performance of the three-dimensional (3D)-printed components [5]. For example, layer thickness influences surface smoothness and dimensional accuracy, as

DOI: 10.1201/9781003686941-11

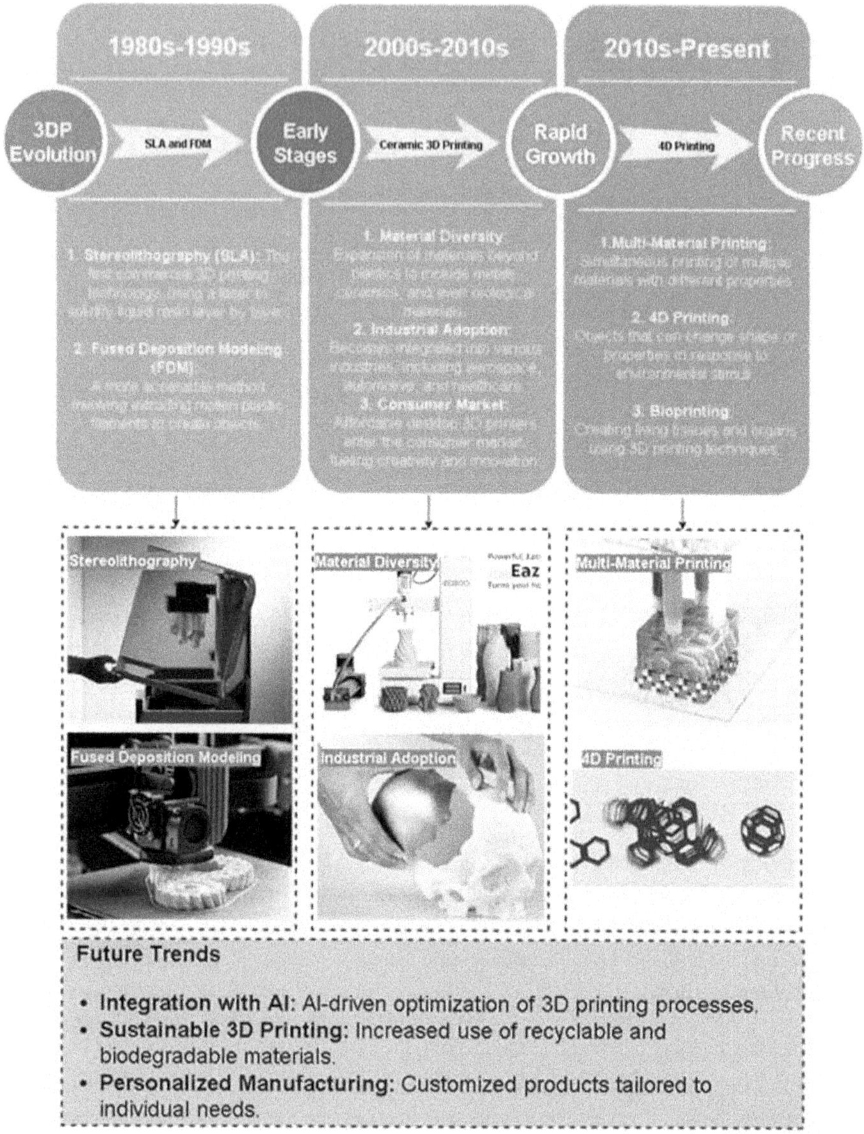

FIGURE 11.1 Evolution of 3D printing technology and future trends.

well as the mechanical strength of the object printed. In general, thinner layers give better detail and layer bonding, although at longer printing times [6].

FDM is among the most popular AM techniques, which creates 3D objects by layer-by-layer extrusion of hot thermoplastic filament. First of all, digital design has to be performed: using computer-aided design (CAD) software, a 3D model of the object to be printed is created [7]. Slicing this digital model into very thin, horizontal layers will instruct the 3D printer [8]. A heated nozzle moves across the build platform during the filament extrusion stage, depositing molten filament layer by layer to create the item, as seen in Figure 11.2. The final structure is progressively built up as each layer is deposited, cooling and solidifying as it goes. After finishing, the item is taken off the build platform and may be post-processed to improve its usefulness or finish [9].

Numerous process variables have a significant impact on the mechanical characteristics of FDM-printed objects. Key parameters include layer height, which influences surface roughness and tensile strength; thinner layers generally give smoother surfaces but longer print times. Infill density defines the strength and weight of the part; higher densities result in stronger parts at the expense of longer print time and more material consumption [11]. Temperature control becomes quite essential, mostly for the nozzle and the build chamber, to attain an accurate flow of filament and dimensional stability. Printing speed is also a major factor: slower speeds generally improve layer adhesion and surface quality but elongate the printing process. The type of thermoplastic filament used, such as PLA or acrylonitrile butadiene styrene (ABS), has a great effect on the mechanical properties of the part, including its strength, flexibility, and temperature resistance. High-quality components with the required mechanical properties can be achieved in FDM-printed objects if the parameters are tuned appropriately [12].

Figure 11.3 shows the workflow of FDM 3D printing. It explains all of the steps needed to make a physical object from a computer design. First, a CAD model is created. Then, it is exported into a printable format (STL, OBJ, AMF, or 3MF), sliced into layers, and G-code instructions are generated. The G-code is then exported to the

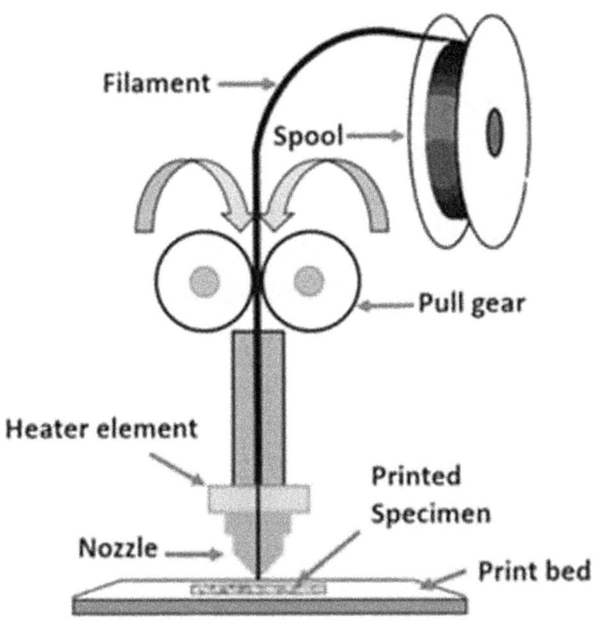

FIGURE 11.2 Diagrammatic illustration of the FDM 3D printing method [10].

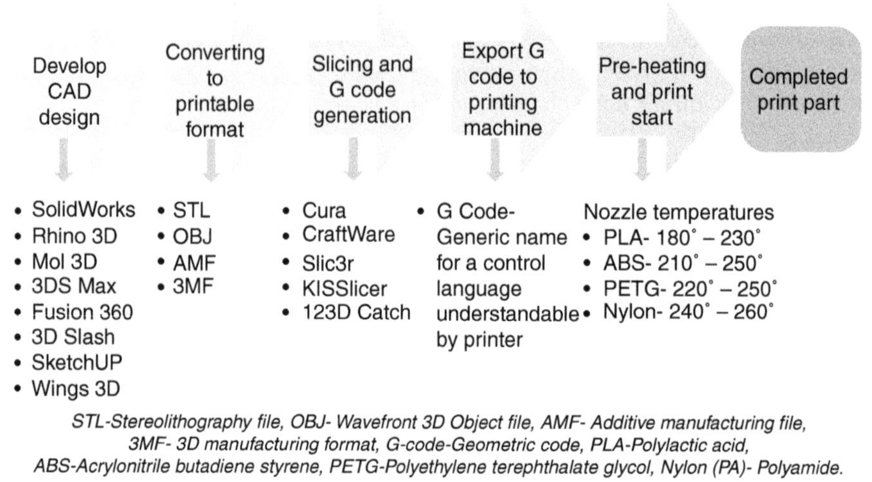

FIGURE 11.3 Process flowchart of the FDM process with a range of phases, beginning from digital design to actual reality, through the variety of software used in the industry [13].

3D printer, which first preheats the nozzle and build platform and then directly starts printing. Finally, the print part is taken out after completion. In the same diagram, there are some common CAD software, slicing software, and nozzle temperature for different materials of filament [13].

This study aims to emphasize how important it is to understand the FDM process factors and how they affect characteristics. Given the importance of these parameters, this chapter will explore their effects on the mechanical properties of build parts produced using FDM. This chapter synthesizes current knowledge, identifies optimal parameter settings, and highlights the challenges faced in achieving desirable mechanical properties through a review of existing literature. This extensive review will provide a guide for engineers, researchers, and practitioners in the optimization of FDM processes for better mechanical performance. This chapter focuses on the mechanical properties of PLA, which is biodegradable and one of the most consumed thermoplastics in FDM [14]. The aim is to determine the optimal settings to raise the hardness and strength of PLA specimens in an attempt to improve efficiency and quality in 3D printing. This research demonstrated how imperative it is to adjust the parameters in FDM processing so that it will be able to meet the special mechanical requirements that parts used in various industries like biomedical engineering, automotive, and aerospace call for [15]. This study also points out the potential of real-time parameter control systems in FDM and calls for further research in this area to raise the consistency and performance level of FDM components. Optimization of these process parameters would undoubtedly have profound effects on improving the reliability of FDM components in demanding applications, paving the way for more widespread use of FDM in high-performance engineering fields [16].

11.2 OVERVIEW OF FDM PROCESS PARAMETERS

Figure 11.4 is a fishbone diagram with the various factors that have an influence on the mechanical properties and general quality of the FDM/FFF 3D-printed parts. The key properties are tensile strength, flexural strength, compressive strength, and build time. All these can be grouped into material selection, process parameters, environmental factors, and part build orientation factors, which, to a large extent, predetermine the final output. The choice of material, such as PLA or ABS, directly impacts the intrinsic mechanical properties of the printed part [17]. Process variables such as layer thickness, raster orientation, raster width, build orientation, and air gap also affect surface quality, dimensional accuracy, and part strength. Other important operational characteristics affecting strength, surface finish, and general performance of the part are print speed, extrusion temperature, infill density, pattern, and nozzle diameter [18]. For instance, print speed is always a trade-off between build time and quality, as the faster the rate, the more it will weaken the bonding between layers, while the smaller layer heights give better surfaces and higher resolution. Other environmental factors such as temperature, humidity, and build orientation along different axes (X, Y, and Z) are very critical in assuring part stability and strength. It requires good management and optimization to acquire the appropriate mechanical characteristics and quality in FDM-printed components. In this way, users can finely adjust these parameters to have the final output meet the requirements, so high-quality products with consistency are realized, performing as intended in manifold applications [19].

11.2.1 Layer Height

The thickness of each deposited layer during the FDM process is referred to as a layer height. It plays a critical role in determining the surface finish, dimensional

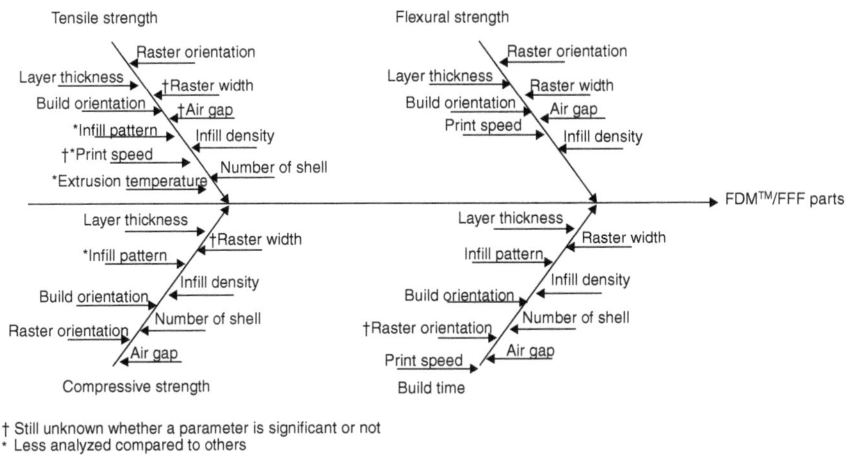

FIGURE 11.4 Fishbone diagram showing how the process parameters mostly affect the FDM/FFF process [18].

accuracy, and mechanical strength of the printed part. Lower layer heights generally enhance surface smoothness and dimensional accuracy but increase build time [20]. Conversely, higher layer heights reduce build time but may compromise mechanical integrity due to weaker interlayer bonding. According to studies, it is critical to strike a balance between build time and layer height; ideal layer heights normally fall between 0.1 and 0.3 mm for PLA parts, as shown in Figure 11.5.

11.2.2 Print Speed

The rate at which the print head moves throughout the deposition process is known as the print speed. It affects the deposition rate, thermal management, quality, and ultimately the mechanical properties of the part as shown in Figure 11.6. High print speeds can lead to insufficient layer adhesion and increased internal stresses, reducing tensile and flexural strength [21]. Low print speeds improve bonding between layers but prolong the manufacturing process [22]. An optimal print speed balances productivity and part strength, typically ranging from 30 to 60 mm/s for PLA materials.

11.2.3 Infill Density and Pattern

The internal structure of the printed part, as depicted in Figure 11.7, is referred to as infill density, whereas the infill's geometric arrangement is indicated by the infill pattern. These parameters significantly influence the part's weight, strength, and impact resistance. Higher infill densities increase mechanical strength and impact resistance, but also increase material usage and print time, as shown in Figure 11.8. The infill pattern affects stress distribution within the part. Common infill densities fall between 20% and 100%, and grid and honeycomb patterns frequently offer a fair trade-off between material efficiency and strength [23].

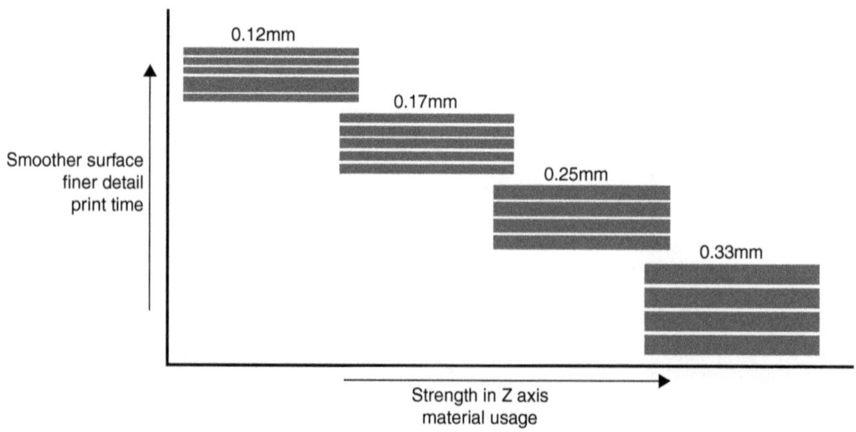

FIGURE 11.5 Influence of layer height on surface roughness and tensile strength.

Source: https://rapidfab.ricoh-europe.com/resources/help/fdm-design-guides/

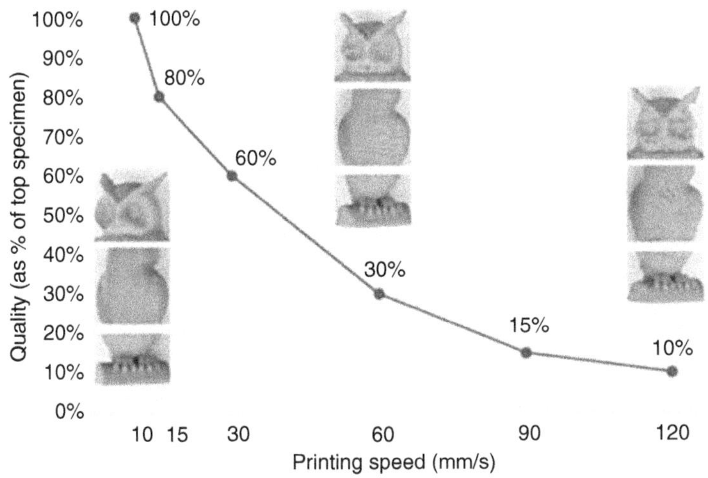

FIGURE 11.6 Studies on print speed and quality of the FDM 3D-printed part.

Source: https://3dprintingindustry.com/news/color-speed-temperature-and-age-how-do-these-factors-influence-the-outcome-57141/

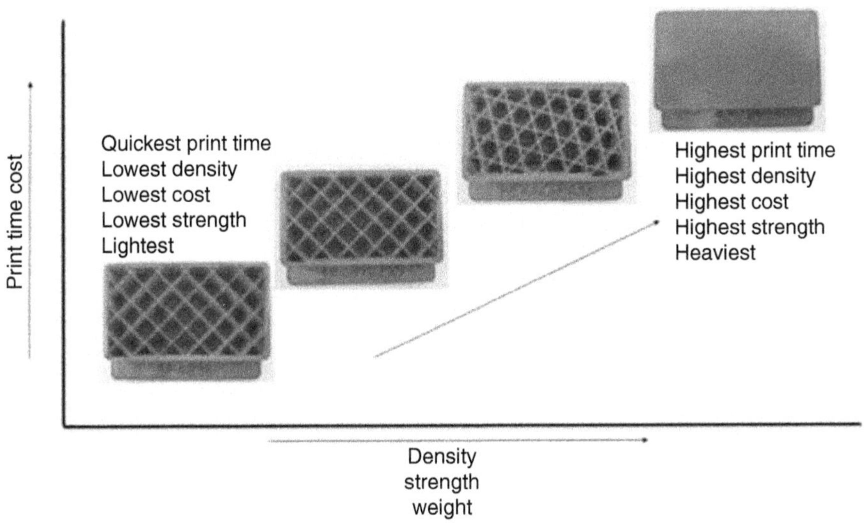

FIGURE 11.7 Density, strength, and weight studies on FDM printing time.

Source: https://rapidfab.ricoh-europe.com/resources/help/fdm-design-guides/

The mechanical characteristics, functionality, and general performance of 3D-printed components are greatly influenced by the infill pattern selection in FDM. Infill patterns influence critical factors such as strength, weight distribution, energy absorption, and even the part's durability under various load conditions. Common

Strength [MPa] — Infill %

Layer Height [mm]	10	30	50	70	80	90	100
0.1	4	9	15	24	29	34	40
0.15	8	13	19	28	33	38	44
0.2	10	15	21	30	35	40	46
0.25	11	16	22	31	37	42	48
0.3	11	16	22	31	36	41	48

Speed [mins] — Infill %

Layer Height [mm]	10	30	50	70	80	90	100
0.1	21	35	47	61	68	74	80
0.15	14	23	32	41	46	50	54
0.2	10	18	24	31	35	38	41
0.25	8	14	20	25	28	31	33
0.3	7	12	16	21	24	26	28

Cost [cents] — Infill %

Layer Height [mm]	10	30	50	70	80	90	100
0.1	5	9	13	16	18	20	22
0.15	5	9	13	17	19	21	23
0.2	6	10	14	18	20	22	23
0.25	6	10	14	18	20	22	24
0.3	6	10	14	18	20	22	24

Quality — Infill %

Layer Height [mm]	10–90	100
0.1	very high	
0.15	high	low
0.2	medium	
0.25	low	
0.3	very low	

BEST ⟷ **WORST**

FIGURE 11.8 Comparative analysis of strength, cost, speed, quality over infill percentage of FDM printing.

Source: https://skmurphy.com/blog/2015/08/08/3d-printing-tradeoffs-and-optimization/

patterns, such as concentric, honeycomb, rectilinear, zigzag, Archimedean chords, Hilbert curve, and octagram spiral, serve different purposes based on the application requirements, as illustrated in Figure 11.9. For example, honeycomb patterns provide excellent load-bearing strength and isotropic properties, while rectilinear or zigzag patterns are lightweight and suitable for reducing material usage and print time [24].

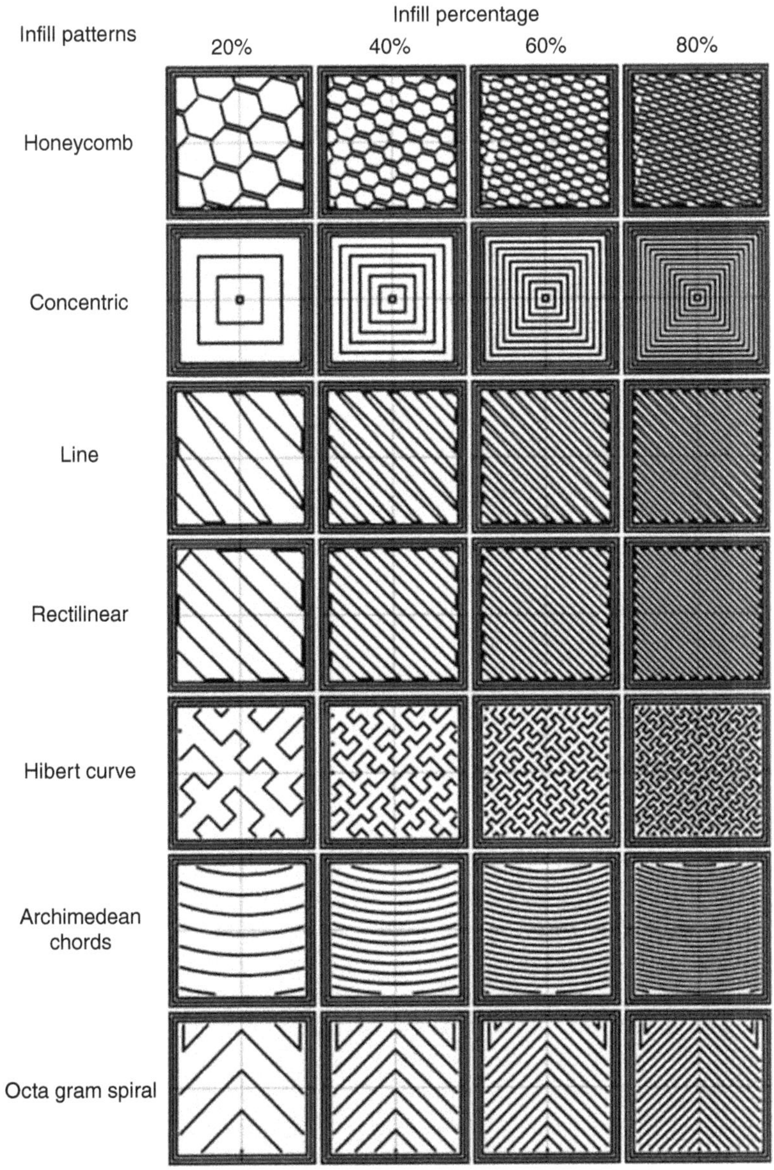

FIGURE 11.9 Illustration of various infill patterns at varying densities.

Source: https://skmurphy.com/blog/2015/08/08/3d-printing-tradeoffs-and-optimization/

Infill patterns have a very important impact on the mechanical properties. Patterns like honeycomb and gyroid excel in distributing stress evenly, making them ideal for load-bearing and impact-resistant applications. In contrast, concentric or linear patterns exhibit anisotropic behavior, meaning their strength varies with direction. The infill pattern also affects the energy absorption capacity of components, with gyroid and honeycomb patterns showing superior resilience in dynamic or crash scenarios [25]. Moreover, patterns like the Hilbert curve are not only visually aesthetic but also improve the internal uniformity of parts, ensuring better bonding and surface finish. Infill patterns also affect other parameters such as material efficiency and sustainability. For instance, lightweight patterns like zigzag or concentric use less material and take less time to print and, therefore, are suitable for prototyping or non-load-bearing applications. On the other hand, denser patterns or combinations of hybrid infills, such as using honeycomb in critical stress zones and rectilinear infill in less critical areas, strike a balance between strength and weight in an optimization of material usage [26].

The new trends of complex infill structures and hybrid patterns further increase the potential of FDM. Complex patterns like gyroid, with their continuous curvature, reduce stress concentration points and therefore improve durability and fatigue resistance. Hybrid infill designs, where more than one pattern is used within the same component, offer a tailored property to suit specific requirements [27]. Gradient density infills, where density varies within the part, offer local reinforcement while maintaining overall weight reduction. By the careful selection of infill pattern and density, it is possible to tune mechanical characteristics, energy absorption, and aesthetics of FDM-printed parts to meet the demands of applications across industries such as automotive, aerospace, medical, consumer, and sports goods. This will result in knowledge that can be used to produce optimized components, balancing strength, weight, cost, and sustainability [28].

11.2.4 Extrusion Temperature

Extrusion temperature is one of the most critical factors affecting the mechanical characteristics of PLA components made using FDM. Some of the important qualities optimized when operating between 190°C and 220°C, especially between 200°C and 210°C, are tensile strength, Young's modulus, impact resistance, and layer adhesion. At optimal temperatures, proper bonding between adjacent layers ensures maximum tensile strength, while lower temperatures result in weak interlayer adhesion, and higher temperatures may degrade PLA, reducing its strength. The stiffness, reflected by Young's modulus, is also enhanced due to uniform filament deposition, minimizing defects [29]. Additionally, impact resistance improves as the optimal temperature reduces brittleness and enhances ductility, making parts less prone to fractures under sudden forces. Layer adhesion benefits significantly, with the molten filament effectively fusing with previous layers, improving structural integrity and reducing anisotropy. Deviations from the optimal range can lead to poor surface finish, warping, or under-extrusion, negatively impacting these properties. Figure 11.10 highlights the connection between extrusion temperature and the mechanical performance of PLA parts, showcasing how precise temperature control enhances their functional attributes [30].

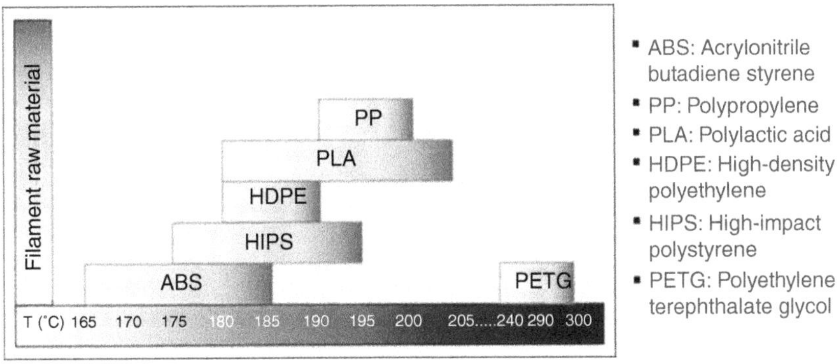

FIGURE 11.10 Illustration of extrusion temperature for various thermoplastic polymers [30].

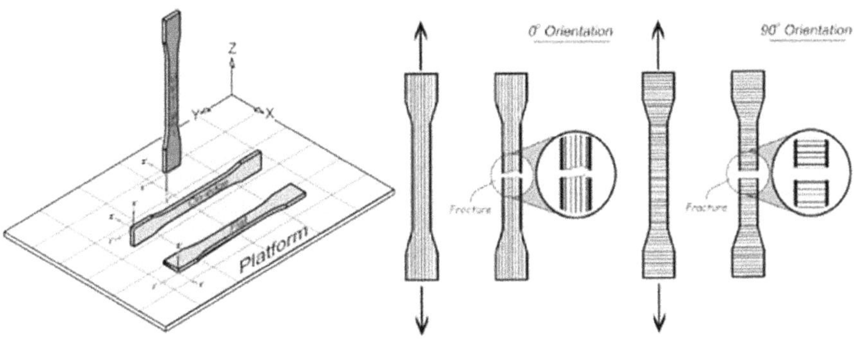

FIGURE 11.11 Illustrates part positioning on the building platform and fracture direction based on orientation shows anisotropic mechanical properties [31–34].

11.2.5 Build Orientation

The alignment of the component with respect to the build platform is known as "build orientation" [31]. Because FDM is layer by layer, it has an impact on the anisotropy of mechanical properties, as seen in Figure 11.11. When printed horizontally, parts frequently have greater tensile strength than when printed vertically. Build orientation can influence dimensional accuracy and surface finish. Selecting the appropriate build orientation based on the intended load-bearing direction can enhance mechanical performance [32].

11.3 FDM SELECTION GUIDE

The FDM selection guide is designed to help users make informed decisions about the appropriate infill density and layer thickness for their 3D printing projects, recognizing that each application has unique requirements, as illustrated in Figure 11.12. The guide starts by determining whether the part being printed is primarily for visual

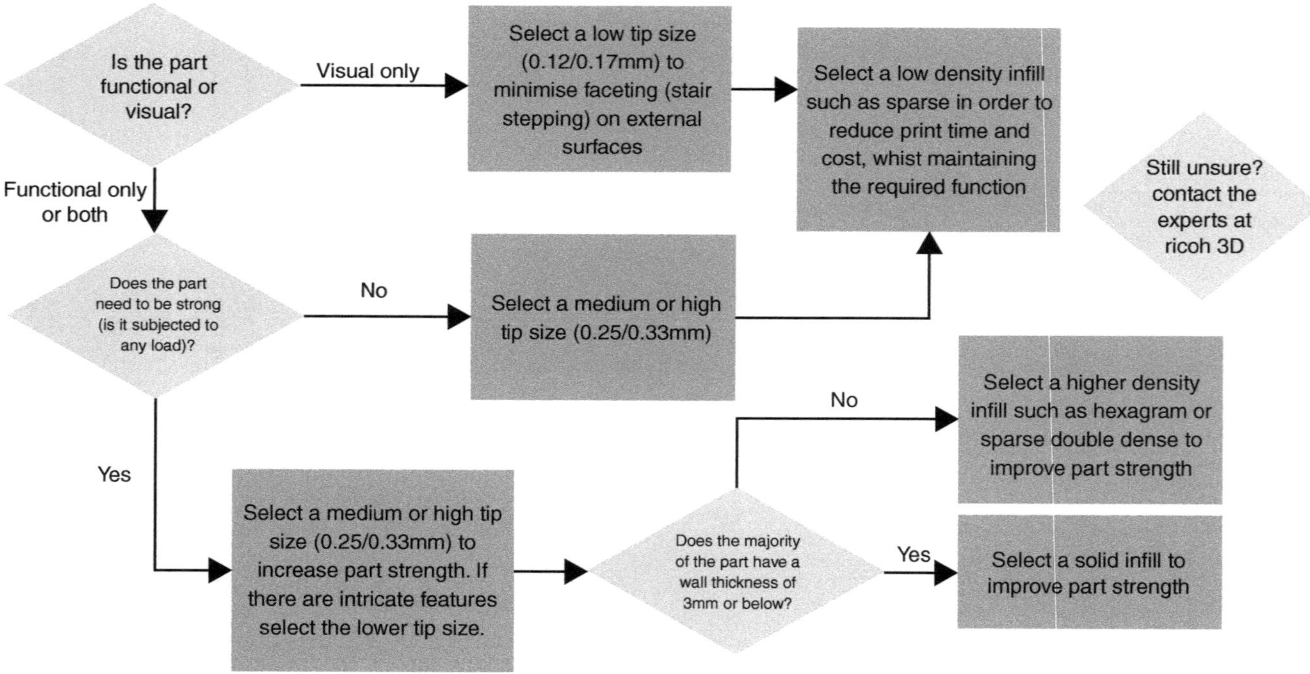

FIGURE 11.12 Illustration of infill and layer height selection guide for FDM 3D printing technology.

Source: https://rapidfab.ricoh-europe.com/resources/help/fdm-design-guides/

purposes or if it also needs to serve a functional role. The diameter of the nozzle opening through which the filament is extruded is referred to as tip size in the context of FDM. It determines the width of the deposited material strand and directly influences the surface finish, resolution, and mechanical properties of the 3D-printed part. Clients must swap out the 3D printer's nozzle for one of the proper diameters to get the required tip size. This involves removing the existing nozzle and installing a nozzle with the required tip size, a process that may require a wrench or specialized tool. Once replaced, the printer's slicer settings must be updated to reflect the new nozzle diameter to ensure accurate filament flow and layer deposition [35]. For parts that are purely visual, a lower tip size (0.12/0.17 mm) is recommended to minimize the faceting effect, also known as stair-stepping, which can occur on external surfaces. This results in a smoother and more aesthetically pleasing finish. Additionally, a low-density infill, such as Sparse, is suggested to reduce both the print time and material costs, while still achieving the necessary visual qualities [36].

If the part is functional or needs to fulfill both functional and visual roles, the next step is to assess whether the part requires strength, meaning whether it will be subjected to any significant loads. If the part does not need to be particularly strong, a medium or high tip size (0.25/0.33 mm) is advised, balancing speed and strength, along with a low-density infill to maintain efficiency. However, if the part does need to be strong, a medium or high tip size is still recommended to enhance part strength. For parts with intricate features, a lower tip size should be selected to ensure detail accuracy. The guide then advises examining the part's wall thickness: if the majority of the part has a wall thickness of 3 mm or less, a solid infill is recommended to further improve strength. If the wall thickness is greater than 3 mm, a higher density infill such as Hexagram or Sparse Double Dense can be used to give the part increased strength with little increase in material usage [37].

Advanced internal structures in FDM are Hexagram and Sparse Double Dense infills, as shown in the flowchart. The hexagram infill pattern is made up of a complex, star-like, interlocking geometry that gives better load distribution and rigidity while maintaining light weight, and thus is suitable for applications where high strength in multiple directions is needed [38]. On the other hand, Sparse Double Dense infill alternates between layers of sparse and denser structures, creating a balanced lattice that provides improved mechanical properties compared to sparse infill alone, while consuming less material and reducing print time compared to solid infill. Such patterns allow for the exact tailoring of internal part structures, enabling manufacturers to optimize strength, weight, and production efficiency based on functional demands [39]. Users who still doubt which settings are best for their project are advised to turn to the experts for tailor-made counseling so that the selected FDM settings conform exactly to the application needs. This integrated decision-making framework will thus help users optimize their 3D printing process and weigh the trade-offs very carefully between appearance, strength, time, and cost.

11.4 PROCESS PARAMETERS IMPACT ON MECHANICAL PROPERTIES

The mechanical characteristics, including tensile strength, impact resistance, flexural strength, and fatigue performance, largely determine the functioning and appropriateness of FDM parts for a wide range of applications. Process optimization for meeting specific performance objectives requires an understanding of how process factors impact these qualities. This synthesis of existing research identifies the optimal settings and common challenges in achieving desired mechanical characteristics. The mechanical performance of FDM parts is directly related to and influenced by some important process parameters utilized during fabrication. Exceptional tensile strength, impact resistance, flexural strength, and fatigue performance can be achieved in the FDM products through careful adjustment of variables such as layer height, print speed, infill density, build orientation, and extrusion temperature. The strong technical expertise involved in the FDM process allows process fine-tuning for differing application requirements; hence reliable, efficient, and good-quality 3D printed parts are provided [40].

11.4.1 Tensile Strength

The tensile strength of a material can be defined as the resistance of a material to breaking under tension and is significantly affected by internal stresses, filament quality, and layer adhesion. Typically, lower layer heights produce stronger parts as the layer adhesion increases, but on the flip side, it results in increased build time. Related to the print speed and tensile strength, faster speeds are expected to have reduced accuracy and mechanical strength because there is not sufficient bonding time between layers. This property is usually measured by means of standard tensile tests following the American Society for Testing and Materials (ASTM) D638 [41]. The graph in Figure 11.13 shows the influence of bed temperature and melt temperature on the tensile modulus and tensile strength of FDM 3D-printed parts. Generally, increasing bed temperature and melt temperature can improve mechanical properties up to a certain point. However, excessive temperatures can lead to material degradation and reduced performance. The optimal temperatures for achieving the best balance of mechanical properties depend on the specific material and printing conditions [42].

11.4.2 Flexural Strength

Temperature, infill density, layer height, and print orientation are some of the variables that affect flexural strength, which is a measurement of a material's capacity to withstand deformation under load. Temperature, particularly bed temperature and melt (extrusion) temperature, significantly impacts flexural properties. Optimal bed and melt (extrusion) temperatures, as shown in Figure 11.14, significantly enhance flexural properties by improving interlayer bonding and filament flow, reducing voids, and increasing structural integrity [43].

For instance, extrusion temperatures between 200°C and 210°C for PLA ensure proper molecular bonding, preventing brittleness and enhancing ductility, while adequate bed temperature minimizes warping and improves adhesion to the build plate. However, excessively high temperatures may degrade the material, negatively impacting performance [42]. Reducing layer height strengthens tensile and flexural

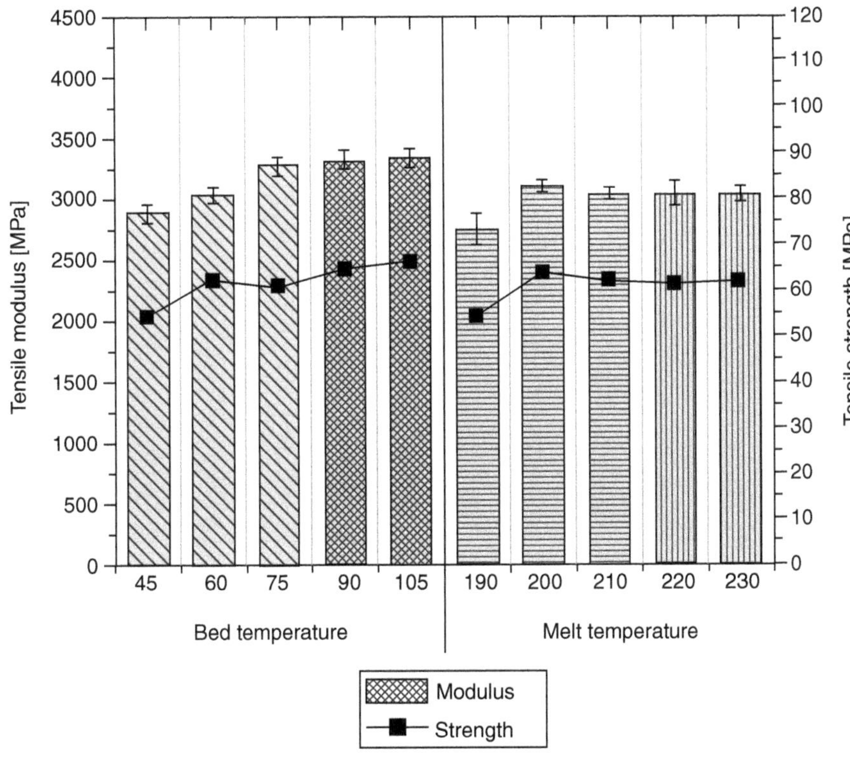

FIGURE 11.13 Effect of bed and extrusion temperatures on tensile strength and modulus [42].

properties by promoting better bonding and minimizing voids. Higher infill densities improve part strength at the cost of increased material usage and print time. Additionally, infill patterns, such as honeycomb, outperform simpler patterns by offering superior strength [44]. Print orientation and speed also play a role; horizontal orientations and slower speeds typically yield better interlayer adhesion and stronger parts. Vertical orientations can result in anisotropic properties, where lower mechanical strength is present along specific axes, and thus, mechanical performance is maximized via maximizing these parameters under material- and application-specific conditions [45].

11.4.3 Impact Resistance

Impact resistance behavior of a material on absorbing energy in sudden impacts without fracturing can be influenced by infill pattern, extrusion temperature, and print speed. Better impact resistance can be achieved with higher infill densities and optimized extrusion temperatures due to fewer internal voids and proper layer adhesion. Standard methods used are the Charpy and Izod impact tests. The following figure shows the effect of bed temperature, melt temperature, and annealing on the Izod notched impact strength of FDM 3D-printed parts. Annealing improves impact

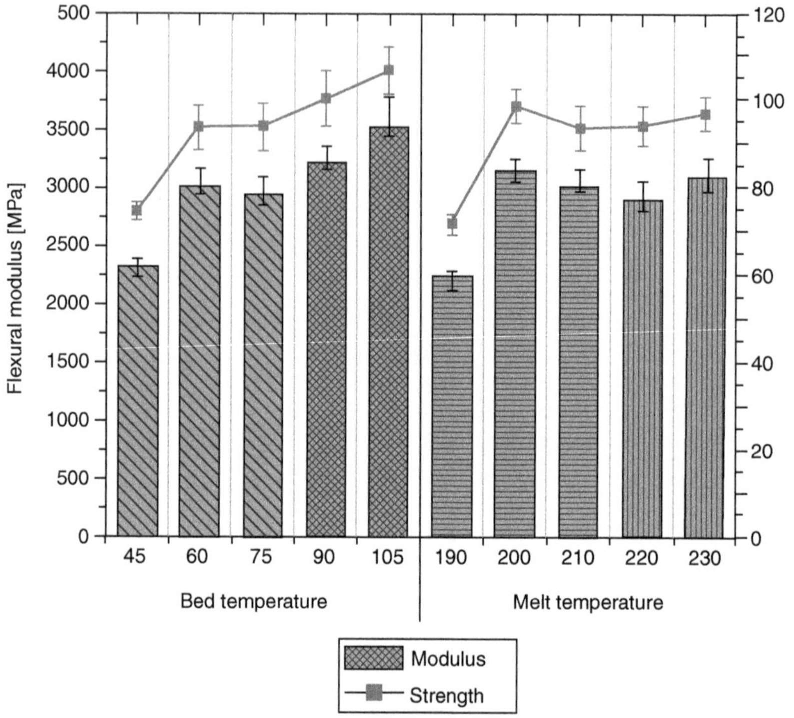

FIGURE 11.14 Effect of bed and extrusion temperatures on flexural strength and modulus [42].

strength, while bed temperature and melt temperature have a rather more complex relationship [42]. The impact strength of materials is increased by high temperatures up to a point, but could degrade at very high temperatures, as shown in Figure 11.15.

11.4.4 Fatigue Performance

Fatigue performance, the durability of a material under cyclic loading conditions, is influenced by layer adhesion, print orientation, and consistency of the material. All these parameters can be changed to improve the fatigue life of FDM parts dramatically to better fit their application, where repetitive stresses are concerned. Fatigue testing involves subjecting samples to repeated loading until failure. Figure 11.16 in the research paper shows fatigue life versus stress amplitude curves for 3D printed ABS parts fabricated by FDM technology. This graph is very important in comparison with the fatigue performance of these materials. From the curves, one can identify which material has better fatigue resistance at what particular stress level. This information can be used by engineers and designers when choosing materials for FDM-printed components loaded cyclically to provide the reliability and longevity of the parts [46].

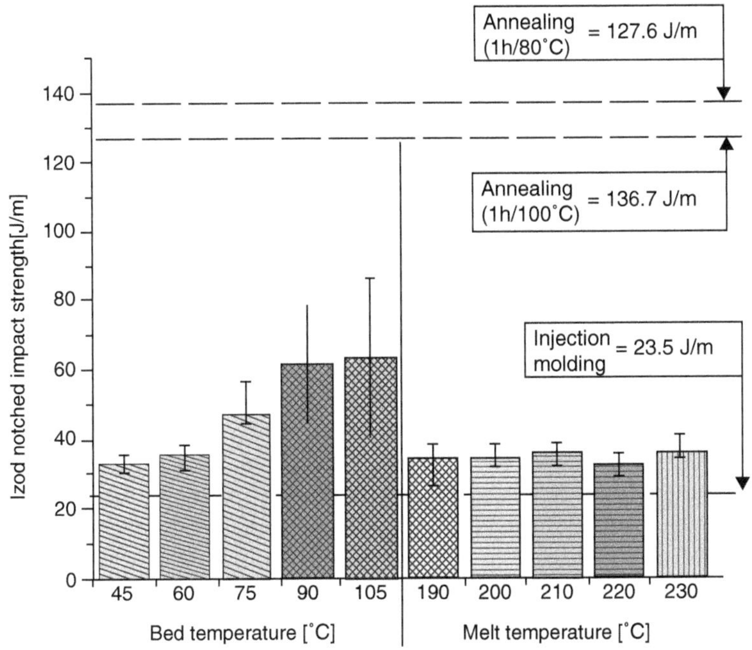

FIGURE 11.15 Illustration of impact resistance across different infill patterns [42].

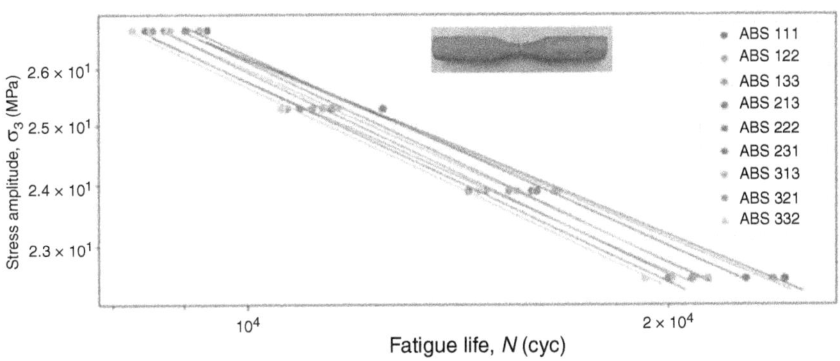

FIGURE 11.16 Fatigue performance of part printed with ABS material using FDM 3D Printer [46].

11.5 OPTIMIZATION OF PROCESS PARAMETERS

In FDM, process parameter optimization is essential to improve mechanical characteristics without sacrificing efficiency. Several optimization strategies have been developed, including the Taguchi method, response surface methodology (RSM), Design of Experiments (DOE), and genetic algorithms (GAs) [47]. The Taguchi method uses orthogonal arrays to find out parameters that cause the most significant influence upon factors such as layer height and infill density. Analysis of variance (ANOVA) analyzes the statistical significance of mechanical properties, such as surface roughness and tensile strength, whereas RSM uses quadratic modeling in the determination of optimum mechanical properties [48]. In parameter tuning, the systematic adjustment is made to one process parameter at a time, such as layer height, extrusion temperature, print speed, and bed temperature, to achieve a set mechanical outcome. By systematic variation and testing, optimum settings can be identified, as has been shown in many studies where parameter tuning has resulted in substantial improvement in mechanical properties [49].

Several instances show that the optimization of FDM parameters can effectively improve the mechanical performance. The Taguchi method optimizes parameters like layer height and infill density to increase the tensile strength and dimensional accuracy by using orthogonal arrays. In some instances, it employs a layer height of 0.1 mm; the surface quality is found to be improved while preserving the structural integrity [50]. Similarly, RSM and ANOVA are adopted for the optimization of printing parameters such as extrusion temperature and print speed. Research has found that within a range, the maximization of tensile and flexural strengths occurs with an extrusion temperature of 210°C and print speed of 40 mm/s. Multi-objective optimization methods, on the other hand including DOE and GA arrive at optimum trade-offs among conflicting goals of maximum strength with minimal material used and minimal time consumed in printing. For instance, the honeycomb infill pattern, at a density of 100%, enhances flexural strength without too much consumption of material. More advanced methods of optimization are machine learning (ML) and simulation tools that can better predict the best parameters to use for each application. Based on empirical studies, for example, a bed temperature of 90°C to 105°C reduces warping while improving adhesion, whereas reduced layer thickness improves the interlayer bonding. Taken together, these methods have shown the importance of systematic parameter tuning to improve the quality and functionality of FDM-printed parts [18]. Table 11.1 shows

TABLE 11.1
Tensile and Compression Specimens Printing Parameters for FDM 3D Printers [51]

	Materials	
Parameter	**PLA**	**ABS**
3D printer	MakerBot Replicator 5th Gen	MakerBot Replicator 2X
Infill density	100%	100%
Layer thickness	0.15 mm	0.15 mm
Number of shells	3	3
Raster orientation	0°/90° crisscross infill	0°/90° crisscross infill
Print speed	90 mm/s	90 mm/s
Support structure	—	—
Rafts	Yes	Yes
Extruder temperature	215°C	230°C

a summary of optimal parameters to maximize the mechanical performance of PLA parts printed via FDM. These settings, adopted from various optimization studies, are a layer height of 0.1 mm, print speed of 40 mm/s, extrusion temperature of 210°C, and 100% infill density with a honeycomb pattern. Such optimization strategies have improved the overall performance of FDM-fabricated parts by successfully balancing the trade-offs between different mechanical qualities [51].

11.6 CHALLENGES IN OPTIMIZING MECHANICAL PROPERTIES OF FDM-PRINTED PARTS

In pursuit of enhancing the mechanical characteristics of components fabricated using FDM, several critical challenges and obstacles persist. These challenges, ranging from material variability to process consistency, notably impact the quality and reliability of FDM-printed parts as mentioned in Table 11.2. To improve the reliability and quality of FDM-printed components, several issues must be resolved. These challenges must be addressed to fully utilize FDM technology's potential for creating dependable, high-performance components for a variety of industrial applications. This will need ongoing research and development in material science, process optimization, and predictive modeling [52].

TABLE 11.2
Comparing AM Technologies: Challenges and Material Properties [53–55]

Fabrication Method	Challenges to Mitigate	Material Variability, Scalability, and Anisotropy
Fused deposition modeling (FDM)	Layer-by-layer deposition, poor surface finish, dimensional accuracy, and limited material choices.	High material variability, limited scalability, and significant anisotropy.
Stereolithography (SLA)	Photopolymerization process, limited build volume, and potential for residual stress.	Moderate material variability, good scalability, and reduced anisotropy compared to FDM.
Selective laser sintering (SLS)	Powder-based process, part porosity, and potential for warping.	Moderate material variability, good scalability, and reduced anisotropy compared to FDM.
Digital light processing (DLP)	Photopolymerization process, limited build volume, and potential for part distortion.	Moderate material variability, good scalability, and reduced anisotropy compared to FDM.
Electron beam melting (EBM)	High-energy electron beam, high cost, and limited material choices.	Low material variability, good scalability, and reduced anisotropy.

11.6.1 MATERIAL VARIABILITY

One of the most pressing challenges in FDM is the variability in filament quality and material composition, which can lead to inconsistencies in mechanical properties. Different materials, even within the same class of thermoplastics like PLA, ABS, and polyethylene terephthalate glycol (PETG), have distinct thermal, mechanical, and rheological characteristics that affect printability and final part quality then exhibit varying responses to process parameters such as nozzle temperature, print speed, and layer height [56]. In that aspect, material choice poses a significant challenge for optimization in FDM due to the inherent variability in material properties and their interaction with process parameters. For example, PLA is easier to print and provides good surface finish, but its brittleness limits its use in applications requiring high impact strength. On the other hand, ABS has better toughness and heat resistance but requires higher bed and extrusion temperatures and, therefore, is prone to warping [57]. Such material-specific behaviors make parameter tuning complicated, as the optimal settings for one material may not work for another. For instance, extrusion temperature and bed temperature, two parameters critical in ensuring proper layer adhesion and reduction of voids, change considerably with a change in material. Further complexity arises with composite filaments that have reinforcements, such as carbon fiber (CF) or metal particles, due to the abrasive nature of the reinforcements, changes in flow dynamics, and requirements for special nozzles and settings [58]. Lastly, even within the same material, different manufacturers can still differ in properties such as diameter consistency, molecular composition, and additives, leading to non-congruous performances. These challenges underscore the need for detailed material characterization and tailored optimization strategies for each material to achieve desired mechanical properties while maintaining print efficiency [59]. This variability complicates efforts to generalize findings and reduces the reproducibility and reliability of printed parts as shown in Figure 11.17. The lack of standardized filament specifications and quality control measures further exacerbates this issue, making it difficult to achieve consistent results across different printing environments and material batches.

The image shown in Figure 11.18 provides a comparative analysis of various FDM 3D printing materials based on their visual quality, ease of printing, maximum stress, impact resistance, elongation at break, heat resistance, and layer adhesion. Each material (PLA, ABS, polyethylene terephthalate (PET), Nylon, thermoplastic polyurethane TPU, and polycarbonate (PC)) is rated on a scale of low, medium, and high for each property. The results show that PLA is generally easy to print and has good visual quality, while TPU excels in impact resistance and elongation at break. ABS and PC offer high strength and heat resistance, but may be more challenging to print. Nylon provides a balance of properties, with medium ratings across most categories. PET is generally easy to print but may have lower strength and impact resistance compared to other materials [60]. In general, the material selection is based on the particular needs of the 3D-printed component, including its mechanical characteristics, processing ease, and intended usage.

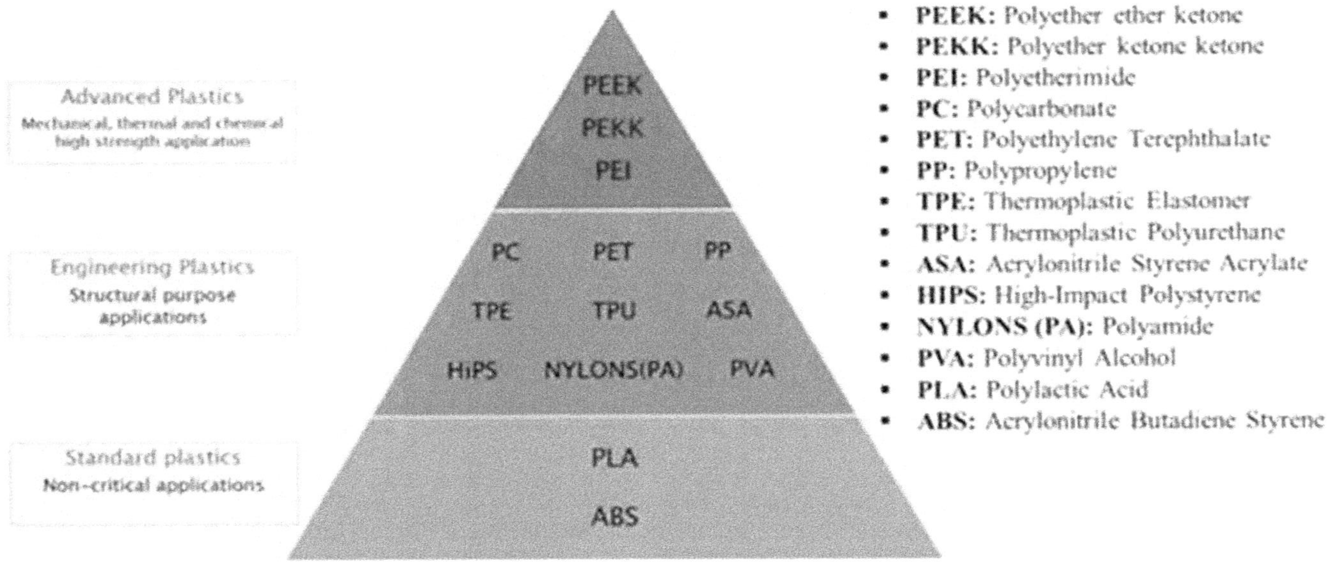

FIGURE 11.17 A pyramid of plastics: from standard to advanced showing the material versatility in FDM 3D printing technology.

Source: https://weerg.com/guides/best-fdm-3d-printing-filament-materials-compared

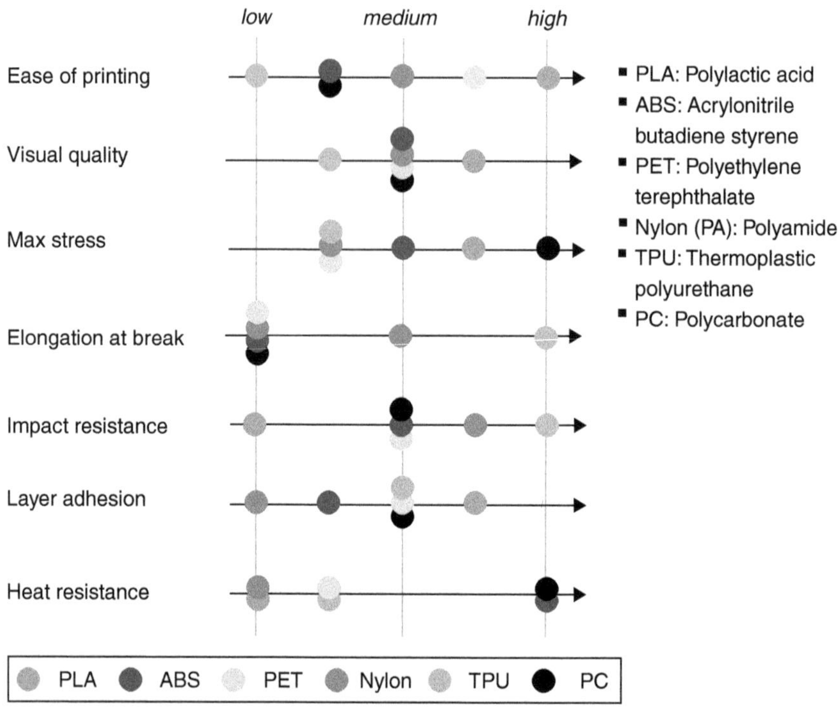

FIGURE 11.18 Effect of material variability on mechanical consistency.

Source: www.hubs.com/knowledge-base/fdm-3d-printing-materials-compared/

11.6.2 Anisotropy

Anisotropy, a property arising from the layer-by-layer deposition process in FDM, poses a significant challenge in achieving uniform mechanical properties. FDM parts tend to exhibit directional dependence of mechanical properties, meaning that their strength, durability, and performance can vary depending on the axis along which they are measured [61]. This anisotropic behavior limits the performance of parts under multi-directional loads and makes it difficult to predict and optimize mechanical properties [62]. While efforts have been made to optimize build orientation and layer adhesion techniques to minimize anisotropy, achieving isotropic properties remains an ongoing area of research.

11.6.3 Scalability

The scaling up of FDM processes to large production and the retention of mechanical integrity is another big challenge. There are intrinsic drawbacks: FDM has lower production speeds, and in larger parts, the strength is quite weaker compared to other techniques. It is much important to develop multi-nozzle systems and improve the process monitoring techniques; on the other hand, assurance of mechanical properties and structural integrity at a larger scale is in itself a big challenge.

11.6.4 Process Consistency

The consistency in the process is of utmost importance to achieve quality and reliability in FDM-printed parts. Variations in machine calibration, material batches, and environmental conditions like humidity and temperature may lead to variations in the quality of parts and their mechanical properties. These lead to repeatability and reliability issues, thus making it very hard to maintain a stable manufacturing process. This is further compounded by a lack of standardized testing methodologies for the determination of mechanical characteristics, as this would hinder the creation of universal parameter recommendations.

11.6.5 Predictive Modeling and Long-Term Performance

Uneven component quality follows from the incomplete knowledge about complex interactions between process parameters and mechanical properties in FDM. The development of more sophisticated predictive models and simulation tools is necessary to have a good prediction of influence of process parameters on the mechanical properties. Another open issue is that there is a lack of information regarding long-term performance, aging effects of FDM-printed parts, durability, and reliability over time.

11.7 CASE STUDIES

FDM has been getting wider recognition because of its versatility in the production of functional parts for so many sectors of industry; improvement in process parameters enhances mechanical properties and surface quality, improving generally the performance of these 3D-printed components [47]. Some case studies discussed below come from various other industry fields, through which again process parameter settings bring out the purposes needed. The section points out real-world applications where optimized FDM process parameters were successfully used in the production of high-performance parts.

11.7.1 Aerospace Applications: Optimizing Layer Height and Infill Density for Lightweight Components

In the aerospace industry, FDM has been an effective means of manufacturing lightweight and complex drone components, particularly through infill pattern optimization to maximize the strength-to-weight ratio, as seen in Figure 11.19. In a case study on the production of drone parts such as frames, propeller mounts, and camera supports, FDM was used to achieve a substantial reduction in weight while maintaining mechanical integrity [63]. Further experiments, using different patterns of the infill, such as grid, honeycomb, and triangular, and playing with the densities of infills, showed that the 50% honeycomb infill gives a good compromise in strength-to-weight ratio [64]. Resulting parts have about a 35% weight reduction but maintain tensile strength at 95% that of the fully solid counterparts. The optimized components of the drone therefore showed a 40% increase in strength-to-weight ratio, which enabled improved

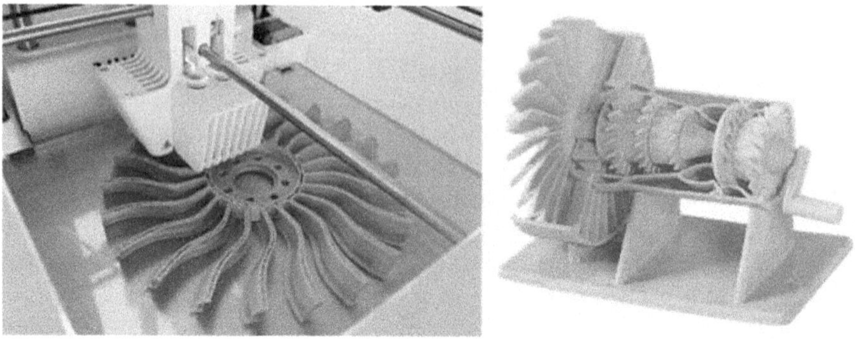

FIGURE 11.19 Aerospace component manufactured via optimized FDM parameters.

Source: https://researchdive.com/blog/Aerospace-3D-Printing and https://dreambot3d.com/3d-printer-for-aerospace/

flight performance with increased flight time by 20% and a payload capacity of 15% [65]. Material consumption decreased by 40%, while printing time dropped by 30% for the infill patterns after optimization. The FDM process thus proves to be not only effective in improving mechanical properties but also cost- and time-efficient [66]. This case study underscores the potential of FDM in aerospace applications, particularly for creating unmanned aerial vehicle (UAV) components, where both performance and efficiency are paramount.

11.7.2 Automotive Industry: Improving Surface Finish and Dimensional Accuracy for Engine Prototype Components

In the automotive industry, FDM has been increasingly utilized to produce prototype components with enhanced surface finish and dimensional accuracy as shown in Figure 11.20. A notable case study involved the production of custom engine covers and intake manifolds, where achieving tight tolerances and smooth surfaces was critical for evaluating design feasibility. Researchers focused on optimizing key process parameters such as layer height, print speed, and extrusion temperature [67]. It was improved by reducing the layer height to 0.1 mm and keeping the extrusion temperature under good control at 210°C. This has greatly reduced the surface roughness, therefore reducing post-processing requirements. In addition, slower print speeds were used to achieve better layer adhesion, which minimized dimensional deviations and ensured that the prototype components were very close to the design specifications [68]. The optimized FDM process has produced prototype engine components with surface roughness values as low as 2.5 μm and dimensional accuracy within ±0.1 mm, which are highly effective for testing and validation in automotive applications. This case study shows that FDM is capable of producing high-quality prototypes that yield meaningful insight into design performance and allow the automotive industry to have faster development cycles [69].

FIGURE 11.20 Automotive parts improved by optimized FDM parameters.

Source: https://rapiddirect.com/blog/3d-printing-for-automotive-applications/

11.7.3 Medical Applications: Enhancing Flexural Strength for Prosthetic Devices

In the medical field, FDM has already been successfully applied to help improve the flexural strength of prosthetic devices so that they could provide durability as well as comfort for users (Figure 11.21). This case study optimized the lower limb prosthetics production process, in which the flexural strength is highly necessary to undergo the repeated stresses experienced by the user throughout daily movements. Researchers studied different combinations of process parameters, infill density, layer height, and print orientation [70]. By increasing the infill density to 80% and reducing the layer height to 0.15 mm, they achieved a significant improvement in the flexural strength of the prosthetics. Additionally, printing with a horizontal orientation further enhanced the mechanical properties by ensuring better layer adhesion and load distribution [71]. The optimized FDM process resulted in prosthetic devices with flexural strength values that met industry standards, providing users with reliable and long-lasting solutions. This case study highlights the potential of FDM in producing high-quality, customized prosthetic devices that combine strength with user-specific design requirements, making it a valuable tool in the medical industry, as illustrated in Figure 11.22.

11.7.4 Consumer Goods: Customization of Wearable Devices

Customization of wearable devices through FDM has been significantly enhanced by optimizing print temperature and filament type, as illustrated in Figure 11.23. A case study focused on creating personalized wearable health monitoring devices, where comfort and durability are critical. By experimenting with different thermoplastic

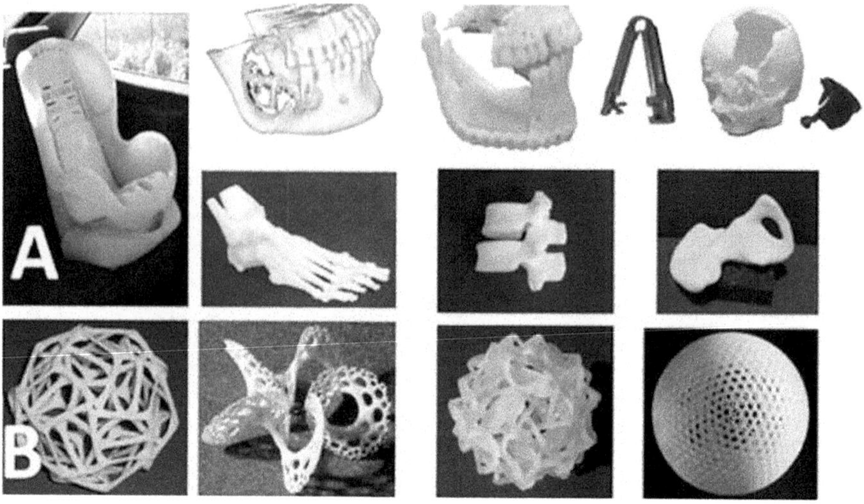

FIGURE 11.21 Illustrates various biomedical applications in FDM 3D printing [72].

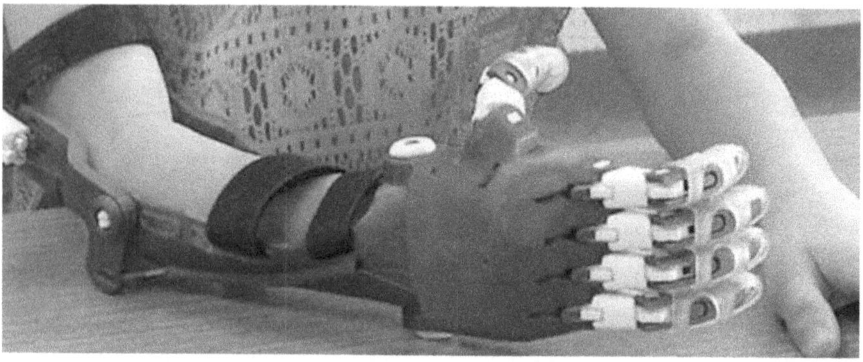

FIGURE 11.22 Customized medical prosthetic manufactured via FDM.

Source: https://sunrise-metal.com/medical-3d-printing/

filaments, such as TPU and PETG, and fine-tuning print temperatures, researchers were able to tailor the mechanical properties of the devices to meet specific user needs [73]. Higher print temperatures ensured better layer adhesion and flexibility, crucial for devices worn close to the skin, while the choice of filament type influenced the overall feel and durability. TPU, known for its flexibility, was optimized at a slightly lower temperature to prevent warping while maintaining the material's elasticity. The result was a range of wearable devices that not only fit comfortably but also exhibited enhanced durability and resilience, meeting the exacting demands of customized medical applications. This approach underscores the importance of process parameter optimization in delivering personalized, high-performance wearable devices through FDM [74].

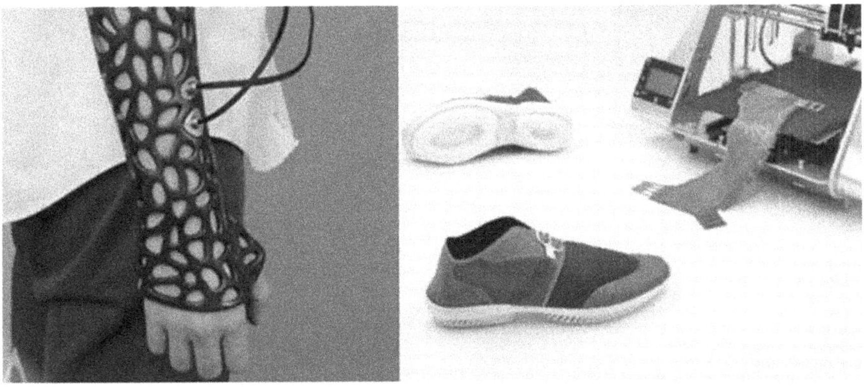

FIGURE 11.23 Consumer goods applications benefiting from FDM parameter optimization.

Source: https://insider-trends.com/top-40-on-demand-3d-printed-products-in-retail/

11.7.5 Architecture and Construction: Structural Integrity of Scaled Models

In the architecture and construction industry, FDM has been leveraged to create scaled models that require a balance between structural integrity and material efficiency. A notable case study involves the application of multi-objective optimization techniques to enhance the mechanical properties of these models while minimizing material usage [75]. By optimizing key FDM process parameters such as layer height, infill density, and print speed, the structural integrity of the models was significantly improved, ensuring they could accurately represent the load-bearing characteristics of the full-scale structures. This approach allowed architects and engineers to produce detailed, robust models that not only served as visual and spatial references but also served as functional prototypes for testing structural concepts. The application of multi-objective optimization in FDM, supported by architectural applications, is of great potential to demonstrate the ability of AM to contribute to more effective and sustainable construction practices [76].

Figure 11.24 shows the 3D concrete printer in action as it lays down successive layers of concrete to build up a wall. It symbolizes how 3D printing can be transformational within the construction industry. Automation in the building process significantly reduces construction time and associated labor costs. It saves on material waste by the deposition of precise quantities and allows creating complex architectural designs that might otherwise be unachievable or very costly to replicate using conventional means [77]. In addition, the use of alternative materials featuring better sustainability, such as recycled aggregates, becomes available. The innovation is exemplified in how AM has changed contemporary construction with more efficiency, flexibility, and sustainability. Figure 11.25 displays a complicated 3D-printed architectural model of a modern building complex. The model has complex details, including multiple levels, a curved facade, and structural elements inside. This level of precision allows architects and engineers to assess the structural integrity of the

FIGURE 11.24 A 3D printer for building creates structural components.

Source: https://news.cornell.edu/stories/2022/09/first-its-kind-3d-printed-home-blends-concrete-wood

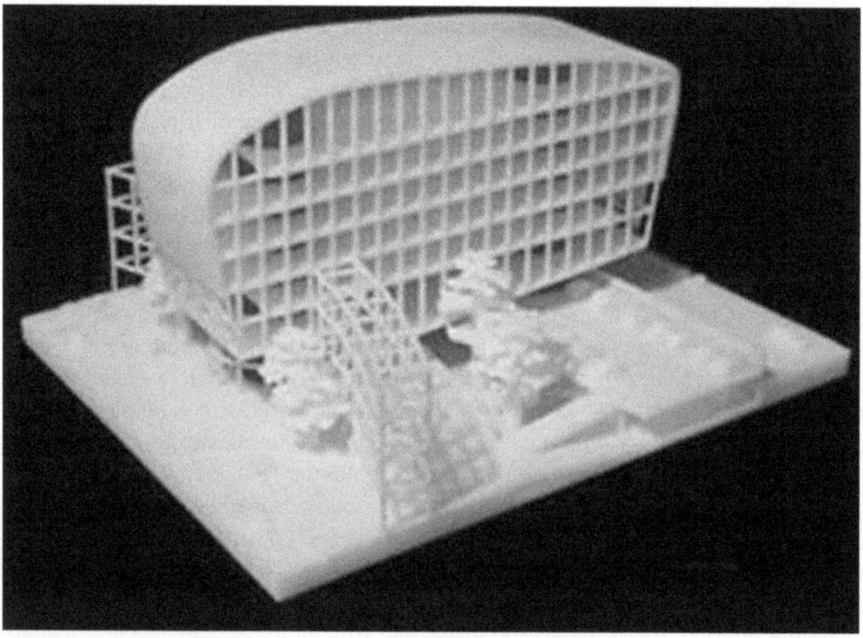

FIGURE 11.25 Visualization of 3D complex building designs.

Source: https://stratasys.com/br/resources/blog/3d-printed-architecture/

design, identify potential issues, and optimize the building's layout and functionality before actual construction begins. 3D printing technology enables the creation of highly accurate and complex models, facilitating informed decision-making and reducing risks in the construction process [78].

11.8 FUTURE DIRECTIONS

The future of FDM technology is set to transcend current limitations through advancements in materials, dynamic process control, hybrid manufacturing techniques, and real-time monitoring systems. By embracing these innovations, FDM is poised to significantly enhance the mechanical properties of printed parts, pushing the boundaries of what this technology can achieve. These developments will collectively enable the production of superior components, opening new avenues for FDM's application in various high-demand industries, from aerospace to medical devices, where precision, strength, and efficiency are paramount.

11.8.1 Advanced Materials

The evolution of FDM will increasingly focus on the integration of advanced composite materials and high-performance thermoplastics, as illustrated in Figure 11.26. The incorporation of materials such as CF-reinforced PLA and metal-infused filaments can drastically improve the mechanical properties of printed parts, offering enhanced strength, thermal stability, and electrical conductivity. Future research will focus on optimizing process parameters such as temperature, print speed, and layer height for emerging materials to fully harness their potential and achieve enhanced mechanical properties and functional performance. This advancement will not only improve the mechanical performance of FDM-printed components but also expand their applicability in high-demand sectors like aerospace, automotive, and medical devices.

Figure 11.27 demonstrates a novel approach to fabricating 3D-printed electrodes for energy storage devices. The process involves creating a slurry of conductive materials (PLA pellets, graphite powder, solvent, and plasticizer), casting it into a thin film, and then extruding the film into filaments for 3D printing. The resulting 3D-printed electrodes exhibit improved performance compared to traditional manufacturing methods. 3D-printed electrodes outperform traditional manufacturing methods by offering enhanced customization, allowing precise control over geometry to increase surface area and improve charge transfer. The incorporation of materials like graphene enhances conductivity, while the process minimizes waste and enables complex designs, such as porous structures, which optimize energy storage and overall efficiency. These features result in superior energy density, better mechanical stability, and cost-effectiveness. This technique holds promise for advancing the development of high-performance batteries and other energy storage devices. Other examples for advanced material 3D printing are CF-reinforced PLA and metal-infused filaments, which enhance strength, thermal stability, and electrical conductivity.

Figure 11.27 illustrates an integrated process for fabricating 3D-printed graphene-based composites, focusing on their potential use in energy storage devices. The

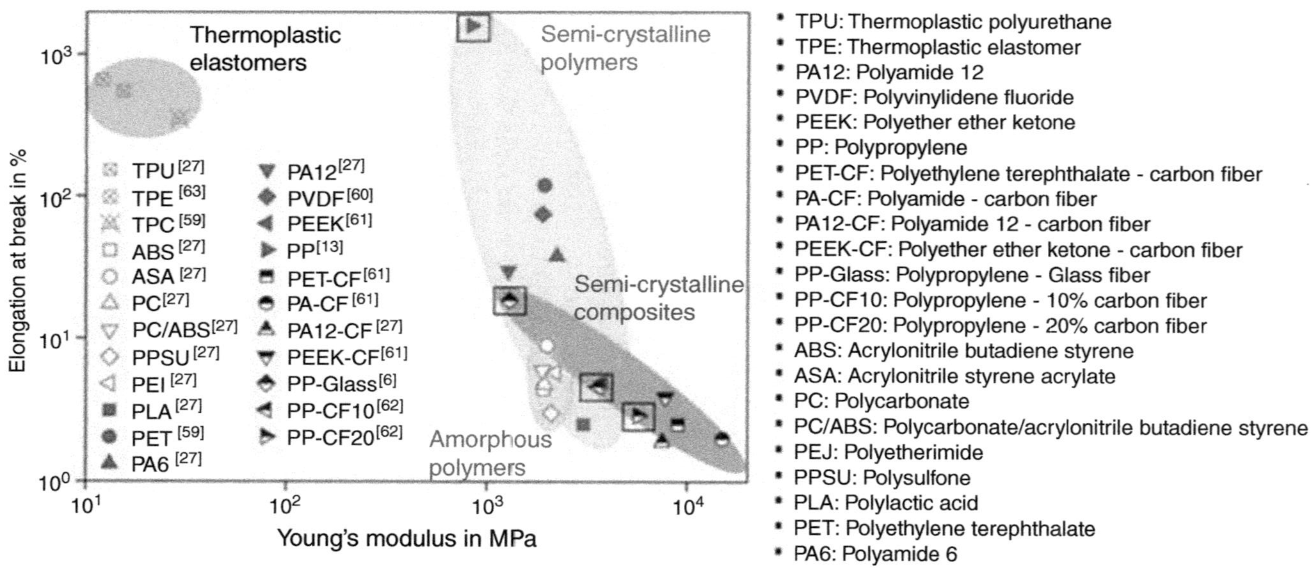

FIGURE 11.26 Illustrating material guide for FDM 3D printing for printing various advanced materials.

Source: https://weerg.com/guides/best-fdm-3d-printing-filament-materials-compared

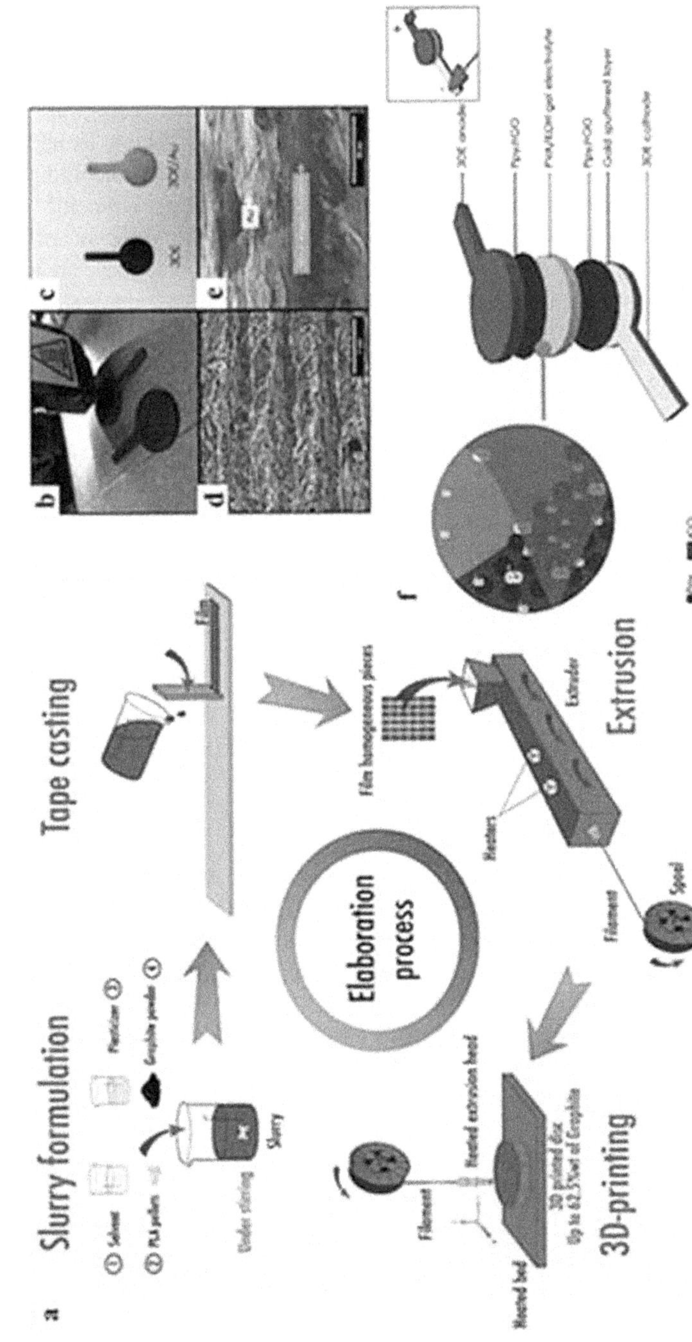

FIGURE 11.27 Advanced composite materials in FDM 3D printing [79].

process begins with slurry formulation, where PLA pellets, graphite powder, a solvent, and a plasticizer are mixed under continuous stirring to ensure uniformity. This homogeneous slurry is then tape-cast into thin films, which are subsequently cut and extruded into filaments suitable for 3D printing. These filaments are used in an FDM-based 3D printer, where a heated nozzle deposits the material layer by layer onto a heated bed to create structures such as a graphite-rich disc. Further post-processing techniques, such as metal sputtering, enhance the conductivity and overall performance of the printed components. The 3D-printed electrodes, including anodes and cathodes, are assembled with a gel electrolyte and a current collector to create functional energy storage devices. This process well demonstrates the versatility and efficiency of 3D printing in developing high-performance graphene-based composites with applications ranging from batteries to supercapacitors. Combination of material tailoring and geometrical flexibility puts the present approach at the forefront of most advanced energy storage applications [79].

11.8.2 Process Monitoring and Control

One step further, this FDM technology integrates real-time monitoring and adaptive control systems. The ML algorithms use sensor-based feedback loops to adjust the process parameters on the go during printing. Such optimization guarantees better consistency of properties and better mechanical performance from one print to another while reducing defects, which, in general, improves the quality of the produced parts by FDM. Thus, it will enable manufacturers to achieve tighter tolerances and more reliable outputs, even for complicated geometries or difficult materials. The actual experimental setup for the data acquisition using the FDM 3D Printer Creality CS-10 and the data acquisition equipment, DAQ MX 6001, is presented in Figure 11.28.

11.8.3 Hybrid Manufacturing

The hybrid manufacturing approaches combining FDM with other processes like computer numerical control (CNC) machining, resin curing, or multi-material printing will redefine the capabilities of AM. With these integrations, makers are able to make parts with better precision and improved mechanical properties. Makers can also use CNC machining to fine-tune the critical dimensions after an initial FDM print, or they can use resin curing to harden and smooth the surface of a part. These hybrid methods will pave the way for the production of industrial standard parts and make FDM a far more viable option for end-use parts where high reliability and durability are demanded [81].

Figure 11.29 shows a hybrid 3D printing system combining vat photopolymerization (VPP) with direct ink writing (DIW) methods. First, a dispensing syringe deposits material A onto a photocurable liquid matrix in a tank, while a scanning laser solidifies the material; the same is repeated with material B. Once one layer is formed, the building platform moves down to transfer the layer to the previously built layers; another light exposure is used for curing the new layer. The system is composed of a multi-material DIW printhead, a laser scanning-based VPP module, and a

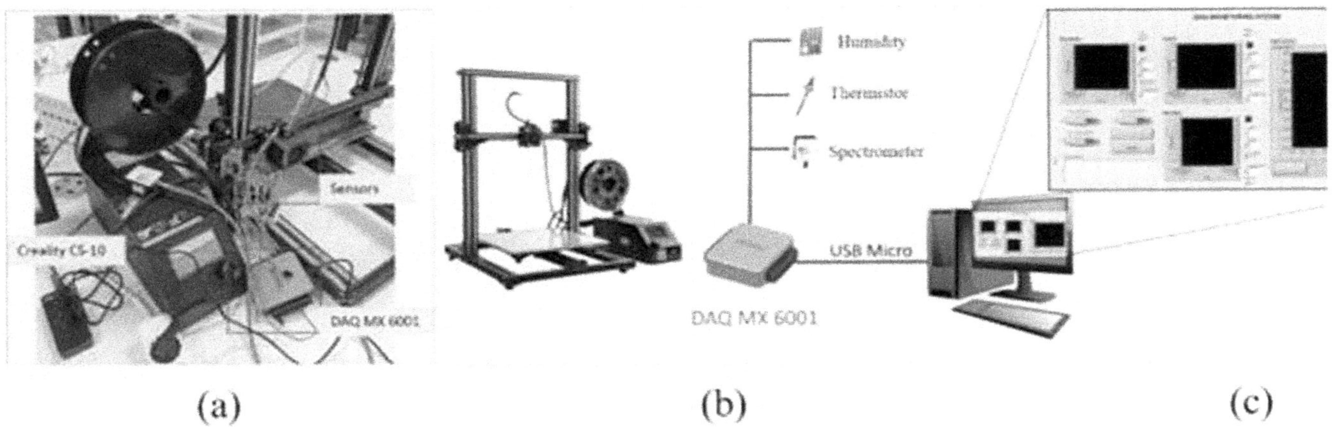

FIGURE 11.28 Emerging technologies in process monitoring for FDM shown for better consistency and improved mechanical performance with ML algorithms and integration of sensors [80]. (a) DOE setup including sensors collected for processing in the proposed MH-ED-TCN system. (b) The graphical user interface of developed monitoring system software.

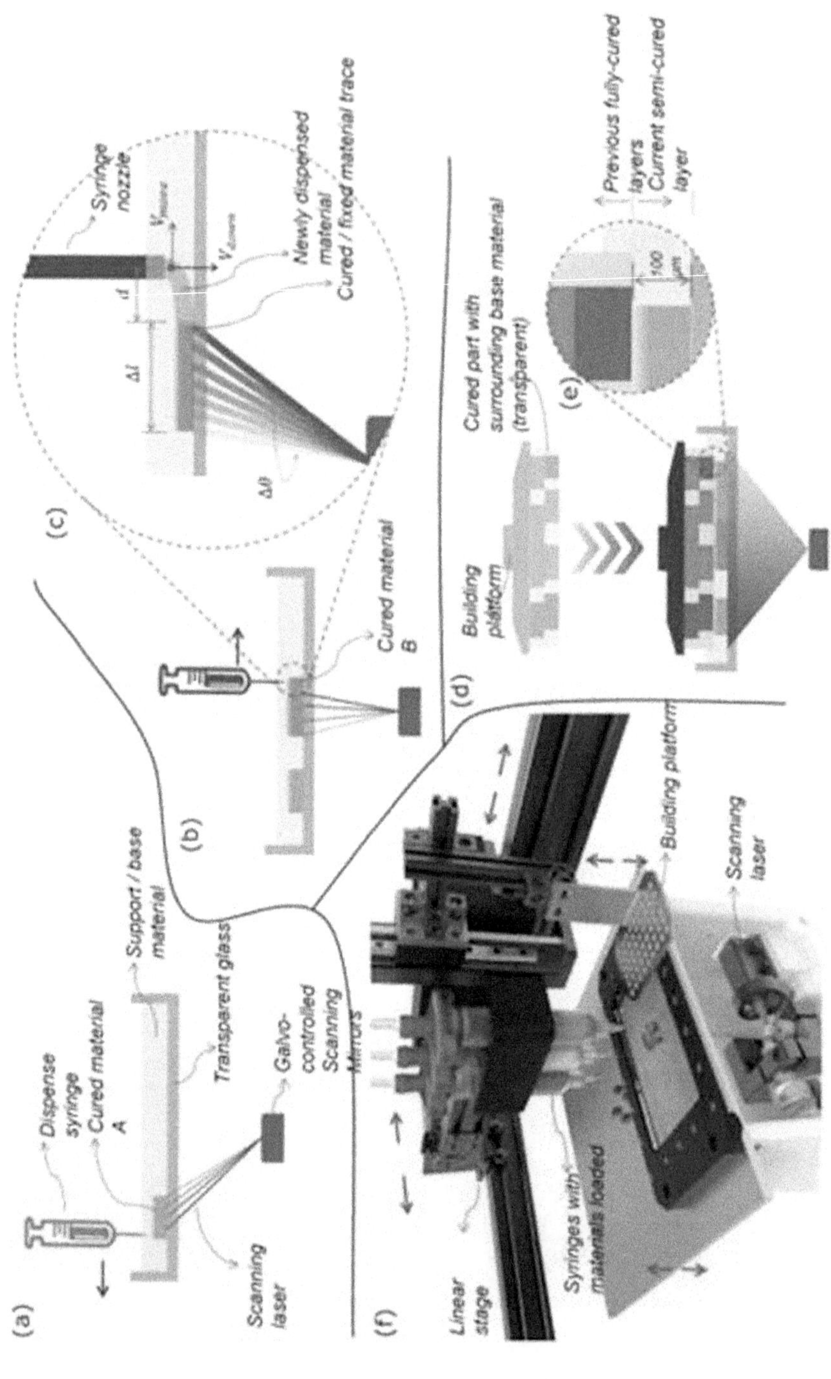

FIGURE 11.29 Multi-material printing for higher precision and improved mechanical performance through hybrid manufacturing processes [82].

three-dimensional movable building platform. Such a hybrid approach allowed the fabrication of complex structures with multiple materials and properties [82].

11.8.4 Real-Time Process Monitoring

One of the critical development areas for the future of FDM is the creation of real-time process monitoring systems. These systems would continuously track and adjust parameters such as temperature, speed, and extrusion rates during the printing process, responding instantly to any deviations from the desired outcomes. Such capability would significantly enhance the consistency, accuracy, and overall quality of FDM parts, making it possible to produce complex, high-performance components with reduced risk of defects. This advancement will be particularly valuable in applications where precision and reliability are critical, such as in aerospace or medical device manufacturing [83].

11.8.5 Smart Materials and Functional Printing

The development of smart or stimuli-responsive materials will expand the scope of FDM, as mentioned in Figure 11.30. These materials can change properties in response to external stimuli such as temperature, light, or moisture, leading to self-healing, shape-shifting, and self-assembling capabilities in printed parts. This will further enhance FDM's applicability in biomedical devices and soft robotics [84].

11.8.6 Multi-Dimensional Printing (4D and 5D Printing)

As shown in Figure 11.31, an extension of 3D printing, 4D printing allows things to change over time in reaction to external stimuli, and 5D printing, which incorporates movement and improved layer alignment, will revolutionize the possibilities for FDM, allowing for more resilient and adaptable parts that can function in complex, real-world environments [85].

11.8.7 Sustainability and Eco-Friendly Materials

The incorporation of biodegradable filaments and recyclable thermoplastics into FDM processes will pave the way for more eco-friendly manufacturing solutions, as shown in Figure 11.32. Research will focus on developing sustainable materials and reducing waste through more efficient printing technologies [86].

11.9 CONCLUSION

This chapter has comprehensively reviewed the critical impact of various FDM process parameters on the mechanical properties of build parts. Both the literature analysis and study findings highlight those parameters such as layer height, print speed, extrusion temperature, infill density, and print orientation, which play pivotal roles in determining tensile strength, flexural strength, impact resistance, and

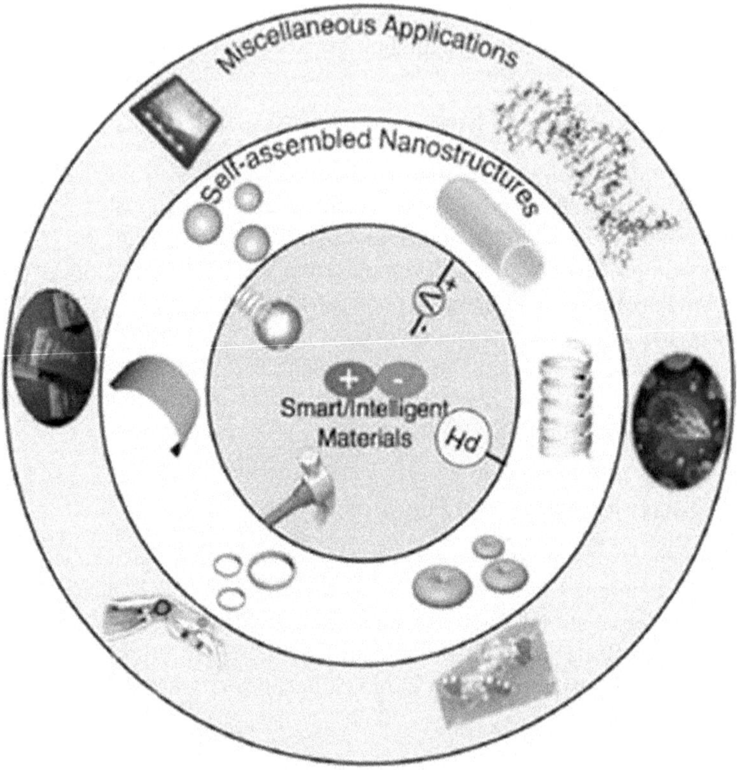

FIGURE 11.30 Intelligent materials and their applications in 3D printing using FDM [84].

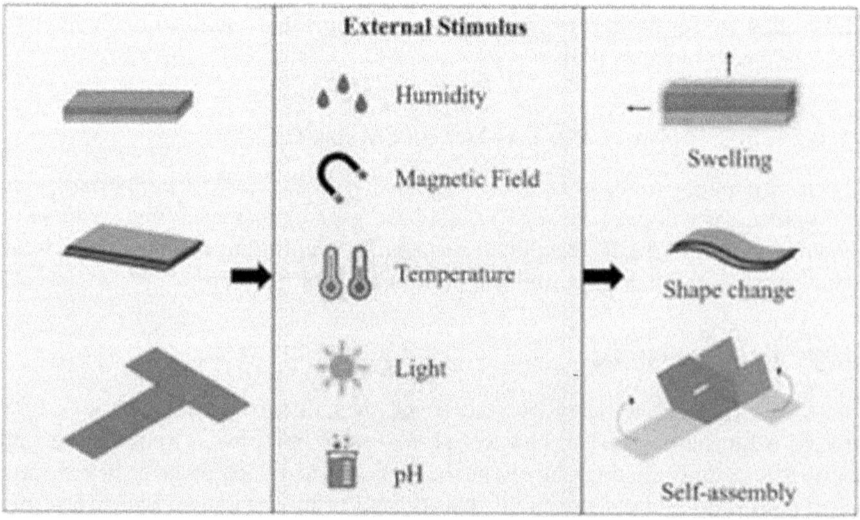

FIGURE 11.31 Different stimuli and reactions seen in materials with many dimensions [85].

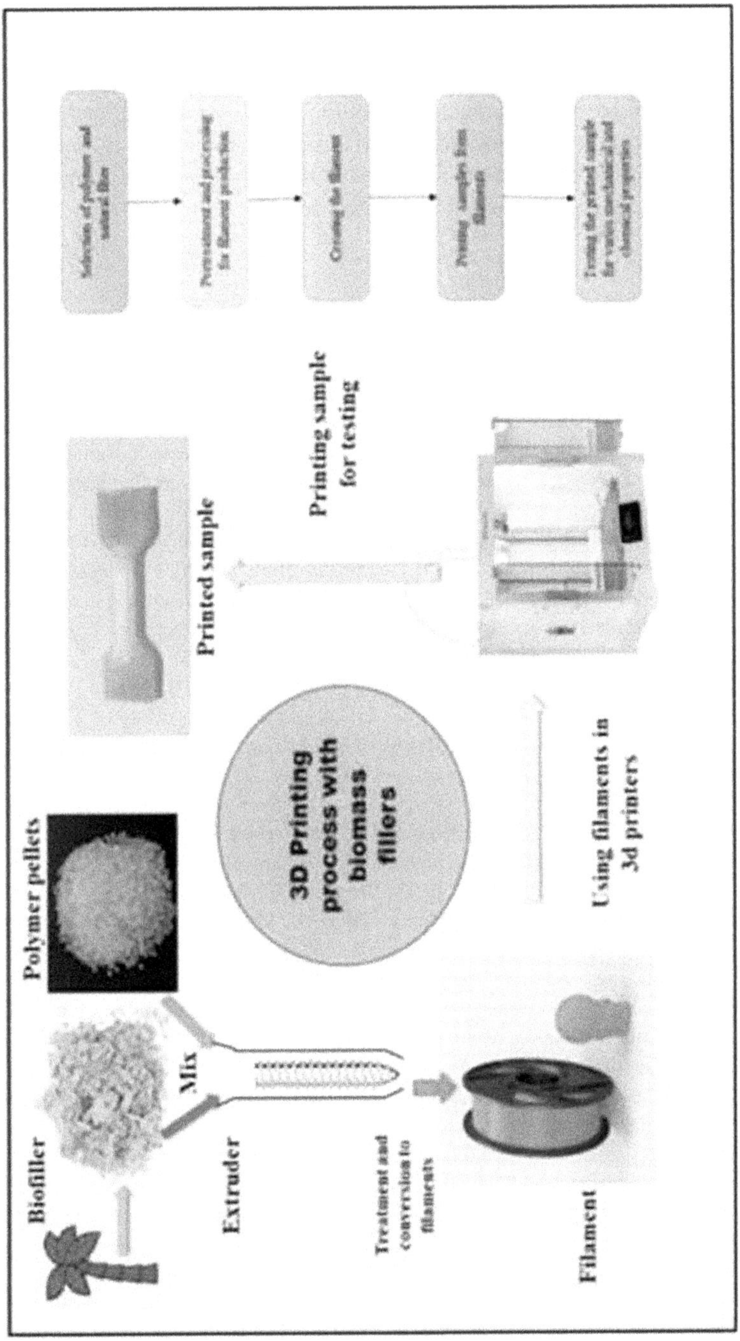

FIGURE 11.32 Using natural fibers in FDM 3D printing techniques to create biomass composite [86].

overall durability of FDM-fabricated components. Thinner layers enhance quality but increase print time, while higher infill densities boost strength at the cost of additional material usage and longer production times. Methods like the Taguchi method, RSM, ANOVA, and ML are instrumental in identifying optimal settings. Despite challenges related to material variability, anisotropy, and scalability, ongoing advancements in real-time process control, mathematical modeling, and hybrid manufacturing show promise for improving FDM consistency and quality. Continued research and development are essential to fully harness FDM technology's potential for producing high-performance, reliable parts across diverse industries.

REFERENCES

1. K. S. Boparai and R. Singh, "Advances in fused deposition modeling," in *Reference Module in Materials Science and Materials Engineering*, Elsevier, 2017.
2. D. K. Bagal, A. Barua, A. K. Pattanaik, S. Jeet, and D. Patnaik. (2020), "Parametric optimization based on mechanical characterization of fused deposition modelling fabricated part using utility concept," in S. Singh, C. Prakash, S. Ramakrishna, and G. Krolczyk (eds), *Advances in Materials Processing, Lecture Notes in Mechanical Engineering*, Springer, pp. 313–325, https://doi.org/10.1007/978-981-15-4748-5_30.
3. S. Sahoo, H. Sutar, P. Senapati, B. Shankar Mohanto, P. Ranjan Dhal, and S. Kumar Baral, "Experimental investigation and optimization of the FDM process using PLA," *Mater. Today Proc.*, vol. 74, pp. 843–847, 2023, doi: 10.1016/j.matpr.2022.11.208.
4. J. Giri, A. Chiwande, Y. Gupta, C. Mahatme, and P. Giri, "Effect of process parameters on mechanical properties of 3D printed samples using FDM process," *Mater. Today Proc.*, vol. 47, pp. 5856–5861, 2021, doi: 10.1016/j.matpr.2021.04.283.
5. M. Kam, A. İpekçi, and Ö. Şengül, "Investigation of the effect of FDM process parameters on mechanical properties of 3D printed PA12 samples using Taguchi method," *J. Thermoplast. Compos. Mater.*, vol. 36, no. 1, pp. 307–325, 2023, doi: 10.1177/08927057211006459.
6. A. D. Tura and H. B. Mamo, "Characterization and parametric optimization of additive manufacturing process for enhancing mechanical properties," *Heliyon*, vol. 8, no. 7, p. e09832, 2022, doi: 10.1016/j.heliyon.2022.e09832.
7. B. N. Dhanunjayarao, N. V. Swamy Naidu, R. Suresh Kumar, Y. Phaneendra, B. Sateesh, J. L. Olajide, and E. R. Sadiku, "3D printing of fiber reinforced polymer nanocomposites: additive manufacturing," in *Handbook of Nanomaterials and Nanocomposites for Energy and Environmental Applications*, Cham: Springer International Publishing, 2020, pp. 1–29, doi:10.1007/978-3-030-11155-7_166-1.
8. D. G. Zisopol, I. Nae, A. I. Portoaca, and I. Ramadan, "A theoretical and experimental research on the influence of FDM parameters on tensile strength and hardness of parts made of polylactic acid," *Eng. Technol. Appl. Sci. Res.*, vol. 11, no. 4, pp. 7458–7463, 2021, doi: 10.48084/etasr.4311.
9. O. Y. Venkatasubbareddy, P. Siddikali, and S. M. Saleem, "Improving the dimensional accuracy and surface roughness of FDM parts using optimization techniques," *IOSR J. Mech. Civ. Eng.*, vol. 16, no. 053, pp. 18–22, 2016, doi: 10.9790/1684-16053041822.
10. A. M. Nazri, I. M. Ridzwan, M. T. Mastura, M. Faizal, L. Zulkiflle, A. Lukista Debby Dyne, I. Ghazali Ihwan, "Application of Taguchi method to optimize the parameter of fused deposition modeling (FDM) using oil palm fiber reinforced thermoplastic

composites," *Polymers (Basel)*, vol. 14, no. 11, p. 2140, 2022, doi: 10.3390/polym14112140.

11. N.V.S.S. Sagar and K. S. Vepa (2019), "Experimental investigations for improving the strength of parts manufactured using FDM process, in: U. Chandrasekhar, L. J. Yang, , and S. Gowthaman (eds), *Innovative Design, Analysis and Development Practices in Aerospace and Automotive Engineering (I-DAD 2018)*, Lecture Notes in Mechanical Engineering, Springer, Singapore, pp. 307–313, https://doi.org/10.1007/978-981-13-2697-4_34.
12. C. Palanisamy, R. Raman, and P. Kumar Dhanraj, "Additive manufacturing: a review on mechanical properties of polyjet and FDM printed parts," *Polym. Bull.*, vol. 79, no. 9, pp. 7065–7116, 2022, doi: 10.1007/s00289-021-03899-0.
13. S. Wickramasinghe, T. Do, and P. Tran, "FDM-based 3D printing of polymer and associated composite: a review on mechanical properties, defects and treatments," *Polymers (Basel)*, vol. 12, no. 7, pp. 1–42, 2020, doi: 10.3390/polym12071529.
14. V. A. Yiga, M. Lubwama, S. Pagel, P. W. Olupot, J. Benz, and C. Bonten, "Optimization of tensile strength of PLA/clay/rice husk composites using Box-Behnken design," *Biomass Convers. Biorefinery*, vol. 13, no. 13, pp. 11727–11753, 2023, doi: 10.1007/s13399-021-01971-3.
15. M. R. Hasan, I. J. Davies, A. Pramanik, M. John, and W. K. Biswas, "Potential of recycled PLA in 3D printing: a review," *Sustain. Manuf. Serv. Econ.*, vol. 3, p. 100020, 2024, doi: 10.1016/j.smse.2024.100020.
16. T. Gajjar, R. Yang, L. Ye, and Y. X. Zhang, "Effects of key process parameters on tensile properties and interlayer bonding behavior of 3D printed PLA using fused filament fabrication," *Prog. Addit. Manuf.*, vol. 10, no. 2, pp. 1261–1280, 2024, doi: 10.1007/s40964-024-00704-y.
17. I. Tharazi, A. B. Sulong, F. Mohd Salleh, A. H. Abdullah, and N. Ismail, "Application of response surface methodology for parameters optimization in hot pressing kenaf reinforced biocomposites," *J. Mech. Eng.*, vol. 17, no. 3, pp. 131–144, 2020, doi: 10.24191/jmeche.v17i3.15318.
18. L. Auffray, P.-A. Gouge, and L. Hattali, "Design of experiment analysis on tensile properties of PLA samples produced by fused filament fabrication," *Int. J. Adv. Manuf. Technol.*, vol. 118, no. 11–12, pp. 4123–4137, 2022, doi: 10.1007/s00170-021-08216-7.
19. L. Wu, K. Yick, S. Ng, and J. Yip, "Application of the Box-Behnken design to the optimization of process parameters in foam cup molding," *Expert Syst. Appl.*, vol. 39, no. 9, pp. 8059–8065, 2012, doi: 10.1016/j.eswa.2012.01.137.
20. M. Ramesh and K. Panneerselvam, K., "PLA-based material design and investigation of its properties by FDM, in: M. Shunmugam and M. Kanthababu (eds), *Advances in Additive Manufacturing and Joining*, Lecture Notes on Multidisciplinary Industrial Engineering, Springer, Singapore, pp. 229–241, 2020, https://doi.org/10.1007/978-981-32-9433-2_20.
21. M. Abouelmajd, A. Bahlaoui, I. Arroub, M. Zemzami, N. Hmina, M. Lagache, and S. Belhouideg, "Experimental analysis and optimization of mechanical properties of FDM-processed polylactic acid using Taguchi design of experiment," *Int. J. Simul. Multidiscip. Des. Optim.*, vol. 12, p. 30, 2021, doi: 10.1051/smdo/2021031.
22. B. N. Dhanunjayarao and N. V. S. Naidu, "Parameter optimization for dimensional accuracy of VAT photopolymerization 3D printed part: Taguchi approach," *AIP Conf. Proc.*, vol. 040004, p. 040004, 2023, doi: 10.1063/5.0186171.
23. K. Kishore and M. K. Sinha, "A state-of-the-art review on fused deposition modelling process," in: R. M. Singari, K. Mathiyazhagan, and H. Kumar (eds), *Advances*

in Manufacturing and Industrial Engineering, ICAPIE 2019, Lecture Notes in Mechanical Engineering, Springer, Singapore, pp. 229–241, 2021, https://doi.org/10.1007/978-981-15-8542-5_75.
24. M. Lalegani Dezaki and M. K. A. Mohd Ariffin, "The effects of combined infill patterns on mechanical properties in FDM process," *Polymers (Basel)*, vol. 12, no. 12, p. 2792, 2020, doi: 10.3390/polym12122792.
25. M. S. Kamer and O. Dogan, "Effects of infill pattern and compression axis on the compressive strength of the 3D-printed cubic samples," *Mater. Test.*, vol. 66, no. 8, pp. 1241–1250, 2024, doi: 10.1515/mt-2024-0037.
26. F. Bandinelli, M. Scapin, and L. Peroni, "Effects of anisotropy and infill pattern on compression properties of 3D printed CFRP: mechanical analysis and elasto-plastic finite element modelling," *Rapid Prototyp. J.*, vol. 30, no. 11, pp. 142–158, 2024, doi: 10.1108/RPJ-11-2023-0385.
27. S. Patel, S. Gupta, H. Saket, K. Bakna, S. S. Patel, S. Kumar, V. Ramakoteswara Rao, R. K. Mandava, "Effect of infill pattern on the mechanical properties of PLA and ABS specimens prepared by FDM 3D printing," *Proceedings of the Institution of Mechanical Engineers, Part E: Journal of Process Mechanical Engineering*. 2024;0(0), doi:10.1177/09544089241258744.
28. M. Daly, M. Tarfaoui, M. Bouali, and A. Bendarma, "Effects of infill density and pattern on the tensile mechanical behavior of 3D-printed glycolyzed polyethylene terephthalate reinforced with carbon-fiber composites by the FDM process," *J. Compos. Sci.*, vol. 8, no. 4, p. 115, 2024, doi: 10.3390/jcs8040115.
29. R. Alkabbanie, B. Aktas, G. Demircan, and S. Yalcin, "Short carbon fiber-reinforced PLA composites: influence of 3D-printing parameters on the mechanical and structural properties," *Iran. Polym. J.*, vol. 33, no. 8, pp. 1065–1074, 2024, doi: 10.1007/s13726-024-01315-8.
30. V. Mishra, S. Negi, S. Kar, A. K. Sharma, Y. N. K. Rajbahadur, and A. Kumar, "Recent advances in fused deposition modeling based additive manufacturing of thermoplastic composite structures: a review," *J. Thermoplast. Compos. Mater.*, vol. 36, no. 7, pp. 3094–3132, 2023, doi: 10.1177/08927057221102857.
31. M. Blaj and G. Oancea, "Fused deposition modelling process: a literature review," *IOP Conf. Ser. Mater. Sci. Eng.*, vol. 1009, no. 1, p. 012006, 2021, doi: 10.1088/1757-899X/1009/1/012006.
32. M. Taufik and P. K. Jain, "Role of build orientation in layered manufacturing: a review," *Int. J. Manuf. Technol. Manag.*, vol. 27, no. 1/2/3, p. 47, 2013, doi: 10.1504/IJMTM.2013.058637.
33. J. M. Chacón, M. A. Caminero, E. García-Plaza, and P. J. Núñez, "Additive manufacturing of PLA structures using fused deposition modelling: effect of process parameters on mechanical properties and their optimal selection," *Mater. Des.*, vol. 124, pp. 143–157, 2017, doi: 10.1016/j.matdes.2017.03.065.
34. B. Caulfield, P. E. McHugh, and S. Lohfeld, "Dependence of mechanical properties of polyamide components on build parameters in the SLS process," *J. Mater. Process. Technol.*, vol. 182, no. 1–3, pp. 477–488, 2007, doi: 10.1016/j.jmatprotec.2006.09.007.
35. F. Kartal and A. Kaptan, "Investigating the effect of nozzle diameter on tensile strength in 3D-printed printed polylactic acid parts," *Black Sea J. Eng. Sci.*, vol. 6, no. 3, pp. 276–287, 2023, doi: 10.34248/bsengineering.1287141.
36. P. Demircioğlu, H. S. Sucuoğlu, İ. Böğrekci, and A. Gültekin, "The effect of three dimensional printed infill pattern on structural strength," *El-Cezeri Fen ve Mühendislik Derg.*, vol. 5, no. 3, pp. 785–796, 2018, doi: 10.31202/ecjse.423915.

37. J. Bonada, M. M. Pastor, and I. Buj-Corral, "Influence of infill pattern on the elastic mechanical properties of fused filament fabrication (FFF) parts through experimental tests and numerical analyses," *Materials (Basel)*, vol. 14, no. 18, p. 5459, 2021, doi: 10.3390/ma14185459.
38. K. Yu, Q. Gao, L. Lu, and P. Zhang, "A Process parameter design method for improving the filament diameter accuracy of extrusion 3D printing," *Materials (Basel)*, vol. 15, no. 7, p. 2454, 2022, doi: 10.3390/ma15072454.
39. L. Racz and M. C. Dudescu, "Numerical evaluation of the infill pattern upon mechanical properties of 3D printed materials," *Procedia Struct. Integr.*, vol. 56, pp. 3–10, 2024, doi: 10.1016/j.prostr.2024.02.030.
40. J. A. Gopsill and B. J. Hicks, "Deriving infill design of fused deposition modelled parts from predicted stress profiles." *Proceedings of the ASME 2016 International Design Engineering Technical Conferences and Computers and Information in Engineering Conference. Volume 2A: 42nd Design Automation Conference.* Charlotte, North Carolina, USA. August 21–24, 2016. V02AT03A033. ASME. https://doi.org/10.1115/DETC2016-59935.
41. H. K. Dave, B. H. Patel, S. R. Rajpurohit, A. R. Prajapati, and D. Nedelcu, "Effect of multi-infill patterns on tensile behavior of FDM printed parts," *J. Brazilian Soc. Mech. Sci. Eng.*, vol. 43, no. 1, p. 23, 2021, doi: 10.1007/s40430-020-02742-3.
42. C. Benwood, A. Anstey, J. Andrzejewski, M. Misra, and A. K. Mohanty, "Improving the impact strength and heat resistance of 3D printed models: Structure, property, and processing correlationships during fused deposition modeling (FDM) of poly(lactic acid)," *ACS Omega*, vol. 3, no. 4, pp. 4400–4411, 2018, doi: 10.1021/acsomega.8b00129.
43. M. S. Kumar, M. U. Farooq, N. S. Ross, C.-H. Yang, V. Kavimani, and A. A. Adediran, "Achieving effective interlayer bonding of PLA parts during the material extrusion process with enhanced mechanical properties," *Sci. Rep.*, vol. 13, no. 1, p. 6800, 2023, doi: 10.1038/s41598-023-33510-7.
44. A. Lepoivre, N. Boyard, A. Levy, and V. Sobotka, "Methodology to assess interlayer quality in the material extrusion process: a temperature and adhesion prediction on a high performance polymer," *Addit. Manuf.*, vol. 60, p. 103167, 2022, doi: 10.1016/j.addma.2022.103167.
45. T. D. Harpool, I. M. Alarifi, A. Alshammari Basheer, A. Aabid, M. Baig, A. R. Malik, S. A. Mohamed, R. Asmatulu, T. M. and A. A. EL-Bagory, "Evaluation of the infill design on the tensile response of 3D printed polylactic acid polymer," *Materials (Basel)*, vol. 14, no. 9, p. 2195, 2021, doi: 10.3390/ma14092195.
46. A. Yankin, G. Serik, S. Danenova, Y. Alipov, A. Temirgali, D. Talamona, and A. Perveen, "Optimization of fatigue performance of FDM ABS and nylon printed parts," *Micromachines*, vol. 14, no. 2, p. 304, 2023, doi: 10.3390/mi14020304.
47. N. D. Borra and V. S. N. Neigapula, "Synthesis and optimization for shape memory behaviour of 4D printed GNPs reinforced shape memory photopolymer composite," *Rapid Prototyp. J.*, vol. 29, no. 6, pp. 1175–1194, 2023, doi: 10.1108/RPJ-08-2022-0254.
48. N. D. Borra and V. S. N. Neigapula, "Parametric optimization for dimensional correctness of 3D printed part using masked stereolithography: Taguchi method," *Rapid Prototyp. J.*, vol. 29, no. 1, pp. 166–184, 2023, doi: 10.1108/RPJ-03-2022-0080.
49. M. Moradi, A. Aminzadeh, D. Rahmatabadi, and S. A. Rasouli, "Statistical and experimental analysis of process parameters of 3D nylon printed parts by fused deposition modeling: response surface modeling and optimization," *J. Mater. Eng. Perform.*, vol. 30, no. 7, pp. 5441–5454, 2021, doi: 10.1007/s11665-021-05848-4.

50. K. Park, G. Kim, H. No, H. W. Jeon, and G. E. O. Kremer, "Identification of optimal process parameter settings based on manufacturing performance for fused filament fabrication of CFR-PEEK," *Appl. Sci.*, vol. 10, no. 13, p. 4630, 2020, doi: 10.3390/app10134630.
51. Banjanin Bojan, Vladic Gojko, Pál Magdolna, Balos Sebastian, Dramicanin Miroslav, Rackov Milan, Knezevic Ivan, "Consistency analysis of mechanical properties of elements produced by FDM additive manufacturing technology," *Matéria (Rio Janeiro)*, vol. 23, no. 4, 2018, doi: 10.1590/s1517-707620180004.0584.
52. E. Asadollahi-Yazdi, J. Gardan, and P. Lafon, "Multi-objective optimization approach in design for additive manufacturing for fused deposition modeling," *Rapid Prototyp. J.*, vol. 25, no. 5, pp. 875–887, 2019, doi: 10.1108/RPJ-07-2018-0186.
53. Q. An, J. Ren, X. Jia, S. Qu, N. Zhang, X. Li, G. Fan, S. Pan, Z. Zhang, and K. Wu, "Anisotropic materials based on carbohydrate polymers: a review of fabrication strategies, properties, and applications," *Carbohydr. Polym.*, vol. 330, p. 121801, 2024, doi: 10.1016/j.carbpol.2024.121801.
54. I. Gibson, D. Rosen, and B. Stucker, *Additive Manufacturing Technologies*. New York, NY: Springer New York, 2015.
55. N. Zohdi and R. (Chunhui) Yang, "Material anisotropy in additively manufactured polymers and polymer composites: a review," *Polymers (Basel)*, vol. 13, no. 19, p. 3368, 2021, doi: 10.3390/polym13193368.
56. J. Quodbach, M. Bogdahn, J. Breitkreutz, R. Chamberlain, K. Eggenreich, A. G. Elia, N. Gottschalk, G. Gunkel-Grabole, L. Hoffmann, D. Kapote, T. Kipping, S. Klinken, F. Loose, T. Marquetant, H. Windolf, S. Geißler, and T. Spitz, "Quality of FDM 3D printed medicines for pediatrics: Considerations for formulation development, filament extrusion, printing process and printer design," *Ther. Innov. Regul. Sci.*, vol. 56, no. 6, pp. 910–928, 2022, doi: 10.1007/s43441-021-00354-0.
57. M. Batista, M. Ramirez-Peña, J. Salguero, and J. M. Vazquez-Martinez, "Optimization of fused filament fabrication for high-performance polylactic acid parts under wear conditions," *Lubricants*, vol. 12, no. 8, p. 281, 2024, doi: 10.3390/lubricants12080281.
58. D. Vieweger, S. Diel, H.-G. Schweiger, and U. Tetzlaff, "Mechanical properties of raw filaments and printed specimens: effects of fiber reinforcements and process parameters," *Polymers (Basel)*, vol. 16, no. 11, p. 1576, 2024, doi: 10.3390/polym16111576.
59. R. Chafik, S. Ebeid, and M. Sayed, "Printing parameters and material characteristics affecting mechanical properties of FDM printed parts," *Eng. Res. J.*, vol. 53, no. 2, pp. 220–225, 2024, doi: 10.21608/erjsh.2023.251818.1252.
60. J. Bochnia, M. Blasiak, and T. Kozior, "A comparative study of the mechanical properties of FDM 3D prints made of PLA and carbon fiber-reinforced PLA for thin-walled applications," *Materials (Basel)*, vol. 14, no. 22, p. 7062, 2021, doi: 10.3390/ma14227062.
61. N. D. Borra, V. S. N. Neigapula, U. K. Sanivada, and R. Fangueiro, "Tailoring the shape memory properties of silica nanoparticle infused photopolymer composites for 4D printing applications: a Taguchi analysis," *Eur. Polym. J.*, vol. 194, p. 112174, 2023, doi: 10.1016/j.eurpolymj.2023.112174.
62. B. N. Dhanunjayarao and N. V. Swamy Naidu, "Assessment of dimensional accuracy of 3D printed part using resin 3D printing technique," *Mater. Today Proc.*, vol. 59, no. 3, pp. 1608–1614, 2022, doi: 10.1016/j.matpr.2022.03.148.
63. P. Venkatesh, P. V. Krishna, B. U. Kiran, B. L. Rao, P. R. V. Kumar, B. N. Dhanunjayarao, N. V. S. Naidu, Y. Phaneendra, and A. B. Beera, "Design and

optimization of 4D printed carbon fiber reinforced poly lactic acid parts using fused deposition modeling for shape memory applications: a Taguchi approach," *Mater. Sci. Forum*, vol. 1110, pp. 143–157, 2023, doi: 10.4028/p-cD5xfu.
64. M. Sonmez, C-E. Pelin, G. Pelin, B. Rusu, A. Stefan, M. D. Stelescu, M. Ignat, D. Gurau, M. Georgescu, M. Nituica, O-C. Oprea, L. Motelica, B. Waśniewski, P. Ortyl, R. D. Trușcă, "Development, testing, and thermoforming of thermoplastics reinforced with surface-modified aramid fibers for cover of electronic parts in small unmanned aerial vehicles using 3D-printed molds," *Polymers (Basel)*, vol. 16, no. 15, p. 2136, 2024, doi: 10.3390/polym16152136.
65. D. D. Lavanya and V. Arulprasanth, "Optimization of an aerospace bracket by additive manufacturing with continuous fiber reinforced plastics," *International J. Sci. Res. Eng. Manag.*, vol. 08, no. 07, pp. 1–10, 2024, doi: 10.55041/IJSREM36381.
66. D. Tarun, S. Sundar, K. C. Kishor Kumar, S. A. Vardhan, and P. S. P. Sai Durga Prasad, "Advancements in fused deposition modeling for aerospace: optimizing lightweight and high-strength components," *Int. J. Innov. Sci. Res. Technol.*, vol. 9, no. 6, pp. 2212–2215, 2024, doi: 10.38124/ijisrt/IJISRT24JUN1751.
67. A. Singh, T. Bhardwaj, P. Saini, N. Gangil, and D. K. Singh, "Investigation on FDM process parameters for the dimensional accuracy with ABS polymer part," *Eng. Res. Express*, vol. 6, no. 2, p. 025553, 2024, doi: 10.1088/2631-8695/ad52ee.
68. N. D. Borra and V. S. N. Neigapula, "Effect of GNPs and resin blend on tear resistance of 4D printed shape memory photopolymer composite," *Rapid Prototyp. J.*, vol. 29, no. 6, pp. 1138–1154, 2023, doi: 10.1108/RPJ-10-2022-0352.
69. H. Rathod, R. Makwana, and G. Acharya, "Optimize process parameter of FDM techniques to improve mechanical properties of aerospace application," *J. Kejuruter.*, vol. 35, no. 5, pp. 1085–1100, 2023, doi: 10.17576/jkukm-2023-35(5)-09.
70. R. P. Kaul, A. Malviya, S. Sagar, S. Kumar, A. Gupta, O. Bhutia, R. Sagar, M. Singhal, and A. Kumar, "An experimental study conducted on material properties used in a FDM printer for medical models." *Fortune J Health Sci*, vol. 7, no. 4, pp.31, 2024, doi: 10.20944/preprints202407.2481.v1
71. A. Casucci, G. Verniani, A. L. Barbieri, N. M. Ricci, E. Ferrari Cagidiaco, and M. Ferrari, "Flexural strength analysis of different complete denture resin-based materials obtained by conventional and digital manufacturing," *Materials (Basel)*, vol. 16, no. 19, p. 6559, 2023, doi: 10.3390/ma16196559.
72. S. Sharma and S. A. Goel, "3D printing and its future in medical world," *J. Med. Res. Innov.*, vol. 3, no. 1, e000141, 2018, doi: 10.15419/jmri.141.
73. Y. Yamamoto, S. Harada, D. Yamamoto, W. Honda, T. Arie, S. Akita, and K. Takei, "Printed multifunctional flexible device with an integrated motion sensor for health care monitoring," *Sci. Adv.*, vol. 2, no. 11, 2016, doi: 10.1126/sciadv.1601473.
74. M. Pulcinelli, D. Lo Presti, C. Massaroni, S. Silvestri, E. Schena, and F. De Tommasi, "Wearable elastic band based on 3D-printed fiber Bragg grating sensor for knee joint monitoring," in *2024 IEEE International Workshop on Metrology for Industry 4.0 & IoT (MetroInd4.0 & IoT)*, May 2024, pp. 597–602, doi: 10.1109/MetroInd4.0 IoT61288.2024.10584188.
75. A. Hutyra, M. Bańkosz, and B. Tyliszczak, "Technology for automated production of high-performance building compounds for 3D printing," *Materials (Basel)*, vol. 17, no. 15, p. 3829, 2024, doi: 10.3390/ma17153829.
76. H. Hassan, E. Rodriguez-Ubinas, A. Al Tamimi, E. Trepci, A. Mansouri, and K. Almehairbi, "Towards innovative and sustainable buildings: a comprehensive review of 3D printing in construction," *Autom. Constr.*, vol. 163, p. 105417, 2024, doi: 10.1016/j.autcon.2024.105417.

77. P. J. S. Cruz, AM perspectives: research in additive manufacturing for architecture and construction. SOAP | Stichting OpenAccess platforms, 2024.
78. M. D. Leles da Silva, L. F. Silva, B. M. Toralles, F. A. R. Cardoso, M. Corso, and L. C. Soto Herek Rezende, "Building a sustainable future: The role of additive manufacturing in civil construction," *Case Stud. Constr. Mater.*, vol. 20, p. e02976, 2024, doi: 10.1016/j.cscm.2024.e02976.
79. H. Guo, R. Lv, and S. Bai, "Recent advances on 3D printing graphene-based composites," *Nano Mater. Sci.*, vol. 1, no. 2, pp. 101–115, 2019, doi: 10.1016/j.nanoms.2019.03.003.
80. G. A. R. Sampedro, D. J. S. Agron, G. C. Amaizu, D.-S. Kim, and J.-M. Lee, "Design of an in-process quality monitoring strategy for FDM-type 3D printer using deep learning," *Appl. Sci.*, vol. 12, no. 17, p. 8753, 2022, doi: 10.3390/app12178753.
81. Y. Yang, Z. Chen, X. Song, B. Zhu, T. Hsiai, I. Wu Pin, R. Xiong, J. Shi, Y. Chen, O. Zhou, and K.K. Shung, "Three dimensional printing of high dielectric capacitor using projection based stereolithography method," *Nano Energy*, vol. 22, pp. 414–421, 2016, doi: 10.1016/j.nanoen.2016.02.045.
82. J. Jin, F. Zhang, Y. Yang, C. Zhang, H. Wu, Y. Xu, and Y. Chen, "Hybrid multimaterial 3D printing using photocuring-while-dispensing," *Small*, vol. 19, no. 50, 2023, doi: 10.1002/smll.202302405.
83. I. Tamaşag, C. Suciu, I. Beşliu-Băncescu, C. Dulucheanu, and D.-A. Cerlincă, "Experimental study on the possibilities of FDM direct colour printing and its implications on mechanical properties and surface quality of the resulting parts," *Polymers (Basel)*, vol. 14, no. 23, p. 5173, 2022, doi: 10.3390/polym14235173.
84. M. Mrinalini and S. Prasanthkumar, "Recent advances on stimuli-responsive smart materials and their applications," *Chempluschem*, vol. 84, no. 8, pp. 1103–1121, 2019, doi: 10.1002/cplu.201900365.
85. D. G. Tamay, T. Dursun Usal, A. S. Alagoz, D. Yucel, N. Hasirci, and V. Hasirci, "3D and 4D printing of polymers for tissue engineering applications," *Front. Bioeng. Biotechnol.*, vol. 7, p. 164, 2019, doi: 10.3389/fbioe.2019.00164.
86. W. Ahmed, F. Alnajjar, E. Zaneldin, A. H. Al-Marzouqi, M. Gochoo, and S. Khalid, "Implementing FDM 3D printing strategies using natural fibers to produce biomass composite," *Materials (Basel)*, vol. 13, no. 18, p. 4065, 2020, doi: 10.3390/ma13184065.

12 Short and Continuous Fibre Composites in Additive Manufacturing

Processability and Affinity with Structural Optimization Methods

Afonso Oliveira and Marco Leite

12.1 INTRODUCTION

In the aerospace industry, the search for the most optimized solution is relentless. With the goal of reducing weight and increasing efficiency, this search allows for fuel savings, cost reduction, and increased aircraft flight autonomy [1]. Since the introduction of composites in the 1970s, aircraft manufacturers have continuously and progressively introduced these materials in their designs [2] and have searched for ways to gain a competitive advantage by introducing new aircraft designs [3, 4]. With additive manufacturing (AM), a layer-by-layer deposition, enables the design of complex parts with added value, minimizing wastes and production costs when the production volume is adequate [5]. Structural optimization software is an enabler of these parts as it allows for the design of lightweight parts with material only where it is needed to support the loads [6].

The use of composite materials in AM for applications in aerospace is not only possible, but also highly desirable, due to the advantages of the technology [7]. One possible advantage is the possibility of complex parts with new shapes with the design freedom to place fibre reinforcements where needed, albeit manufacturing constraints [8]. For maximum results, the tripod of material, shape, and process, shown in Figure 12.1, should be balanced as the shape optimizers seldom have results that cannot be manufactured [9].

In this work, a discussion on processability and the affinity with structural optimization methods of short and continuous fibres in AM material extrusion is presented. A case study of unmanned aerial vehicle (UAV) wing ribs with topological optimization (TO) and use of short and continuous fibres in fused filament fabrication (FFF) is detailed. Several geometries are designed and numerical and experimentally tested.

DOI: 10.1201/9781003686941-12

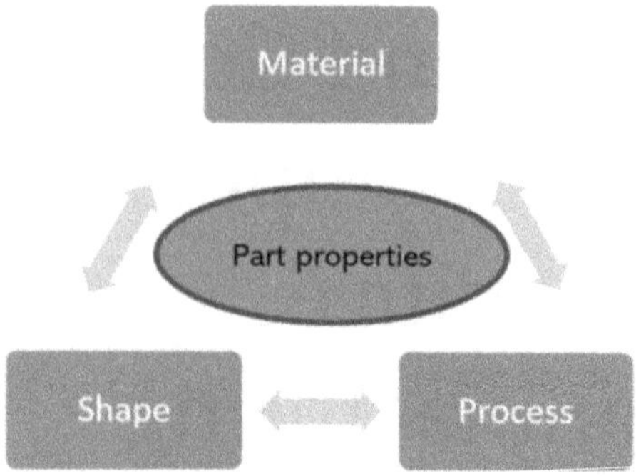

FIGURE 12.1 Part properties dependence on material, shape, and process. (Adapted from ref. [9].)

12.2 SHORT AND CONTINUOUS FIBRE-REINFORCED FFF

Polymeric parts manufactured by three-dimensional (3D) printing lack strength and functionality as load-bearing elements, due to their mechanical properties. These are on par with injection-moulded polymers in the X–Y directions but are much frailer in the Z direction. The low mechanical properties exhibited by neat polymers are not suitable for high-performance engineering applications. As such, they still tend to be used mostly as conceptual prototypes, restricting their industrial applications [10, 11]. 3D printing of polymer composites addresses these problems as these have enhanced mechanical properties over existing polymeric materials [12]. A composite material is made from two or more materials with different physical or chemical properties that, when combined, produce a material that offers superior overall performance when compared to its individual constituents [13]. In the case of AM-reinforced polymers, the polymer is the matrix of the composite, where the reinforcements are introduced. Different types of reinforcements are chosen depending on the desired properties of the final product [14].

Love et al. [15] theorized that the merging of polymer AM with carbon fibre (CF) would revolutionize the AM industry, as AM is used to make complex shapes in opposition to traditional CF technology, which is excellent at manufacturing strong, simple, lightweight structures. This can be extended to glass fibre (GF) and aramid fibre (AF) reinforcements as well. Among the AM methods, both FFF and selective laser sintering (SLS) have been reported to use CF to fabricate composites [16].

During the FFF process, the polymers can be reinforced by either short or continuous fibres [10]. For short fibre, the method involves embedding short fibres SLS inside a polymer filament by premixing them into the polymer matrix before extrusion. As such, the printing process is similar to a neat polymer FFF extrusion. The fibres end up being oriented in the same direction as the raster movement, resulting in increased mechanical properties [11]. The characteristics of both fibres and polymer matrices, including fibre length, orientation, type, and matrix–fibre bonding strength, dictate the increase in mechanical properties [17].

In this work, an UltiMaker S5 printer and the UltiMaker™ Cura slicer are used to slice and manufacture the parts. The Ultrafuse™ from BASF company, PAHT CF15, a polyamide (PA) filament reinforced with 15% of short CFs is used.

Another fibre reinforcement method for FFF-printed composites is continuous fibres. The continuous fibre reinforcement method is characterized by having a separate fibre filament, which gets deposited continuously throughout the layer. This way it is possible to control the fibre orientation, unlike in the short fibres, where these are only partially aligned [18].

Continuous fibre reinforcement can be produced by two different methods:

- A co-extrusion process where there is only one nozzle responsible for deposition, and the fibre filament and matrix filament are incorporated in the injector. After deposition, the material solidifies rapidly and adheres to the previous layer, so that the fibre can be pulled out continually by the foregoing fibre inside the part.
- A dual extrusion method where one nozzle is dedicated to the deposition of the matrix and a second nozzle is dedicated to the fibre deposition. In this work, a dual extrusion method is used, where two nozzles, one responsible for the polymer and other for the fibre deposition, can deposit independently.

To print these composites, a Markforged™ Mark II printer and its PA filament are used. For the reinforcement, both CF- and GF-reinforced filaments are used. The Eiger slicer software is used, which allows some control over fibre deposition quantities and patterns.

12.3 TOPOLOGICAL OPTIMIZATION

TO is one of the three main structural optimization categories, together with size and shape optimization [19].

In size optimization, the structure layout is already defined, but its dimensions are variable. For example, a truss structure whose overall dimensions are known, but the cross sections of each bar element are unknown. Shape optimizations add the shape change to the size change. As such, the boundary design geometry of the structure is modified to meet the required objectives. If a beam with holes were to be shape optimized, the geometry of the holes would change to better support the load.

In contrast to size and shape optimization, where components of a structure can be manipulated but must always be present, TO goes on step above by allowing topology changes. TO allows sections of the parts to have null thickness, thereby determining the number and shape of the holes. As such, it is beneficial to obtain an initial design, when there is no prior knowledge about the geometry of the part. Shape and size optimizations are usually used afterwards to update current designs.

TO optimizes material layout by using a mathematical algorithm to minimize or maximize chosen performance parameters. It is necessary to define a design space, the boundary conditions, applied loads, and constraints. The compliance of the structure is usually the most common objective function to be minimized, which is equivalent to maximizing its stiffness. In some cases, maximizing the strength may also be of importance, meaning the algorithm will give a solution that can withstand maximum load without reaching the failing conditions.

After proven use in the automotive industry as an optimization method capable of weight saving and efficient design, recent trends have seen TO being used in aerospace industry as well. Airbus has been implementing this technology in civil aircraft design in its newer models [20], having taken it to a new level for the A380 design [21]. TO is performed on a never-seen-before scale to save weight in leading edge ribs, which amounted to a reported 700 kg saved. The main software used by Airbus is Altair Hyperworks™, being that both companies work closely together to develop a framework for topology, size, and shape optimizations for aircraft components.

TO optimizes material layout by using a mathematical algorithm to minimize or maximize chosen performance parameters. The formulation of this problem is explained in detail in ref. [22]. It is necessary to define a design space, the boundary conditions, applied loads, and constraints. The compliance of the structure is usually the most common objective function to be minimized, which is equivalent to maximizing its stiffness. In some cases, maximizing the strength may also be of importance, meaning the algorithm will give a solution that can withstand maximum load without reaching the failing conditions. This problem may be solved for set conditions such as maximum displacement or stress but, in this case, it is chosen to be solved for a set volume fraction (V_f).

12.4 CASE STUDY

For this case study, wing ribs for a UAV with a single spar are designed and manufactured. An aerofoil is chosen for the wing and therefore for the ribs. Later, pressure loads are translated from the aerofoil to the rib so that structural loads can be used. For those loads and geometry, the topological parameters are studied.

For many aircraft, the wings and other aerodynamic surfaces are comprised of the following main components: spars, ribs, stringers, and a shell. The ribs are responsible for transmitting the aerodynamic loads from the shell of the wing to the spar. This includes shear and bending loads due to the pressure difference between the upper and lower surface of the wing. Additionally, the ribs also ensure the transmission of hinge reactions from the control surfaces and other concentrated loads coming from the undercarriage and engine thrust attachments. In the presence of stringers, the ribs provide end restraint, which increases the buckling stress of the skin panels. The ribs are also responsible for maintaining the wing shape throughout its span, especially near the wing tips [13].

To calculate the loads that the ribs support, it is necessary to select the appropriate aerofoil profile. A National Advisory Committee for Aeronautics (NACA) 2412 aerofoil is chosen for the ribs, as it has been extensively studied and has large relative thickness allowing more design space for the TO process. For illustration purposes, the initial geometry is shown in Figure 12.2a, where the two holes will be used to fix the manufactured ribs in place while testing. The pressure distribution on the aerofoil, obtained using XFLR5, for a chosen angle of attack of 10° and velocity of 25 m/

s, is presented in Figure 12.2b. A total of 100 points are used to create the aerofoil, with cosine spacing, which allows for a highly refined representation of its geometry towards the leading and trailing edges of the aerofoil, the most critical areas. Also, in Figure 12.2c, the pressure forces in each of the 100 points are presented. The values of the force in each of the points are then imported to HyperWorks™. The thickness of each rib is chosen to be 2 mm. As the ribs are thin, only 2D optimizations were made, being that the optimized aerofoil shape is constant in the Z axis.

The material selection is complex as the composite materials used in this study do not have a digital twin in the software. However, as the nature of the optimization is to distribute material where it is needed and stress distributions on a part are material

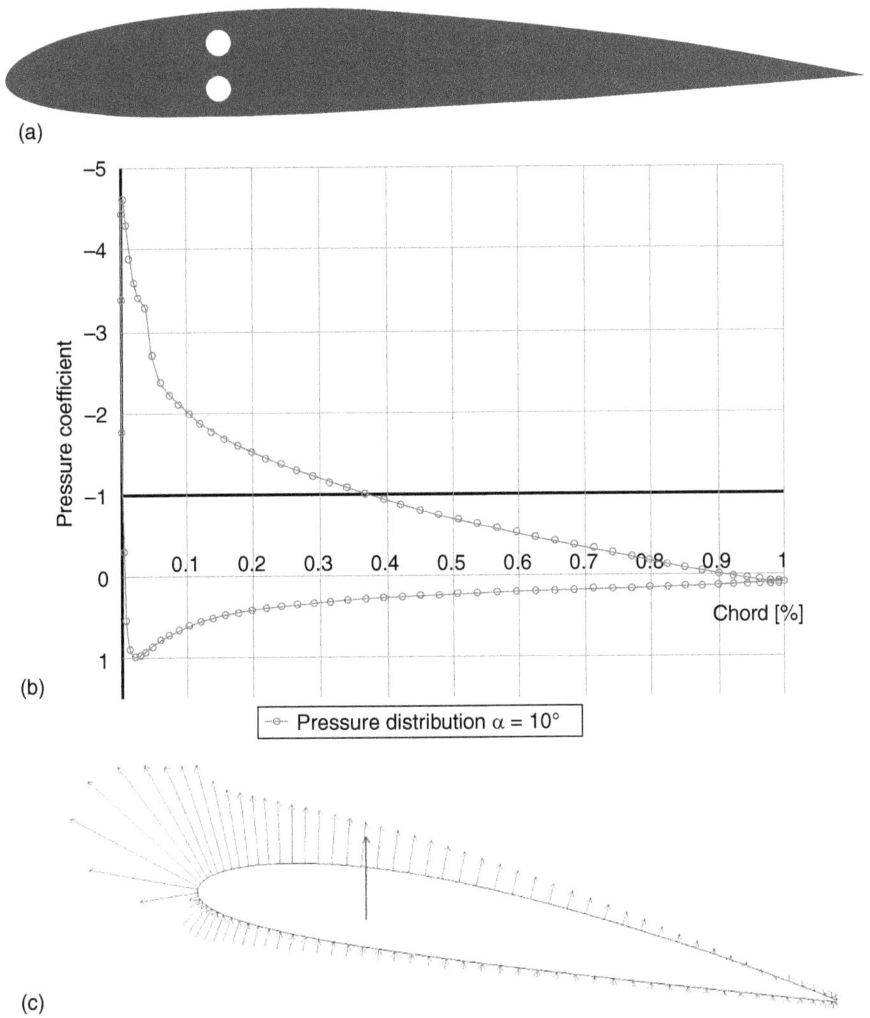

FIGURE 12.2 (a) NACA 2412 aerofoil initial geometry, (b) pressure distribution at α = 10°, (c) pressure force in each point.

independent, the use of aluminium (any isotropic material would be the same) is therefore an acceptable approximation.

With geometry, loads and material defined, the next step is to create a mesh. Throughout every optimization described, mixed thin shell elements are used. HyperWorks™ prioritizes the creation of quadrilateral elements and fills the mesh with triangular elements where needed.

The boundary conditions are then defined. The nodes in the two circular holes are set to fixed, having all 6° of freedom constrained. To simulate the distributed pressure on the surface corresponding to an $\alpha = 10°$, a linear interpolation is applied to the forces displayed in Figure 12.2. If this is not done, the forces would be applied at their coordinates, which does not equate to a distributed load. By using a linear interpolation, the software interpolates the forces and applies a force in each node of the surface, simulating a continuous pressure distribution.

The objective of the optimization is to minimize the compliance, which is equivalent to maximizing the stiffness, and the constraint is the maximum volume fraction (V_f). The algorithm used for the optimization is the Method of Feasible Directions. A detailed review of this method and its mathematical formulations can be consulted in ref. [14].

Without the proper tuning of the optimization parameters, the output geometry is not viable. In Figure 12.3, the relative densities, between 0 and 1, of the geometry obtained without parameter tuning are shown. A phenomenon named "checkering" is clearly visible in this geometry. This happens when the software produces a geometry with voids between higher density elements. As there is no minimum element dimension defined, the software sees this structure as the most efficient one, despite not being manufacturable.

To avoid the "checkering," some parameters must be defined. The most important parameter in this case is the Minimum Member Size (MINDIM). This dimension is recommended to be set at 3–12 times the average mesh element size, which, for 2D elements, is calculated as the average of the square root of the area of the elements.

Another important parameter is the penalty factor. As the optimization is seeking discrete values of 0 or 1 for the element densities, this value is used to penalize the intermediate densities so that discrete solutions may be reached.

Finally, to ensure the aerofoil surface is only composed of solid elements, an offset of 1.2 mm is added to the surface. The output geometry after the changes is shown in Figure 12.4.

Further mesh convergence tests are conducted, settling in a mesh size of 0.3 mm. Several different volume fraction values are tested, being that 0.3 and 0.6 are chosen to be printed, shown in Figure 12.5.

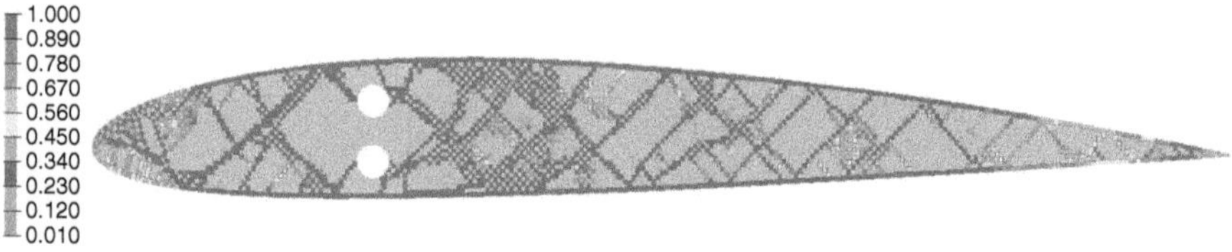

FIGURE 12.3 Relative densities of the output geometry without parameter tuning.

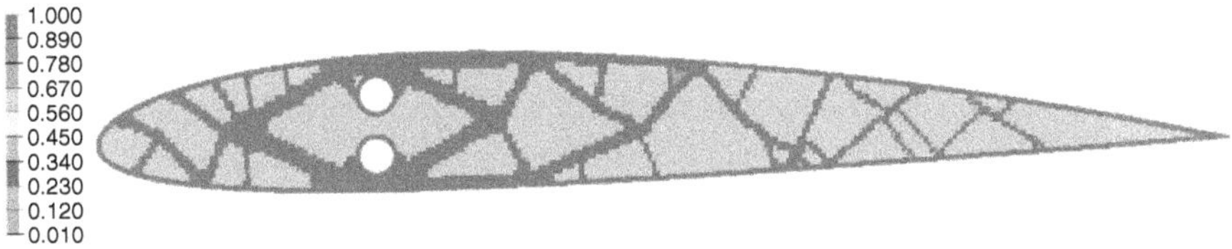

FIGURE 12.4 Relative densities of the output geometry after parameter tuning.

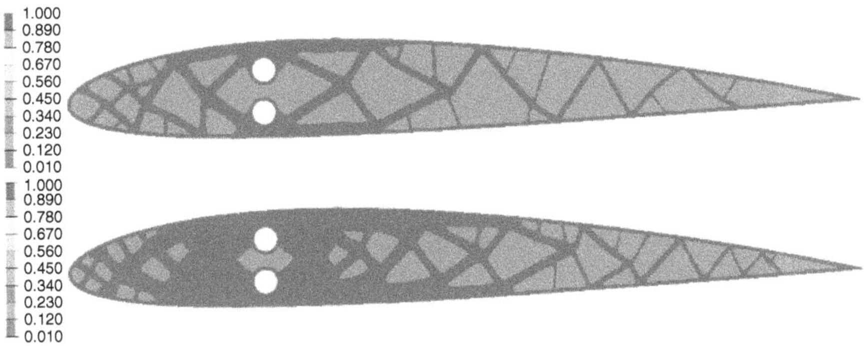

FIGURE 12.5 Relative densities of the final output geometries: $V_f = 0.3$ (top) and $V_f = 0.6$ (bottom).

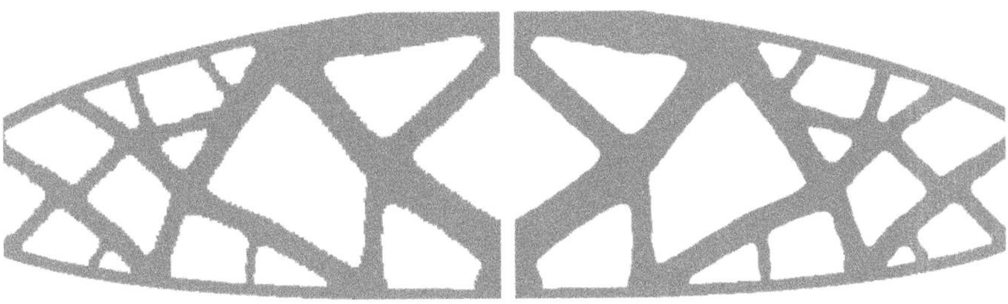

FIGURE 12.6 Geometry before (left) and after (right) smoothing.

The geometries obtained still need post processing, as the arm edges are serrated which does not translate well to 3D printing. To do so, OSSmooth™ is used, a post-process software built in HyperWorks™. This software remeshes the optimized geometry obtained to smooth the serrated edges, as shown in Figure 12.6.

12.4.1 Continuous Fibre Manufacturing Constraints

Contrary to printing with short fibre reinforcement, printing with continuous fibres introduces constraints related to part geometry and fibre placement. Most importantly, there is a minimum width of 2 mm, necessary for fibre to be placed.

As such, additional topology optimizations are performed for a V_f of 0.3 and 0.6, but with an offset of 2 mm, in opposition to the 1.2 mm offset used previously for short fibres. Due to this change in offset value, the outputs of the TO are slightly different geometries than previously shown. While maintaining similar topologies, the designs with an offset of 2 mm show less internal reinforcements in the form of organic arms in the leading and trailing edge areas. These rib designs are capable of withstanding stress with less organic arms because they contain a greater amount of material on their surfaces.

While maintaining similar topologies, the designs with an offset of 2 mm show less internal reinforcements in the form of organic arms in the leading and trailing edge areas. These rib designs are capable of withstanding stress with less organic arms because they contain a greater amount of material on their surfaces.

12.5 EXPERIMENTAL TESTING

To evaluate the performance of each of the designs, an experimental testing methodology is developed. The objective of these is to simulate the pressure distribution on the rib and observe the displacements and maximum force of each rib design before failure. To simulate this pressure distribution, without resorting to a wind tunnel and with controllable variables, an adaptation of the solution used by Carneiro et al. [23] is employed. This solution converts the distributed load due to the pressure differentials into two equivalent point forces. By applying these equivalent forces in specific points, a similar stress distribution is achieved to the distributed pressure in the ribs, having the same highest stress areas, as shown in Figure 12.8.

To do this experimental procedure, a jig is built.

The main components of the jig are:

- A vertical 8-mm-thick wooden base, with epoxied supports for stability;
- Two 8-mm-diameter carbon tubes serving as spars. Contrary to ref. [24], two smaller tubes were used instead of a larger one, to prevent the rotation of the rib. The ribs are slid into these tubes;

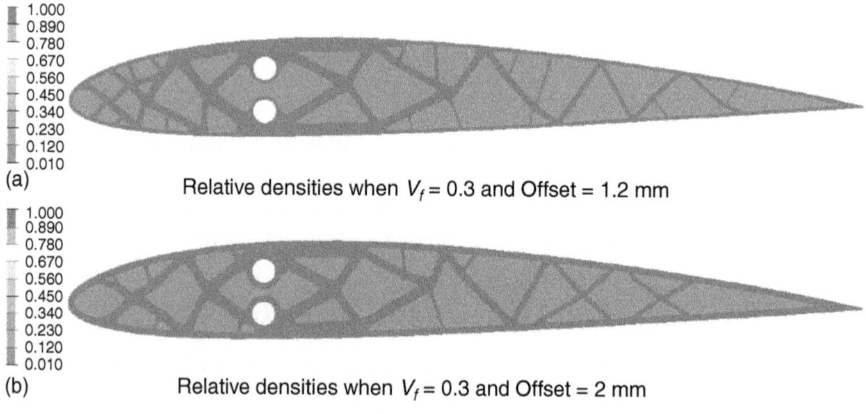

FIGURE 12.7 Comparison between 0.3 volume fraction geometries with different offset values.

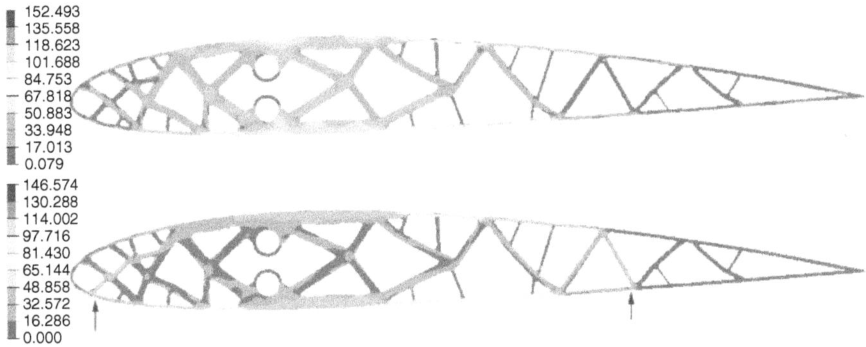

FIGURE 12.8 Stress distributions for the distributed load when $\alpha = 10°$ (top) and for the point forces represented in darker (bottom) in the 0.3 V_f geometry.

- Two 3-mm acrylic pieces, present to prevent the ribs from buckling and allow the observation of the deformation of the ribs;
- Three pieces of 3-mm-thick plywood, which space the acrylic from the base. A thickness of 3 mm is chosen to allow the ribs 0.5 mm clearance in each side, to prevent unnecessary contact and friction;
- Two metal hooks that transfer the load from the ropes to the rib;
- Two pieces of rope, capable of withstanding 75 kg, connecting the metal hooks to the wooden distributing forces bar;
- A plywood bar with specifically placed holes, so as to distribute a single force applied into the two forces applied to the rib.

The jig is then set up on two elevated surfaces (two equally tall tables in this case), allowing the force to be applied below, between the tables. Each test consisted in measuring the displacement of the aluminium bar with a digital calliper between two different loads, to determine the rib rigidity. Afterwards, each rib is loaded until failure, to determine the maximum load supported. For each test, three specimens were built and tested to increase the quality of the results.

In Figure 12.9, the jig is shown in a computer-aided design (CAD) environment and after being built.

12.5.1 Short Fibre-Reinforced Ribs

Four different designs with a 1.2-mm offset are printed with short fibre reinforcement. Rib 1 is printed from the optimized geometry with a V_f of 0.3, with 100% concentric infill. Ribs 2–4 are printed from the 0.6 V_f optimized geometry, with 100% concentric infill, 80% triangular infill, and 20% triangular infill, respectively, shown in Figure 12.10. These are manufactured to compare the effect of the introduction of infill in the stiffness, strength, and weight. To maximize the impact of these changes, only one wall is used on the prints, else they would mimic the behaviour of

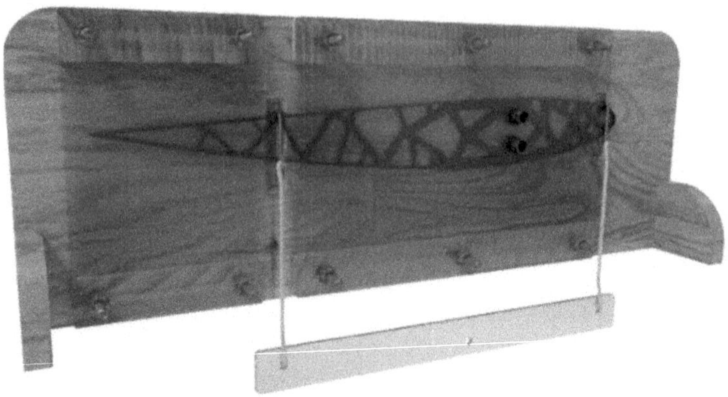

(a)

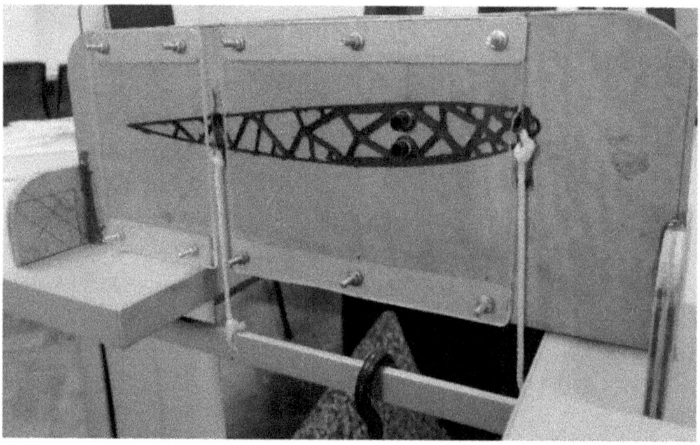

(b)

FIGURE 12.9 (a) CAD model, (b) built experimental setup.

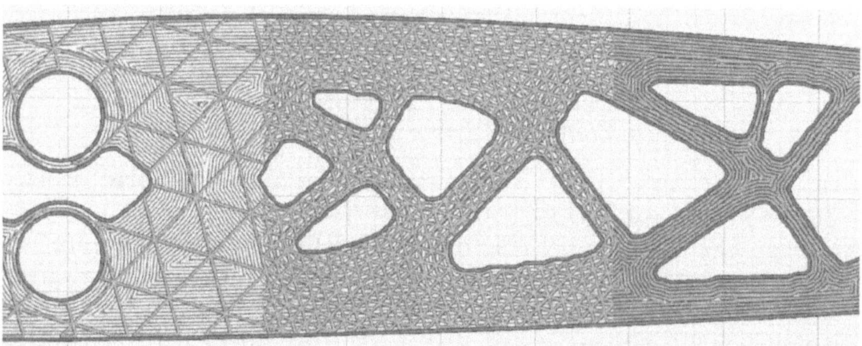

FIGURE 12.10 Bead orientation in the 0.6 V_f rib when using different infills: 20% triangular (left), 80% triangular (middle), and 100% concentric (right). The red lines represent the walls, the yellow ones the bottom layer, and the orange ones the infill.

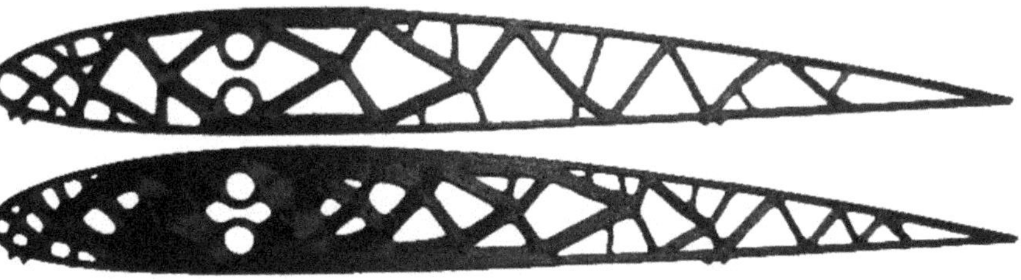

FIGURE 12.11 Comparison between the 0.3 V_f (top) and 0.6 V_f (bottom) short fibre-reinforced ribs.

the concentric infill and would bear most of the load, diminishing possible performance differences.

In Figure 12.11, the manufactured short fibre-reinforced ribs are shown.

12.5.2 Continuous Fibre-Reinforced Ribs

For the continuous fibre-reinforced ribs, both 0.3 and 0.6 V_f geometries with a 2-mm offset are printed, while reinforcing the outside surface of the rib and the largest organic arms with fibre. The 0.3 V_f geometry (A) is printed with a solid infill, whereas the 0.6 V_f one (B) is printed with 37% triangular infill, as in Figure 12.12. The inputs allowed by Eiger™ in relation to the fibre placement are minimal, being that the software places the fibres automatically in each layer according to some manufacturing restrictions.

Besides the minimum width restriction aforementioned, there is also a restriction regarding the minimum fibre length, which exists due to the distance between the servo that cuts the fibre and the fibre nozzle. This value is 45 mm, making it impossible to place fibres shorter than that, meaning there are some features which may be too small to be reinforced. One example of this are the spar holes.

Due to the minimum dimension of the fibre, the minimum diameter for a hole to be reinforced all around is 12.16 mm (D1) [25]. This means the spar holes, which are 8 mm in diameter, are not fully reinforced on their own, but may be reinforced by a continuous fibre that comes from other geometries.

The option to expand thin features in Eiger is used, which transforms geometries that are too thin for the printer to print, and automatically converts them into printable geometries. Naranjo-Lozada et al. [26] stated that it is a good practice to avoid having the same starting position for all the fibre depositions. By distributing these starting points, it is possible to obtain stronger parts. For these specific parts, however, this is disregarded and the initial points of all the fibres are set close to the trailing edge. This is done as the section behind the second point force has little to no stresses. As such, this is deemed a better solution than having the starting point of the fibre changing from layer to layer.

The fibre placement for both the different V_f optimized geometries (0.3 and 0.6) is displayed in Figure 12.12. As it can be seen, the 0.3 V_f geometry has one contour of fibre around all the rib surface and at least one fibre contour in each of the larger organic arms. Despite the spar holes not being fully reinforced all around, they are

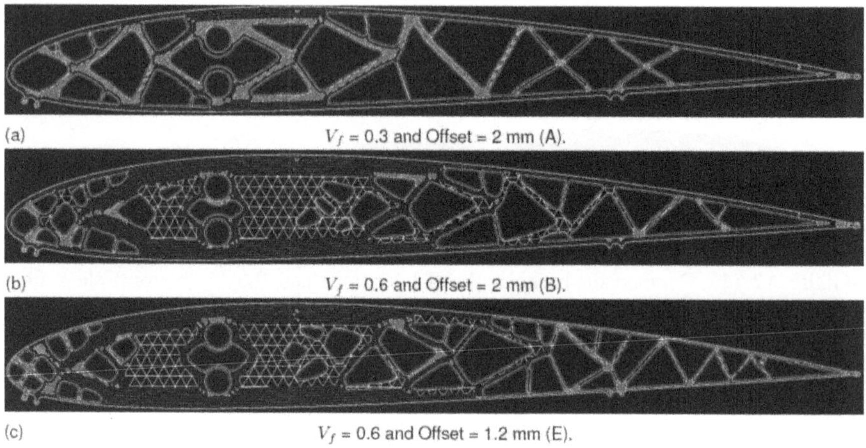

FIGURE 12.12 Fibre placement in the different optimized geometry. The dark blue lines represent the fibre, the light blue ones the walls of the print, and the white ones the infill.

TABLE 12.1
Experimental Properties of Each Long Fibre-Reinforced Rib

Rib No.	A	B	C	D	E
Print time (h:m)	1h 51m	2h 18m	1h 30m	1h 56m	1h 54m
Price estimated by Eiger ($)	2.88	5.59	4.04	10.02	9.16
Plastic volume (cm^3)	4.30	4.81	4.53	5.04	4.92
Fibre volume (cm^3)	1.23	2.85	1.02	2.98	2.7
Fibre coefficient (%)	22.24	37.21	18.38	37.16	35.43
Weight (theo.) (g)	6.34	9.48	6.07	9.35	8.82

still partially reinforced in all the rib designs, as the fibre passes through some of the walls of the holes. Due to the higher space available in the 0.6 V_f geometry, it is possible to apply a larger amount of fibre in the organic arms and, most importantly, in the areas with more stress concentrations.

Finally, one additional case is added, which consists of the previous 0.6 V_f geometry with the initial offset of 1.2 mm. This rib will not have fibre reinforcing the surface and is manufactured to examine the influence of the presence of fibre in the surfaces. This case still has plenty of fibre in the zones of maximum stress but may be compromised as there are no fibres near the loading point behind the spars.

The two first cases are printed with both glass (A and B) and carbon (C and D) and the last one (E) is printed only with CF reinforcement, resulting in a total of five different cases, described in Table 12.1. The case when there is no fibre all around the surface (E) is only reinforced with CF as the conclusions would be the same as if they were reinforced with GF, meaning there is no need to print twice.

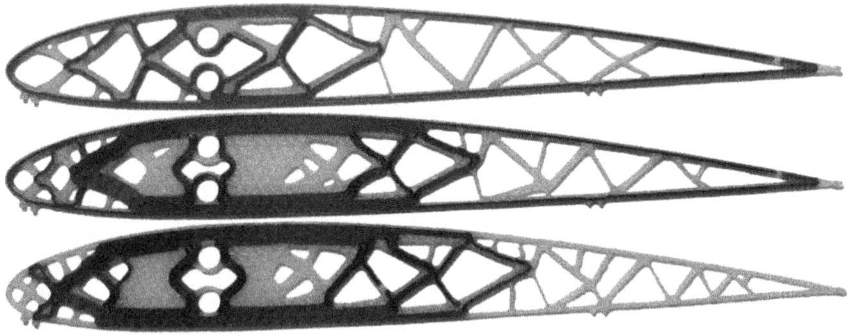

FIGURE 12.13 Comparison between the 0.3 V_f (top) and 0.6 V_f (middle and bottom) continuous fibre-reinforced ribs.

The fibre placement for both the V_f optimized geometries (0.3 and 0.6) is displayed in Figure 12.13.

12.6 EXPERIMENTAL RESULTS

As mentioned above, the objective of these tests is to compare the rib's performance in regard to stiffness and maximum strength. After putting a rib in the jig and securing it in place with the carbon spars, the metal hooks are attached to the supports in the rib. The wooden base is excavated to ensure that in no part of the tests are the metal hooks in contact with the base, which would lead to friction. If they get caught in the wood, it might even compromise the validity of the test. Slotted calibration weights are used to load the rib.

These are stacked on a hook which is placed on the aluminium bar. The assembly of the bar, ropes, and hooks weighs 169 g, which are taken into account for in the final calculations. A crevice is made in the aluminium bar, to keep the hook in place.

To determine the maximum strength of each rib in newton, calibration weights are added until failure or critical deformation occurs. With regard to stiffness, this property is defined as how much does the rib deform between two certain applied loads. This deformation is measured by the displacement of the aluminium bar between two loads with the aid of a calliper rule. The first load guarantees that the ropes are fully stretched so as to eliminate any errors due to slack. The stiffness, in newton per millimetre can then be obtained by dividing the difference in applied loads by the difference in distances between the aluminium bar and the base for the two load cases. Due to occurrence of buckling, certain areas of the ribs come into contact with the limiting acrylic plates, resulting in some friction resistance that may increase the load at which failure occurs. However, the normal forces applied on the limiting plates are relatively low, and the coefficients of friction for acrylic fall within the low range of 0.2–0.4 [24]; therefore, the uncertainty due to the friction forces lies within reason.

To validate the approach, a test rib is printed with polylactic acid (PLA) and tested. The broken rib is shown in Figure 12.14, showing that the failure occurred in the area where the highest stress was predicted (Figure 12.8). This corroborates the results

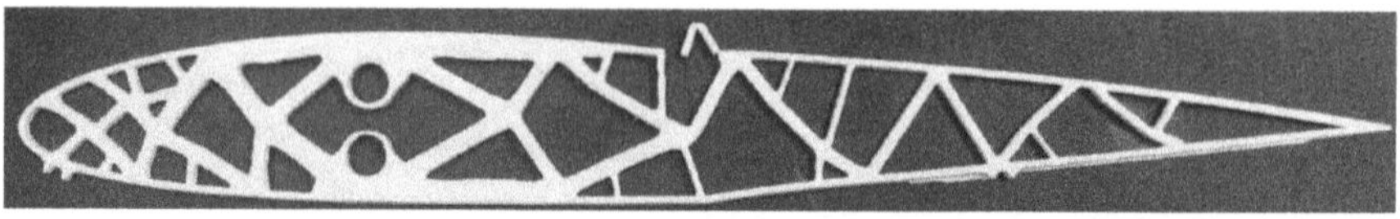

FIGURE 12.14 PLA test rib showing failure in the rib that are predicted by the finite element analysis stress analysis.

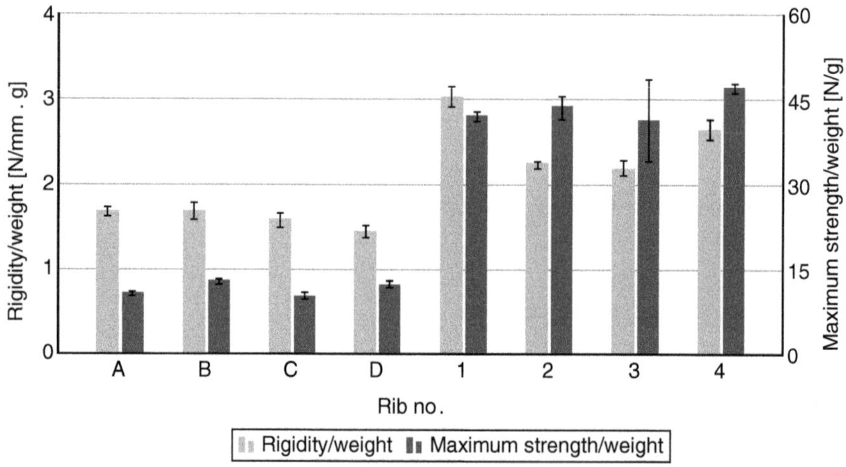

FIGURE 12.15 Properties per weight ratio of the ribs.

TABLE 12.2
Experimental Properties of Each Rib

Rib No.	A	B	C	D	1	2	3	4
Weight (g)	5.74	8.51	5.74	8.51	5.60	8.88	8.23	6.74
Rigidity (N/mm)	9.61	14.28	9.61	14.28	17.01	19.81	18.01	17.81
$CV_{Rigidity}$ (%)	3.555	4.597	3.555	4.597	3.196	0.562	1.345	4.766
Rigidity/weight (N/mm.g)	1.676	1.678	1.676	1.678	3.036	2.231	2.189	2.642
Maximum strength (N)	60.92	108.75	60.92	108.75	235.68	386.48	339.29	317.34
$CV_{Strength}$ (%)	3.489	4.091	3.489	4.091	1.444	4.331	14.717	2.079
Max. strength/weight (N/g)	10.62	12.78	10.62	12.78	42.07	43.52	41.24	47.08

obtained in the stress analysis previously, thus confirming the effectiveness of the testing procedure.

The properties obtained through the experimental testing for each rib are presented in Figure 12.15 and Table 12.2. The low coefficient of variation (CV) values obtained indicate homogeneity in the results and the high representativeness of the average values. The rigidity-to-weight and strength-to-weight ratios are plotted in a bar graph shown in Figure 12.15.

In the case of the continuous fibre-reinforced ribs, due to poor PA properties, the areas that are not reinforced deformed significantly. This behaviour, displayed in all the ribs, means that despite not breaking, the ribs reach a deformation state that is deemed not acceptable for a wing rib. This effect, shown in Figure 12.16, is

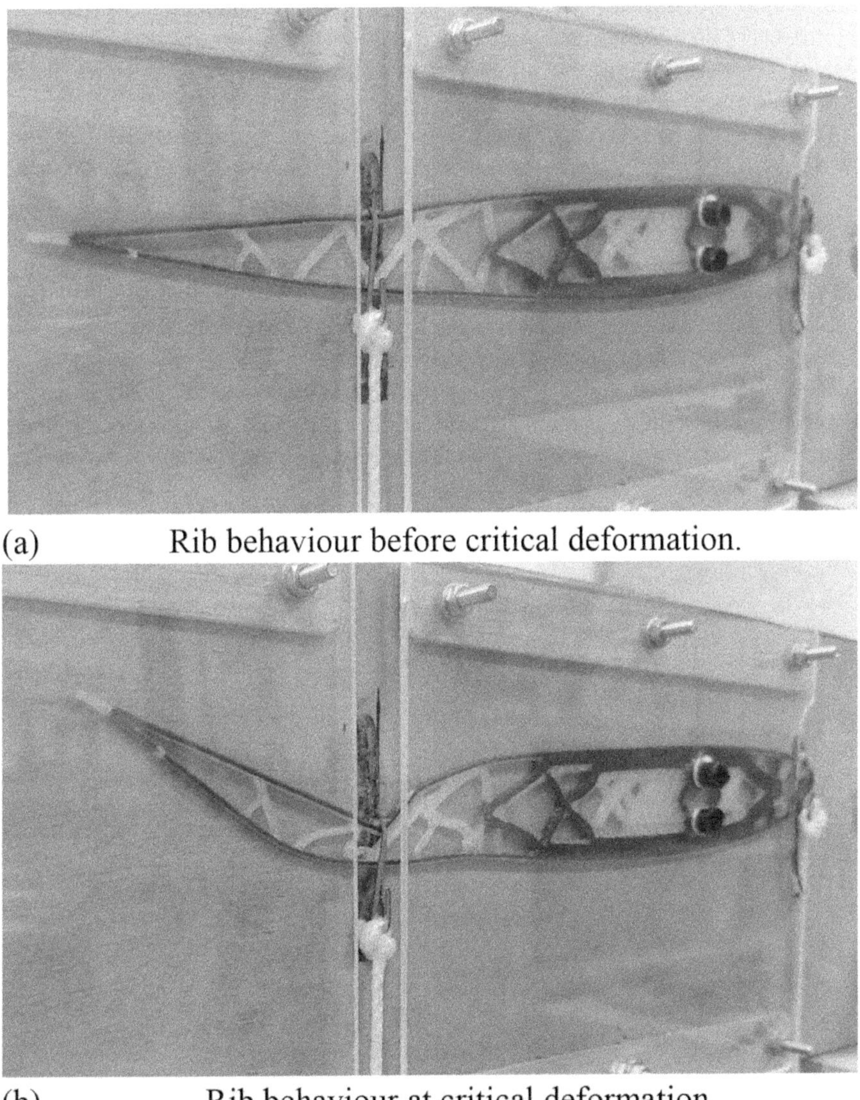

FIGURE 12.16 Critical deformation of continuous fibre-reinforced ribs. (a) Rib behaviour before critical deformation. (b) Rib behaviour at critical deformation.

accentuated where the point loads are applied, but better results are predicted in a more realistic case where distributed loads are applied.

The short fibre-reinforced ribs exhibited a more predictable behaviour as there are no disparities in the material properties in the rib. All the ribs failed at the maximum stress areas of the simulations and supported the necessary load easily, even when considering possible safety factors. When considering the weight, the 0.3 V_f rib has the better results, as it is the lightest and takes less time and material to print.

On the other hand, the short fibre-reinforced ribs have a more predictable behaviour, as there are little disparities in the material properties in the rib. The ribs with a V_f of 0.3 (1) failed in the same place as the PLA one. However, this fracture is much more aggressive as the material is able to store more energy, being that the rib shatters in multiple places, as shown in Figure 12.17.

Regarding the ribs with a V_f of 0.6, the expected results are observed, being that the ribs with 100% infill (2) performed better than the ones with 80% infill (3), which in turn performed better than the 20% ones (4). It is worth noting the 80% infill ribs show a dispersion in maximum force supported, which can easily be seen in Table 12.2 and in the graph in Figure 12.15. The ribs with 80% and 20% infill both fail near the carbon spars, which is in accordance with the simulations, which show high stress concentrations near both spars. In Figure 12.18, a comparison between the breaking points of these ribs and their stress distribution is done.

Finally, the ribs with 100% concentric infill are the only ones that break in the load points. The failure in this area is logically coherent though, as with the increase in infill, the area near the spars that would fail is now stronger. This allows the addition

FIGURE 12.17 Breaking points of the 0.3 V_f ribs.

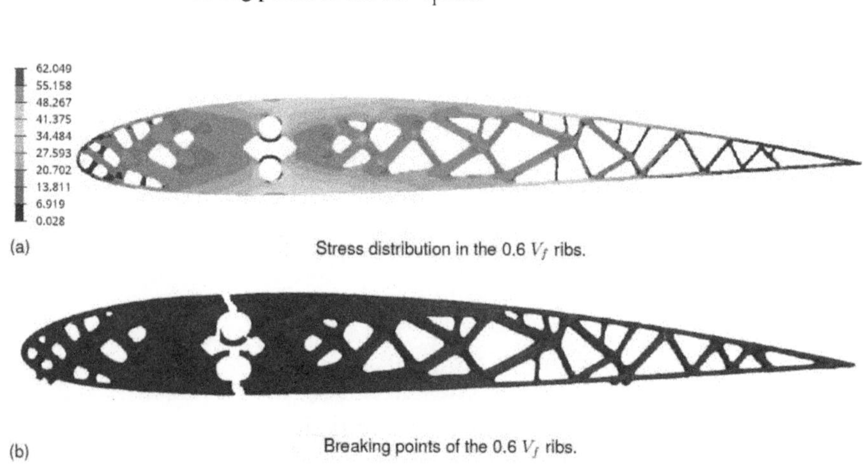

FIGURE 12.18 Correlation between the maximum stress and breaking points of the 0.6 V_f ribs.

FIGURE 12.19 Breaking points of the 0.6 V_f 100% infill ribs.

of more weight, which in turn forms a stress concentration in the zone where the forces are applied, higher than in the other cases, which induces a local failure. These failure points are shown in Figure 12.19.

When taking into account the weight of the ribs, the 0.3 V_f rib (1) has the best stiffness-to-weight ratio, whereas the 0.6 V_f 20% infill rib (4) has the highest maximum strength-to-weight ratio. Being that all the ribs tested are able to support loads vastly greater than the one needed for an extreme situation, even with the inclusion of a safety factor of 2 or 3, the 0.3 V_f rib is more advantageous than the others. Not only is it lighter, but it also needs less material, being cheaper and faster to manufacture. Despite the remaining short fibre-reinforced ribs having supported higher loads and having better stiffness, they are over-dimensioned for the study case at hand. Potentially, ribs with even smaller V_f can still withstand the necessary load in this case.

12.7 CONCLUSIONS

The difficulty in reinforcing specific areas on the continuous fibre-reinforced prints is evident, especially in complex geometries, showing little synergy with topology optimization. Due to this, the ribs had some unreinforced areas that compromised their performance. This would not be as noticeable in a more realistic case with distributed loads.

On the other hand, short fibre reinforcement shows great synergy with topology optimization. The short fibre-reinforced ribs performed as expected. For the case at hand, the ribs can be considered over-dimensioned but can perfectly transition to other cases requiring higher load-bearing ribs.

Continuous fibre-reinforced ribs fail at low loads due to deformations in unreinforced areas, highlighting issues with fibre placement in complex parts. The short fibre-reinforced ribs have a more predictable behaviour, being able to withstand greater forces than the continuous fibre-reinforced ones. All the short fibre-reinforced ribs can support loads far greater than the necessary one for the case at hand, making the 0.3 volume fraction rib favourable due to its lighter weight.

This investigation helps bridge the gap in the literature regarding specific case studies comparison between continuous- and short-reinforced 3D-printed polymers. As such, it makes a stride to move this technology beyond its current status as a research tool and specific use method and establish it as a commercially viable manufacturing option in fields such as aerospace engineering.

Finally, with the successful validation of the rib optimization and design methodology, and the conclusion that these reinforcements can be employed in wing ribs, this study paves the way for the creation and optimization of 3D printable reinforced wing ribs for mass-produced UAVs. Despite the successful outcome of this work, it is important to acknowledge some of its limitations. Overall, this study was performed regarding a general case study to develop a design and testing methodology for 3D-printed wing ribs. A more specific real-world wing design would probably introduce more restraints in terms of airfoil and weight constraints.

During the topology optimization phase, both the materials used could have been more intrinsically characterized, which would result in different output rib geometries that consider anisotropies in the parts. Not taking these anisotropies into account may undermine the integrity of the optimized geometries of the ribs, which happened in the case of the continuous-fibre reinforced ones.

Furthermore, even though the ribs manufactured can withstand the load corresponding to an angle of attack of 10°, other angles of attack may arise difficulties. While in flight, the loads will also be distributed and more complex. However, the methodology for designing and manufacturing 3D-printed carbon-reinforced wing ribs can be considered validated after this study. Regarding possible future works, ribs for a more complex wing could be designed and manufactured with this method. Larger parts can favour long fibres, thereby allowing for more arms to be reinforced; however, there are still limitations concerning print size. Application to a specific aircraft with a defined flight envelope would also be beneficial.

ACKNOWLEDGEMENTS

This work was supported by Fundação para a Ciência e Tecnologia (FCT), through IDMEC, under LAETA project, UIDB/50022/2020. This work has been supported by the European Union under the Next Generation EU, through a grant of the Portuguese Republic's Recovery and Resilience Plan (PRR) Partnership Agreement, within the scope of the project PRODUTECH R3 – "Agenda Mobilizadora da Fileira das Tecnologias de Produção para a Reindustrialização," aiming the mobilization of the production technologies industry towards the reindustrialization of the manufacturing industrial fabric (project ref. no. 60 – C645808870–00000067).

REFERENCES

1. H. Shakhatreh, A.H. Sawalmeh, A. Al-Fuqaha, Z. Dou, E. Almaita, I. Khalil, N.S. Othman, A. Khreishah, M. Guizani, Unmanned aerial vehicles (UAVs): a survey on civil applications and key research challenges, *IEEE Access* 7 (2019): 48572–48634.
2. R.A. Witik, F. Gaille, R. Teuscher, H. Ringwald, V. Michaud, J.A.E. Månson, Economic and environmental assessment of alternative production methods for composite aircraft components, *J. Clean. Prod.* 29–30 (2012): 91–102.
3. E.A. Calado, M. Leite, A. Silva, Selecting composite materials considering cost and environmental impact in the early phases of aircraft structure design, *J. Clean. Prod.* 186 (2018): 113–122.
4. W.J. Marx, D.N. Mavris, D.P. Schrage, Effects of alternative wing structural concepts on high speed civil transport life cycle costs, in: *37th AIAA/ASME/ASCE/AHS/ASC Structure, Structural Dynamics and Materials Conference*. American Institute of Aeronautics and Astronautics, 1996, pp. 562–582.

5. C.M.S. Vicente, M. Sardinha, L. Reis, A. Ribeiro, M. Leite, Large-format additive manufacturing of polymer extrusion-based deposition systems: review and applications, *Prog. Addit. Manuf.* (2023): 1–24.
6. S. Junk, B. Klerch, U. Hochberg, Structural optimization in lightweight design for additive manufacturing, *Procedia CIRP* 84 (2019): 277–282.
7. R.T.L. Ferreira, I.C. Amatte, T.A. Dutra, D. Bürger, Experimental characterization and micrography of 3D printed PLA and PLA reinforced with short carbon fibers, *Compos. Part B Eng.* 124 (2017): 88–100.
8. T.A. Dutra, S.F.M. de Almeida, Composite plate stiffness multicriteria optimization using lamination parameters, *Compos. Struct.* 133 (2015): 166–177.
9. M. Leite, A. Silva, E. Henriques, J.F.A. Madeira, Materials selection for a set of multiple parts considering manufacturing costs and weight reduction with structural isoperformance using direct multisearch optimization, *Struct. Multidiscip. Optim.* 52, 635–644 (2015).
10. J. M. Gardner, G. Sauti, J. W. Kim, R. J. Cano, R. A. Wincheski, C. J. Stelter, B. W. Grimsley, D. C. Working, and E. J. Siochi, "Additive manufacturing of multifunctional components using high density carbon nanotube yarn filaments," in: *International SAMPE Technical Conference*, 2016. [Online]. Available: https://ntrs.nasa.gov/citations/20160009176
11. A. Gebhardt, *Understanding Additive Manufacturing*. Hanser, 2011. [Online]. Available: https://doi.org/10.3139/9783446431621
12. V. Shanmugam, D. J. J. Rajendran, K. Babu, S. Rajendran, A. Veerasimman, U. Marimuthu, S. Singh, O. Das, R. E. Neisiany, M. S. Hedenqvist, F. Berto, and S. Ramakrishna, "The mechanical testing and performance analysis of polymer-fibre composites prepared through the additive manufacturing," *Polymer Testing*, vol. 93, 2021. [Online]. Available: https://doi.org/10.1016/j.polymertesting.2020.106925
13. G. D. Goh, V. Dikshit, A. P. Nagalingam, G. L. Goh, S. Agarwala, S. L. Sing, J. Wei, and W. Y. Yeong, "Characterization of mechanical properties and fracture mode of additively manufactured carbon fiber and glass fiber reinforced thermoplastics," *Materials and Design*, vol. 137, pp. 79–89, 2018. [Online]. Available: http://dx.doi.org/10.1016/j.matdes.2017.10.021
14. K. Goda, M. S. Sreekala, S. K. Malhotra, K. Joseph, and S. Thomas, *Advances in Polymer Composites: Biocomposites – State of the Art, New Challenges, and Opportunities*. John Wiley and Sons, Ltd, 2013, ch. 1, pp. 1–10. [Online]. Available: https://onlinelibrary.wiley.com/doi/abs/10.1002/9783527674220.ch1
15. L. J. Love, V. Kunc, O. Rios, C. E. Duty, A. M. Elliott, B. K. Post, R. J. Smith, and C. A. Blue, "The importance of carbon fiber to polymer additive manufacturing," *Journal of Materials Research*, vol. 29, no. 17, pp. 1893–1898, 2014. [Online]. Available: https://doi.org/10.1557/jmr.2014.212
16. X. Zhang, W. Fan, and T. Liu, "Fused deposition modeling 3D printing of polyamide-based composites and its applications," *Composites Communications*, vol. 21, no. June, p. 100413, 2020. [Online]. Available: https://doi.org/10.1016/j.coco.2020.100413
17. K. K. Chawla, *Ceramic Matrix Composites*. Boston, MA: Springer US, 1993, pp. 1–10. [Online]. Available: https://doi.org/10.1007/978-1-4757-2216-1_1
18. M. Mohammadizadeh and I. Fidan, "Tensile performance of 3D-printed continuous fiber-reinforced nylon composites," *Journal of Manufacturing and Materials Processing*, vol. 5, no. 3, 2021. [Online]. Available: https://doi.org/10.3390/jmmp5030068

19. K. Ghabraie, *Structural Seismic Design Optimization and Earthquake Engineering: Formulations and Applications*. IGI Global, 01: 2012. [Online]. Available: http://dx.doi.org/10.4018/978-1-4666-1640-0.ch010
20. D. H. Lukaszewicz, C. Ward, and K. D. Potter, "The engineering aspects of automated prepreg layup: history, present and future," *Composites Part B: Engineering*, vol. 43, no. 3, pp. 997–1009, 2012. [Online]. Available: https://doi.org/10.1016/j.compositesb.2011.12.003
21. M. Iragi, C. Pascual-González, A. Esnaola, C. S. Lopes, and L. Aretxabaleta, "Ply and interlaminar behaviours of 3D printed continuous carbon fibre-reinforced thermoplastic laminates; effects of processing conditions and microstructure," *Additive Manufacturing*, vol. 30, 2019. [Online]. Available: https://doi.org/10.1016/j.addma.2019.100884
22. F. S. Senatov, K. V. Niaza, A. A. Stepashkin, and S. D. Kaloshkin, "Low-cycle fatigue behavior of 3D-printed PLA-based porous scaffolds," *Composites Part B: Engineering*, vol. 97, pp. 193–200, 2016. [Online]. Available: http://dx.doi.org/10.1016/j.compositesb.2016.04.067
23. P. Carneiro, M. Cardoso, and P. Gamboa, "Structural analysis of wing ribs obtained by additive manufacturing," *Rapid Prototyping Journal*, vol. 25, 2019. [Online]. Available: https://doi.org/10.1108/RPJ-02-2018-0044
24. S. Kasmi, G. Ginoux, S. Allaoui, and S. Alix, "Investigation of 3D printing strategy on the mechanical performance of coextruded continuous carbon fiber reinforced PETG," *Journal of Applied Polymer Science*, vol. 138, no. 37, 2021. [Online]. Available: https://doi.org/10.1002/app.50955
25. S. R. Rajpurohit and H. K. Dave, "Flexural strength of fused filament fabricated (FFF) PLA parts on an open-source 3D printer," *Advances in Manufacturing*, vol. 6, no. 4, pp. 430–441, 2018. [Online]. Available: https://doi.org/10.1007/s40436-018-0237-6
26. J. Naranjo-Lozada, H. Ahuett-Garza, P. Orta-Castañón, W. M. Verbeeten, and D. Sáiz-González, "Tensile properties and failure behavior of chopped and continuous carbon fiber composites produced by additive manufacturing," *Additive Manufacturing*, vol. 26, no. August 2018, pp. 227–241, 2019. [Online]. Available: https://doi.org/10.1016/j.addma.2018.12.020

Index

Note: Page numbers in *italics* and **bold** denote figures and tables, respectively.

G

H

I

Q

R

T

For Product Safety Concerns and Information please contact our EU representative GPSR@taylorandfrancis.com
Taylor & Francis Verlag GmbH, Kaufingerstraße 24, 80331 München, Germany

www.ingramcontent.com/pod-product-compliance
Ingram Content Group UK Ltd.
Pitfield, Milton Keynes, MK11 3LW, UK
UKHW022218190926
475702UK00003B/420
* 9 7 8 1 0 3 2 7 9 8 3 9 4 *